Birkhäuser

Computer Science Foundations and Applied Logic

Computer Science Foundations and Applied Logic is a growing series that focuses on the foundations of computing and their interaction with applied logic, including how science overall is driven by this. Thus, applications of computer science to mathematical logic, as well as applications of mathematical logic to computer science, will yield many topics of interest. Among other areas, it will cover combinations of logical reasoning and machine learning as applied to AI, mathematics, physics, as well as other areas of science and engineering. The series (previously known as *Progress in Computer Science and Applied Logic*, https://www.springer.com/series/4814) welcomes proposals for research monographs, textbooks and polished lectures, and professional text/references. The scientific content will be of strong interest to both computer scientists and logicians.

Shuichi Yukita

Category Theory Using Haskell

An Introduction with Moggi and Yoneda

Birkhäuser

Shuichi Yukita
Department of Digital Media, Faculty
of Computer and Information Sciences
Hosei University
Koganei, Japan

ISSN 2731-5754 ISSN 2731-5762 (electronic)
Computer Science Foundations and Applied Logic
ISBN 978-3-031-68537-8 ISBN 978-3-031-68538-5 (eBook)
https://doi.org/10.1007/978-3-031-68538-5

This book is published under the imprint Birkhäuser, www.birkhauser-science.com by the registered company Springer Nature Switzerland AG
The registered company address is: Gewerbestrasse 11, 6330 Cham, Switzerland

Preface

In the first half of this Preface, we will see the outline of this book, for some readers, with unfamiliar technical terms. But just take it easy since the whole details follow soon in the coming chapters in a gentle manner. In the latter half of this Preface, the intended readers will be precisely defined.

Algebraic topology gave birth to category theory in the 1940s. Category theory became a working language in algebraic geometry and number theory in the 1950s. Algebraic analysis of Kyoto school introduced homological techniques systematically in the study of partial differential equations, and further pushed forward their theory to a more advanced level with category theory. Eilenberg and Mac Lane sought for a systematic way of describing the situations where two things are *naturally* or essentially the same, or something is *universal* among others. The first choice of textbook on category theory might be Mac Lane's famous book [1] if you are familiar with various notions in undergraduate mathematics such as groups, rings, modules, homomorphisms, tensor products, continuous mappings in general topology, homology groups in algebraic topology, and so on. Do not worry. If you are familiar with Haskell programming, this monograph is for you.

In the nineteenth century, before category theory got formal foundations, Galois had actually established an *equivalence of categories* between the category of fields and the category of groups, known today as the *Galois connections.*

Category theory began to spread to logic and computer science soon after it was born. It is also widely used in physics. In the functional programming language Haskell, libraries are defined, designed, or redesigned over revisions, from a category-theoretic point of view. In a computer algebra system Axiom, it is possible to program directly the language of category theory in a way that mathematicians do with pencil and paper. Many other popular programming languages are ready to take in category-theoretic ideas.

Under such circumstances, if you go to a bookstore, you will find many books on category theory, from introductory books to highly technical ones. A quick search on the Internet will show you that there are many publications on category theory related to mathematics, logic, computer science, and physics. The number of research papers is orders of magnitude higher.

This book is an introduction to category theory. The goal is to understand Yoneda's lemma. The Yoneda lemma is often referred to as "the hardest trivial thing in mathematics." Mathematicians often feel frustrated and ask why we can't just say it's right even though it's self-evident. "Lemma" means usually a helper proposition for proving a big theorem, but, in some special cases, it's an honorific title for a great theorem, which is the mother of daily proved theorems by professional mathematicians. For example, Nakayama's lemma and Zorn's lemma are very well known by mathematicians all over the world and are used once for a while, or touched on once in their schooldays. Sorry, they are not mentioned in this book. In category theory, if you can understand Yoneda's lemma properly, you can consider you graduated from your introductory category theory course. No, is that too much? I must confess that it took me decades to get the Yoneda lemma as a life-long companion just like the formula for the quadratic equations. It might be better to restate that if you can get to the chapter of Yoneda, you've graduated from the introductory course on category theory.

I stuck to the explicitness of the data structure and the *grammar* of diagram tracing. The reason is as follows:

- Most beginners have a hard time in grasping the concept of natural transformation, the first hurdle of abstraction, unless they are conscious of types. I feel a bit awkward to take the following example, but let me try. We use expressions like "function $f(x)$" in a math class or daily technical conversation. But, when we bring it into a computer program, the function must be f and not $f(x)$. The latter expression $f(x)$ is a function application of the function f to the data x. And even for us humans, when discussing category theory, there are times when we are required to abandon our human minds and think of ourselves as a computer. When the calculation is completed and the meaning is to be reconsidered, the human mind is allowed to reenter.
- The word "chasing" appearing in the technical phrase "diagram chasing" refers to trying to see something that cannot be seen at a glance where one has no idea in what order the focus should be moved. Such a situation occurs everywhere in category theory.

Taking all these into account, I adopted the following policy:

- For concepts such as functors, natural transformations, cones, and limits, we give them concrete data structures, aiming that those data can be input to a computer.
- Contains many short examples of Haskell code. The compiler scolds you if it fails to match the types, so you don't have to waste your time trying to reach a goal that you never succeed in. Running the code is not a proof of the theorem, but it does convince you that the inferences here and there in the proof are correct.
- Samples of Haskell code are roughly classified into two types. One consists of functors and natural transformations found in the wild in the Haskell standard

libraries. On the other hand, Haskell lets you create and play with small worlds of toy categories, functors, natural transformations, and limits. I won't make the distinction between the two types explicit, but you can easily distinguish them.
- In presenting diagram chasing, the drawer's intentions are described as clearly as possible; why the arrow should be drawn at this time and at this place.
- Diagrams have a grammar and a drawing order. I have not been able to formulate this clearly yet, but I will try to make a presentation with that necessity in mind.

I will describe the features of this book as a precaution to the reader:

- This is an introductory book on category theory aiming at understanding *Yoneda's lemma.* I would also recommend it to those who want to re-challenge category theory. I would like you to experience that the theory will look different when you are strongly aware of the "grammar" of diagram chasing and the data structure of the category.
- All proofs of the theorems are given. You should be falimiliar with the concepts of sets, maps, injections, surjections, Cartesian products, relations, symmetricity, reflexivity, transitivity, and so on. It is assumed that you have the basic knowledge or the skill to acquire new knowledge that you have a proper understanding of equivalence relations and equivalence classes, or that you can quickly recall or supplement them by looking them up in textbooks or on the Internet.
- In some chapters, applications to linear algebra, groups, rings, and fields are given. However, you can safely skip those examples. It just a bit narrows down the way you can enjoy it.
- Haskell example code is everywhere. You can skip it and follow the discussion of category theory itself, but using Haskell is a major feature of this book, so I guarantee you'll enjoy reading it. See [2] for a quick or in-depth introduction to Haskell. You can find many other friendly books on Haskell on the Internet. You should be able to enjoy category theory even more if you do so. In principle, the sample code should be "one-shot," and with some exceptions, mutual references (import, export) are not employed. It means that you can pick up and read sample code independently. There is no versatility at all. However, if you read and compare them, you will find that there is a consistent design pattern.
- Mathematics students will find this book useful as a companion to a standard introductory text.
- I won't mention any of the beautiful libraries for category theory that are actively being developed in the Haskell community. They are for intermediate-level category theory students and above, and I do not have enough competence and experience to introduce them to the reader.

I highly recommend Awodey [3] to readers interested in logic and computer science. For the technical terms, I referred to Mac Lane's book [1], the bible on category theory, and Leinster's masterpiece [4]. The Haskell processor used GHCi version 8.6.5.

I am grateful to all those who reviewed the manuscript.

Koganei, Japan Shuichi Yukita

References

1. Mac Lane S (1998) Categories for the working mathematician, Graduate texts in mathematics, vol 5, 2nd edn. Springer
2. Lipovaca M (2011) Learn you a Haskell for great good! PDF on-line available
3. Awodey S (2006) Category theory. Oxford University Press
4. Leinter T (2014) Basic category theory, Cambridge studies in advanced mathematics, vol 143. Cambridge University Press

Contents

1 Category, Functor, Natural Transformation 1
1.1 Sets and Functions 1
1.2 Category, Object, Morphism 7
1.3 Data Structures of Categories 10
1.4 Functor and Contravariant Functor 12
1.5 Faithful Functors and Full Functors 18
1.6 Natural Transformations 21
1.7 Subcategories of Hask 30
References 33

2 Equivalence of Categories 35
2.1 Functor Category 35
2.2 Equivalence of Categories 39

3 Universality and Limits 51
3.1 Initial and Terminal Objects 51
3.2 Products 53
3.3 Coproducts 57
3.4 Limits 61
3.5 Colimits 69
3.6 Existence of Limits 80
3.7 Existence of Colimits 91

4 Functors and Limits 103
4.1 Functors Map Cones to Cones 103
4.2 Hom Functors and Limits 105
4.3 Hom Functors and Colimits 108
4.4 Executable Examples 111
4.5 Limit Functors 114
4.5.1 Limits as a Functor 115
4.5.2 Colimits as a Functor 116

5 Adjoints 119
5.1 Adjunctions 119
5.2 Unit and Counit 121
5.3 Triangle Identities 124
5.4 Universal Arrows and Adjunctions 126
5.5 Equivalent Formulations of Adjunction 128
5.6 Adjunction and Category Equivalence 138
5.7 Global Naturality of Adjunction 145
5.8 Adjunctions and Limits 147

6 Monads 151
6.1 Kleisli Triples 151
6.2 Moggi' Theory 154
6.2.1 Non-determinism and the List Functor 154
6.2.2 Computation with Output 156
6.2.3 Continuation Passing 159
6.2.4 Computation that May Fail and the Maybe Functor 163
6.3 Monads 165
6.4 Monad Instances in Haskell 174
6.4.1 Computation with Output(revisited) 174
6.4.2 Computation with Continuation(revisited) 178
6.5 Functor, Applicative, and Monad 183
6.6 Monads and Adjoints 189
6.6.1 Adjoints to Monads 189
6.6.2 List Monad as Adjoint 191
Reference 196

7 Representable Functors 197
7.1 Representation of Covariant Functors 197
7.2 Representation of Contravariant Functors 202
7.3 The Yoneda Lemma 204
7.3.1 Covariant Case 204
7.3.2 Contravariant Case (Presheaves) 210
7.4 Milewski's "Understanding Yoneda" 215
7.4.1 Specification Determines Implementation 216
7.4.2 A Function That Has the Same Signature as the Identity Function 217
7.5 Reverse Engineering by the Yoneda Lemma 218
7.5.1 A Machine That Has a Hidden Parameter 218
7.5.2 A Machine That Hides a List Inside 220
7.5.3 A Machine That Has a Secret Morphism 222
7.5.4 The Design Pattern with Yoneda 223
7.6 Adjoints Preserve Limits 224
References 226

8 Monoidal Categories and Coherence 227
8.1 Categories with Tensor Products 227
8.2 Coherence—Part One 238
8.3 Monoidal Functors 245
8.4 Coherence—Part Two 246
References 263

Epilogue 265

Appendix A: Sets and Universe 267

Appendix B: Calculations with Output 269

Appendix C: Continuation Passing 283

Index 295

Category, Functor, Natural Transformation

1

Beginning with sets, functions, and relations, we will see how we are tempted to introduce category theory. After defining the data structures for categories, we will examine some examples. A functor is a function from one category to another that preserves the data structures of the source in its image. Given two categories and two functors between them, a natural transformation is used to compare how the two functors preserve data structures. We will see polymorphic functions in Haskell are natural transformations.

1.1 Sets and Functions

The purpose of the first half of this section is to tempt the reader to use a new device that can handle both functions and their variants such as partial functions and relations. The new device is called category. We will first introduce the *category of sets and functions* or simply the *category of sets*, then the *category of sets and partial functions* and the *category of sets and relations*.

Given two sets A and B, we can form their Cartesian product $A \times B$. A subset $\mathrm{R} \subset A \times B$ is called a *relation* between A and B.

We often write as

$$a \,\mathrm{R}\, b \Leftrightarrow (a, b) \in \mathrm{R}\,.$$

Some authors like to say R is the graph of a relation.

When the relation R satisfies the following condition, it is called a *function* or *mapping*.

$$Condition : \forall a \in A \ . \ \exists! b \in B \ . \ a \,\mathrm{R}\, b$$

Here, $\exists!$ means there exists exactly one thing that satisfies the condition.

S. Yukita, *Category Theory Using Haskell*, Computer Science Foundations and Applied Logic, https://doi.org/10.1007/978-3-031-68538-5_1

Let us consider the case where $A = B = \{1, 2, 3\}$ and three relations:

$$\begin{aligned} \mathrm{R}_1 &= \{(1, 1), (2, 1), (3, 3)\}, \\ \mathrm{R}_2 &= \{(1, 1), (1, 2), (2, 2), (3, 3)\}, \text{ and} \\ \mathrm{R}_3 &= \{(1, 1), (2, 3)\}. \end{aligned}$$

Pictorially, we have the following.

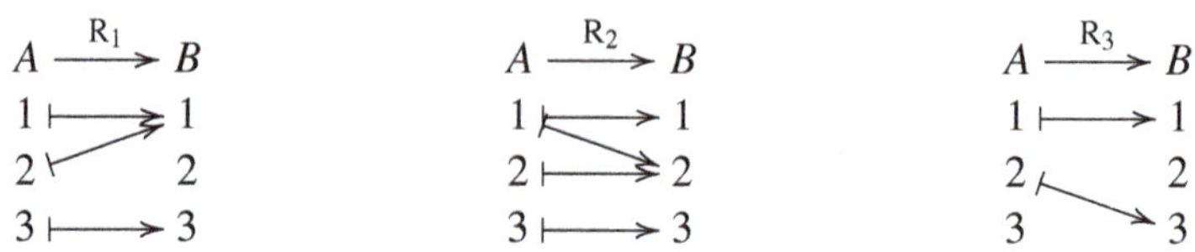

$\mathrm{R}_1, \mathrm{R}_2$, and R_3 are all relations. Which of them are functions? While R_1 is a function, neither R_2 nor R_3 are functions.

Roughly speaking, all the points of the source set emit exactly one arrow $\mapsto$ in order for the relation to be a function. In other words, no two arrows $\mapsto$ start at the same point in the case of a function. Notice that the notion of function is extremely non-symmetric between input and output.

R_3 is not a function, but it is called a *partial function* that represents a computation that may fail or a computation that may not terminate forever, which are a primary object of study in computer science. Rational functions, functions that are formed by a fraction of two polynomials, are examples of a partial function. The values are undefined when a given input makes the denominator null. We can write a logic formula for the condition for a relation to be a partial function.

$$\textit{Condition} : \forall a \in A \ . \ (a \,\mathrm{R}\, b \wedge a \,\mathrm{R}\, b' \Rightarrow b = b')$$

In the case of a partial function, every source point emits at most one arrow $\mapsto$.

Not to mention, the notion of relation is the basis of database theory, where relational algebra plays the central role.

To sum up, we have the following inclusions:

- A function is a partial function.
- A partial function is a relation.

Relation R_2 is neither a function nor a partial function. However, in the theory of functions of a complex variable, we are familiar with the notion of *multi-valued function* that allows several, possibly infinite, values for a single input value. A typical example is a function of square root $f(z) = \sqrt{z}$, which allows twofold values.

The desire to make those multi-valued functions a single-valued function led mathematicians to topology. The Riemann surfaces were thus introduced.

Given two functions $f : A \to B$ and $g : B \to C$, we define their composition $g \circ f : A \to C$ as

$$(g \circ f)(a) = g(f(a)).$$

We know that *function composition* is *associative*. But, it is not obvious.

Proposition 1.1 *Given three functions $f : A \to B$, $g : B \to C$, and $h : C \to D$, we have*

$$(h \circ g) \circ f = h \circ (g \circ f).$$

Proof To prove the statement, we have to show that

$$((h \circ g) \circ f)(a) = (h \circ (g \circ f))(a)$$

holds for all $a \in A$.

Let us calculate the LHS (left-hand side).

$$\begin{aligned} &((h \circ g) \circ f)(a) \\ =&(h \circ g)(f(a)) \\ =&h(g(f(a))) \end{aligned}$$

Next, calculate the RHS (right-hand side).

$$\begin{aligned} &(h \circ (g \circ f))(a) \\ =&h((g \circ f)(a)) \\ =&h(g(f(a))) \end{aligned}$$

We have established that both sides coincide. □

Besides function composition, associative compositions can be found everywhere in mathematics and mathematical sciences. This makes us include associativity in the axioms of the category in what follows.

The composition of partial functions is also associative. We omit the proof. The reader might check each of the following three cases:

- $f(a) = \perp$,
- $f(a)$ exists, but $g(f(a)) = \perp$, and
- $g(f(a))$ exists, but $h(g(f(a))) = \perp$,

where we wrote $f(a) = \perp$ when $f(a)$ does not exist.

Remark 1.1 We used the symbol $\perp$ (bottom) to mean that the thing is not defined. If we add $\perp$ as a member to every set, we can make all the partial functions into total ones. Further, if we require $f(\perp) = \perp$ for every function, we are led to the world of denotational semantics, which is out of the scope of this exposition.

We have just established that the composition of functions and partial functions is associative. Let us consider relations. Given two relations $F \subset A \times B$ and $G \subset B \times C$, we define their composition $G \circ F \in A \times C$ as

$$G \circ F = \{(x, z) \in A \times C \mid \exists y \in B.(x, y) \in F \text{ and } (y, z) \in G\}.$$

Proposition 1.2 *Given three relations $F \subset A \times B$, $G \subset B \times C$, and $H \subset C \times D$, we have*

$$(H \circ G) \circ F = H \circ (G \circ F).$$

Proof We show that

$$\begin{aligned} &(a, d) \in ((H \circ G) \circ F) \\ \Leftrightarrow &(a, d) \in (H \circ (G \circ F)) \end{aligned}$$

holds for any $a \in A$ and $d \in D$.

Let us calculate the LHS, applying the definition of $\circ$ from the outermost (top level) to the inner part.

$$\begin{aligned} &(a, d) \in ((H \circ G) \circ F) \\ \Leftrightarrow &\exists b \in B \ . \ aFb \text{ and } b(H \circ G)d \\ \Leftrightarrow &\exists b \in B \ . \ aFb \text{ and } \exists c \in C \ . \ bGc \text{ and } cHd \\ \Leftrightarrow &\exists b \in B \ . \ \exists c \in C \ . \ aFb \text{ and } bGc \text{ and } cHd \end{aligned}$$

Next, we calculate the RHS, applying the definition of $\circ$ again from the outermost to the inner part.

$$\begin{aligned} &(a, d) \in (H \circ (G \circ F)) \\ \Leftrightarrow &\exists c \in C \ . \ a(G \circ F)c \text{ and } cHd \\ \Leftrightarrow &\exists c \in C \ . \ (\exists b \in B \ . \ aFb \text{ and } bGc) \text{ and } cHd \\ \Leftrightarrow &\exists c \in C \ . \ \exists b \in B \ . \ aFb \text{ and } bGc \text{ and } cHd \end{aligned}$$

We can exchange the positions of $\exists b \in B$ and $\exists c \in C$. Thus, we have proved the equality. □

If we carefully reflect on the discussion above, we find that associativity of function composition comes from that of relations. We have only to show that function composition is the restriction of relation composition to the subclass (of relations) consisting of functions and that the subclass is closed under relation composition.

Note that it is a lucky situation in which a certain binary operation is associative. There are many important non-associative algebras. The most familiar ones everyone can imagine are Lie algebras, but we find an example at the elementary school level.

Example 1.1 Is subtraction associative? Does the following hold?

$$(100 - 50) - 50 = 100 - (50 - 50)$$

Many people carry out mathematical work under various frameworks. Common features are categories consisting of objects and morphisms.

Category	Objects	Morphisms
Category of sets	Sets	Functions
Category of sets and partial functions	Sets	Partial functions
Category of sets and relations	Sets	Relations

What's the good of this? Well, it is common to switch the framework of discussion saying, for example, "so far we have talked about everything in the category of sets and functions, but from now on we will talk about the category of sets and relations."

Let us see further examples that strongly motivate to introduce categories.

Sets in the wild that we encounter everyday are no more than a rabble. Usually, they have structures and structure preserving morphisms. We sometimes want to forget such structures intentionally to gain insight within the framework of *adjunctions*. Such activities are conducted by *forgetful functors*. All such emphasized key words will be defined later. Anyway, we must collect examples of such structures that are forgettable.

We first see an algebraic system *group*. The reader should consult with standard textbooks on algebra. We do not give the definitions here.

Example 1.2 Let $\mathbb{R}_{\mathrm{add}}$ be a group of real numbers with the addition operation. And let $\mathbb{R}^{>0}_{\mathrm{mult}}$ be a group of positive real numbers with the multiplication. The (natural) logarithm function

$$\log : \mathbb{R}^{>0}_{\mathrm{mult}} \to \mathbb{R}_{\mathrm{add}}$$

is called a *group homomorphism* from $\mathbb{R}^{>0}_{\mathrm{mult}}$ to $\mathbb{R}_{\mathrm{add}}$. This function observes the group structures of the two groups involved. An awkward reader should consult with one's high school textbook. There, everyone find a formula like this.

$$\log(x \times y) = \log x + \log y$$

In this example, we have the inverse homomorphism $\exp : \mathbb{R}_{\mathrm{add}} \to \mathbb{R}^{>0}_{\mathrm{mult}}$ that also keeps the group structure via the formula

$$\exp(x + y) = \exp(x) \times \exp(y).$$

We say, in this case, the logarithm function is an *isomorphism*.

Homomorphisms in algebraic systems can be viewed as just a function between two sets if we ignore algebraic structures. When we introduce a new algebraic system, we should check if the composition of morphisms as a plain function yields a structure preserving morphism.

Example 1.3 Let us consider two real vector spaces A and B with linear mapping $f : A \to B$. We may regard linear mapping as just a homomorphism of additive groups. However, f, as a linear mapping, has the property of preservation of scalar multiplication. Namely, we require that for all $\vec{u}, \vec{v} \in A$ and $a, b \in \mathbb{R}$

$$f(a\vec{u} + b\vec{v}) = af(\vec{u}) + bf(\vec{v})$$

should hold.

We may replace "real numbers" with other fields such as "complex numbers." To sum up, we have the following.

Category	Objects	Morphisms
Category of groups	Groups	Group homomorphisms
Category of real vector spaces	Real vector spaces	Real linear mappings
Category of complex vector spaces	Complex vector spaces	Complex linear mappings

So far, we examined various cases where sets play the central role. They have elements and structures that force elements to behave consistently under operations the particular category imposes. Anyway, we treated objects that disclose their internal structures to the world.

Information hiding is one of the key concepts in the object oriented programming paradigm. We are reluctant to accept information that we can do without it. Not knowing anything without which we can do, we can keep a certain level of security.

Now, we proceed to tackle situations where, at first, we have no clue to imagine the internal structures.

Example 1.4 Let us consider the set $A = \{1, 2, 3\}$. We can see the internal structure of A, knowing what the elements of A are. We introduce a function $f : A \to A$:

$$\begin{array}{ccc} A & \xrightarrow{f} & A \\ 1 & \longmapsto & 2 \\ 2 & \longmapsto & 3 \\ 3 & \longmapsto & 1 \end{array}$$

When someone does not know the internal structure of A, what kind of thing can he/she do at best? They are provided with only information about f. They only know that it yields nothing if f is iterated three times. With this observation, they might suppose A has three elements and these are circulated by f. This is exactly what physicists of elementary particles are doing in their daily life.

So far, we have seen examples without giving definitions. In the next sections, we discuss various mathematical phenomena based on a rigorous basis.

1.2 Category, Object, Morphism

The reader is familiar with the term "set" and with the fact that naïve set theory leads to paradoxes.

Russel's paradox arises when we consider a set like this:

$$R = \{x \mid x \notin x\}.$$

If $R \in R$, then by the definition of R we can deduce that $R \notin R$. If $R \notin R$, again by the definition of R we can deduce that $R \in R$. Thus, we get a contradiction.

To avoid such an absurdity, we must use the term carefully. One solution is to introduce the notion of *universe* originally proposed by Grothendieck. We roughly follow the style of Mac Lane [1]. Some details are given in Appendix A. We assume the existence of a sufficiently large set U called *universe*. U has all the needed things as its elements. If a set x is an element of U, we say x is *small* or x is a small set. Universe U is not a small set. Subsets of U are called a *class*. U is a subset of itself, so U is a class.

Remark 1.2 In the rest of this exposition, we often talk about *type class* in Haskell. Note that this concept of class and the class we have defined above have nothing to do with each other.

Definition 1.1 A *category* consists of a set of *objects* $A, B, C, \cdots$ and a set of *morphisms* $f, g, h, \cdots$. These constituents are subject to the following conditions.

1. For every morphism f we have the unique object *domain* $\operatorname{dom} f$ and the unique object *codomain* $\operatorname{cod} f$. When $\operatorname{dom} f = A$ and $\operatorname{cod} f = B$, we write $f : A \to B$ or $A \xrightarrow{f} B$.
2. Given two morphisms f and g with $\operatorname{dom} g = \operatorname{cod} f$, there exists their unique composite $g \circ f : \operatorname{dom} f \to \operatorname{cod} g$.
3. The composition operation is associative, namely, given three morphisms
$$A \xrightarrow{f} B \xrightarrow{g} C \xrightarrow{h} D,$$
we have
$$h \circ (g \circ f) = (h \circ g) \circ f.$$
4. For all object A, there exists a morphism $1_A : A \to A$ called the *identity morphism* that satisfies the following conditions:

for all $f : A \to B$ and $g : B \to A$,

$$f \circ 1_A = f \quad \text{and} \quad 1_A \circ g = g.$$

We denote by Obj($\mathscr{C}$) the set of all objects of $\mathscr{C}$ and by Mor($\mathscr{C}$) the set of all morphisms of $\mathscr{C}$. We denote by $\mathrm{Hom}_{\mathscr{C}}(A, B)$ the set of all morphisms of $\mathscr{C}$ that have domain A and codomain B.

Remark 1.3 The identity morphism 1_A is uniquely determined for each A. Given another morphism $1'_A$ that satisfies the same condition above, we can easily deduce that $1_A = 1'_A$.

There are morphisms that have a special property. When this kind of morphism exists between two objects, we can say they are essentially the same.

Definition 1.2 We say that $f : A \to B$ is an isomorphism if there exists a morphism $g : B \to A$ such that $g \circ f = 1_A$ and $f \circ g = 1_B$. If this is the case, g is called an inverse morphism (or simply an inverse) of f and written as $g = f^{-1}$. The relation inverse is easily shown to be symmetric. So, we also have that f is an inverse morphism of g.

Proposition 1.3 *If f has an inverse morphism, then the inverse morphism is uniquely determined.*

Proof Let g' be another inverse to f, namely g' satisfies the same conditions for g. Let us begin with the condition for g.

$$g \circ f = 1_A$$

We can compose g' from right to both sides. Then, we have

$$g \circ f \circ g' = g'$$

Since $f \circ g' = 1_B$, we have $g = g'$. □

Let $\mathscr{C}$ is a category. If Obj($\mathscr{C}$) and Mor($\mathscr{C}$) are both small, then $\mathscr{C}$ is called *small* or a *small category*. A category that is not small is called *large* or a *large category*.

Let $\mathscr{C}$ is a category. If $\mathscr{C}(A, B)$ is small for all pairs of objects A and B, then $\mathscr{C}$ is called *locally small* or a *locally small category*. $\mathscr{C}$ is not necessarily small when it is locally small.

Remark 1.4 We often say "A is not a set" to mean that it is not a small set. This rough usage is seen everywhere, and we can easily notice the exact meaning of such a statement.

We will see several examples of a large category. All of them happen to be locally small.

Example 1.5 Category **Set** is defined as follows. The set of objects is the set of all small sets. The set of morphisms is all the functions between those small sets. For all pairs of small sets A and B, we know $\mathbf{Set}(A, B)$ is small. Thus, **Set** is locally small. See Appendix A.

All of the algebraic systems we are going to see have small sets as their underlying set.

Example 1.6 Category **Grp** is defined as follows. The set of objects is the set of all groups. The set of morphisms is all the homomorphisms between those groups. Category **Grp** is locally small.

Example 1.7 Category **Ab** is defined as follows. The set of objects is the set of all abelian groups. The set of morphisms is all the homomorphisms between those abelian groups. **Ab** is locally small. Category **Ab** is a subcategory of **Grp**. We will give the definition of a subcategory shortly.

Example 1.8 Category **Top** is defined as follows. The set of objects is the set of all topological spaces. The set of morphisms is all the continuous functions between those topological spaces. Category **Top** is locally small.

Definition 1.3 We say that category $\mathscr{A}$ is a *subcategory* of $\mathscr{B}$ if the following conditions are all satisfied:

- $\mathrm{Obj}(\mathscr{A})$ is a subset of $\mathrm{Obj}(\mathscr{B})$.
- $\mathscr{A}(X, Y)$ is a subset of $\mathscr{B}(X, Y)$ for all pairs of $X, Y \in \mathrm{Obj}(\mathscr{A})$.
- Equality $g \circ_{\mathscr{A}} f = g \circ_{\mathscr{B}} f$ holds for all the pairs of composable morphisms $f, g \in \mathrm{Mor}(\mathscr{A})$, where $\circ_{\mathscr{A}}$ and $\circ_{\mathscr{B}}$ are composition of morphisms in $\mathscr{A}$ and $\mathscr{B}$, respectively.

Example 1.9 Let k be a field, such as $\mathbb{Q}$, $\mathbb{R}$, $\mathbb{C}$, $\mathbb{F}_p$, or others. Vect_k is a category of finite-dimensional k-vector spaces; the set of objects consists of all finite-dimensional k-vector spaces, and the set of morphisms consists of all k-linear mappings between these vector spaces. Vect_k is large, but locally small.

Example 1.10 We introduce some finite categories that are frequently used in the rest of this exposition.

- **0** is a category with no objects and no morphisms.
- **1** is a category with one object and no morphisms other than the identity morphism.

- **2** is a category with two objects A, B and one morphism $f : A \to B$ other than the identity morphisms $1_A, 1_B$

Example 1.11 Any partially ordered set (poset) can be seen as a category. The set of objects is the underlying set. When $a \leq b$, there is exactly one morphism from a to b. The reader should check morphisms so that they satisfy all the axioms of the category.

Example 1.12 Category Hask is defined as follows. Objects are Haskell types. Morphisms from object A to B are functions of type $A \to B$.

Remark 1.5 Many authors accuse Hask of not being even a category. See [2] for more information. The author takes the position that viewing Hask as a category helps us to understand various phenomena in computation. Let us compromise, expecting a theoretical breakthrough is coming someday. We will use Hask instead of **Set** when we present code for Yoneda machines.

If we reverse all the arrows in a discussion, it is very often the case that the discussion is still valid. Such phenomena are formulated as the *principle of duality*.

Definition 1.4 Given a category $\mathscr{A}$, we can construct an *opposite category* as follows. $\mathscr{A}^{\mathrm{op}}$ is a category, where $\mathrm{Obj}(\mathscr{A}^{\mathrm{op}}) = \mathrm{Obj}(\mathscr{A})$ and morphisms are all reversed.

1.3 Data Structures of Categories

We have seen examples of categories in the wild of mathematics. All of them are based on sets and functions. Transitivity and associativity of composition are obvious in those cases. In this section, we will examine artificial toy examples where such properties are not clear.

Example 1.13 Let us consider a category with one object A and one morphism $f : A \to A$ other than mandatory 1_A as in Fig. 1.1. We want a composition table for the composition of morphisms. f is composable with itself. The result will be either 1_A or f. There are no other options.

Fig. 1.1 One morphism other than the identity

In the case of $f \circ f = 1_A$, we have

	1_A	f
1_A	1_A	f
f	f	1_A

In the case of $f \circ f = f$, we have

	1_A	f
1_A	1_A	f
f	f	f

Both cases observe associativity. Therefore, we can define two different categories out of a *generator* f. Equations $f \circ f = 1_A$ and $f \circ f = f$ are called *relations*.

Example 1.14 Let us consider a category with one object A and the set of morphisms

$$\{1_A, f, f^2, f^3, \ldots, f^n, \ldots\}$$

as in Fig. 1.2.

The composition table is given as follows.

	1_A	f	$\cdots$	f^n	$\cdots$
1_A	1_A	f	$\cdots$	f^n	$\cdots$
f	f	f^2	$\cdots$	f^{n+1}	$\cdots$
$\vdots$	$\vdots$	$\vdots$	$\ddots$	$\vdots$	$\cdots$
f^m	f^{m+1}	$\cdots$	$\cdots$	f^{m+n}	$\cdots$
$\vdots$	$\vdots$	$\vdots$	$\vdots$	$\vdots$	$\ddots$

This is a free monoid with generator f (Fig. 1.2).

Fig. 1.2 A free monoid

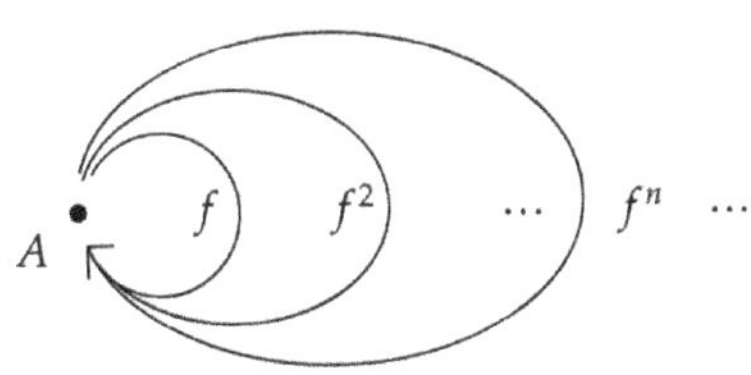

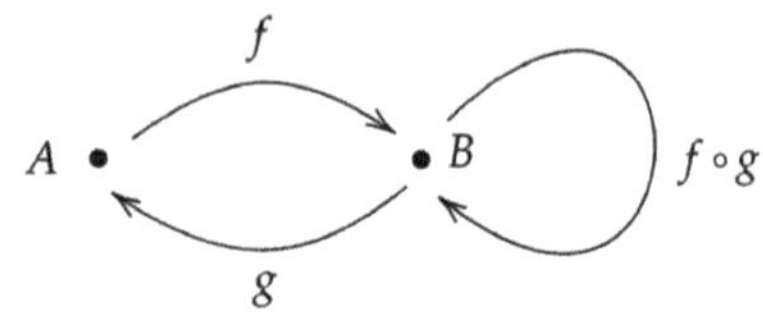

Fig. 1.3 A non-symmetric relation

Example 1.15 Let us consider a category with two objects A and B, and with two morphisms $f : A \to B$ and $g : B \to A$ other than identity morphisms. f and g generate all the morphisms. We say that $\{f, g\}$ is the *system of generators*.

We impose one relation on this category: $g \circ f = 1_A$. If we examine the table carefully, we see that the table is closed under the binary operation of composition. The empty slots mean that the case is not composable. To establish associativity, we must check in all the cases of triples (a, b, c), where a, b, c are morphisms, that $(a \circ b) \circ c = a \circ (b \circ c)$ holds.

	1_A	1_B	f	g	$f \circ g$
1_A	1_A			g	
1_B		1_B	f		$f \circ g$
f	f			$f \circ g$	
g		g	1_A		g
$f \circ g$		$f \circ g$	f		$f \circ g$

Examples 1.13–1.15 are narrowly tractable with pencil and paper. We will provide Haskell programs for checking the consistency of a table to be that of a category.

We can construct a new category out of two categories in various ways. The most important one is the product category.

Definition 1.5 Let $\mathscr{A}$ and $\mathscr{B}$ be categories. The product category $\mathscr{A} \times \mathscr{B}$ is given by:

- Objects are all pairs (A, B) where $A \in \mathrm{Obj}(\mathscr{A})$ and $B \in \mathrm{Obj}(\mathscr{B})$.
- Morphisms are all pairs (f, g) where $f : A_1 \to A_2 \in \mathscr{A}(A_1, A_2)$ and $g : B_1 \to B_2 \in \mathscr{B}(B_1, B_2)$.

1.4 Functor and Contravariant Functor

Given two linear spaces, we naturally compare their structures via linear mappings. Given two groups, we compare their group structures via group homomorphisms. Functors are homomorphisms between two categories.

Definition 1.6 The data for a category consists of objects, morphisms, table of morphism composition. A *functor* from category $\mathscr{A}$ to $\mathscr{B}$ is a collection of functions that is united to be one big function as follows:

- A function $F_0 : \mathrm{Obj}(\mathscr{A}) \to \mathrm{Obj}(\mathscr{B})$ called a *function on objects*,
- A collection of functions

$$F_1(A_1, A_2) : \mathscr{A}(A_1, A_2) \to \mathscr{B}(F_0(A_1), F_0(A_2))$$

  for all the pair $A_1, A_2 \in \mathrm{Obj}(\mathscr{A})$. Since there are no overlapping of domains of definition, we can collectively call the collection as a *function on morphisms.*

From now on, we omit suffixes and parentheses in $F_0(A)$ and write as FA. We omit suffix, object parameters, and parentheses in $F_1(A_1, A_2)(f)$ and write as Ff.

We impose the following conditions on the function of morphisms:

- For all $A \in \mathrm{Obj}(\mathscr{A})$, equality $F(1_A) = 1_{F(A)}$ holds.
- For all pairs of morphisms $f : A_1 \to A_2$ and $g : A_2 \to A_3$, equality $F(g \circ f) = F(g) \circ F(f)$ holds.

Figure 1.4 may convey the feeling of this formality.

Example 1.16 Homology functors H_n from **Top** to **Ab** return a group homomorphism $H_n(f) : H_n(X) \to H_n(Y)$ for a given continuous function $f : X \to Y$.

Example 1.17 Let X and Y be posets and $f : X \to Y$ be order preserving function. We say f is order preserving if $f(x_1) \leqq f(x_2)$ for all pairs $x_1 \leqq x_2$ in X. When we regard X and Y as categories, f is a functor.

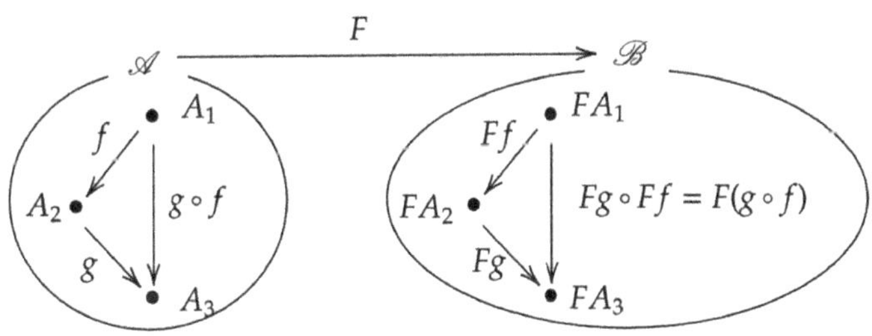

Fig. 1.4 Morphism composition is preserved

Example 1.18 Figure 1.5 depicts a functor between posets. When we draw a graph representation of a category, we often omit obvious transitive closure and identities. The reader should check if this example satisfies all the axioms.

Example 1.19 Let us examine the List functor in Hask. Given a type `A` in Haskell, we can construct a type `[ ] A` or `[A]`. Both are syntactically correct and have the same meaning. `[ ]` is called a type constructor. Figure 1.6 shows that `map` plays the role of functor's function on morphisms. `map` satisfies the condition:

```
map (g . f) = map g . map f .
```

Example 1.20 Let us examine the Maybe functor in Hask. Given a type `A` in Haskell, we can construct a type `Maybe A`. `Maybe` is a type constructor as well as a List. Figure 1.7 shows that `fmap` plays the role of functor's function on morphisms. `map` satisfies the condition:

```
fmap (g . f) = fmap g . fmap f .
```

Function `fmap` is shared by all the functor instances in Haskell. Programmers do not explicitly write which functor is intended. In most cases, the compiler precisely infers which one to take. If this is not the case, programmers must add an annotation to `fmap`.

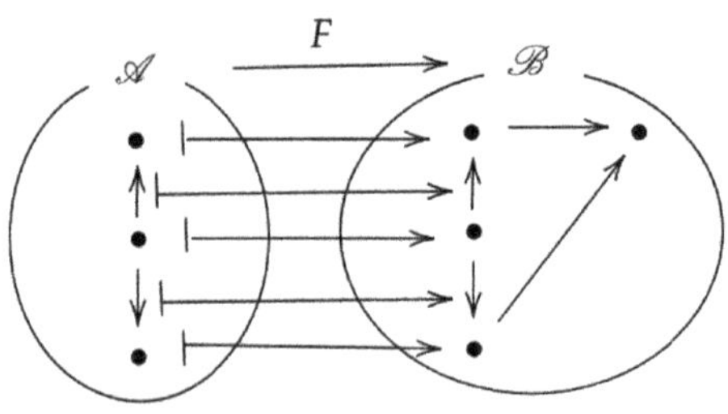

Fig. 1.5 A functor between posets

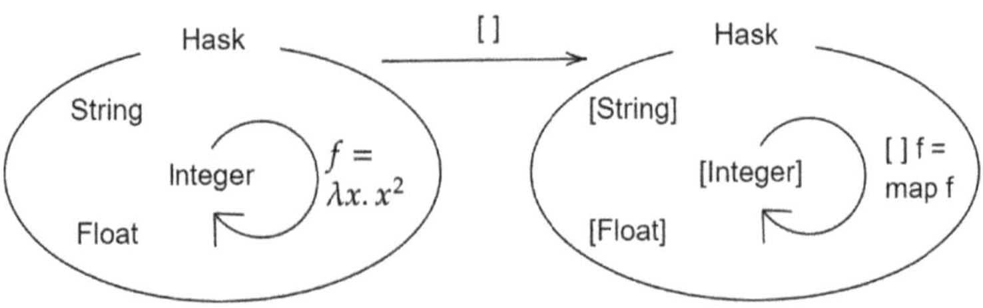

Fig. 1.6 List functor

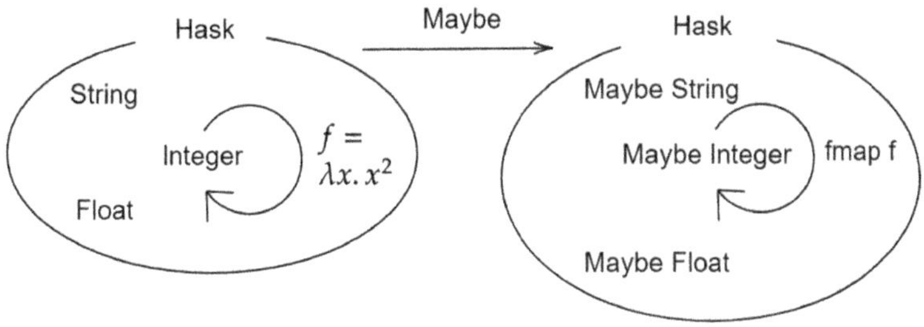

Fig. 1.7 Maybe functor

Example 1.21 Tree structures are naturally seen as an instance of functor. Let us consider binary trees.

Listing 1.1 Tree.hs

```
module Tree where
import Data.Char

data Tree a = Empty | Node a (Tree a) (Tree a)

instance (Show a) => Show (Tree a) where
  show x = show1 0 x

show1 :: Show a => Int -> (Tree a) -> String
show1 n Empty = ""
show1 n (Node x t1 t2) =
  show1 (n+1) t2 ++
  indent n ++ show x ++ "\n" ++
  show1 (n+1) t1

indent :: Int -> String
indent n = replicate (n*4) ' '

instance Functor Tree where
  fmap f Empty = Empty
  fmap f (Node x t1 t2) =
    Node (f x) (fmap f t1) (fmap f t2)

-- test data
tree2 = Node "two" (Node "three" Empty Empty)
                   (Node "four" Empty Empty)
tree3 = Node "five" (Node "six" Empty Empty)
                    (Node "seven" Empty Empty)
tree1 = Node "one" tree2 tree3

string2int :: String -> Int
string2int "one"   = 1
string2int "two"   = 2
string2int "three" = 3
string2int "four"  = 4
string2int "five"  = 5
```

```
37 string2int "six"   = 6
38 string2int "seven" = 7
39 string2int _       = 0
40
41 tree0 = fmap string2int tree1
42
43 {- suggested tests
44 fmap (map toUpper) tree1
45 fmap string2int tree1
46 fmap length tree1
47 -}
```

Line 4 defines the binary tree data structures. Lines 6–17 implement an instance of the Show type class. Lines 19–22 implement a functor instance for the binary tree.

Some tests are shown in Fig. 1.8.

Definition 1.7 A functor from category $\mathscr{A}^{\mathrm{op}}$ to $\mathscr{B}$ is often called a *contravariant functor*. When we want to distinguish functors that we saw so far from contravariant functors, we use the term *covariant functor*.

Example 1.22 Figure 1.9 illustrates a contravariant functor. If we redraw Fig. 1.9 by replacing $\mathscr{A}^{\mathrm{op}}$ with $\mathscr{A}$ while keeping the correspondence, we get Fig. 1.10.

```
*Tree> tree1
        "seven"
    "five"
        "six"
"one"
        "four"
    "two"
        "three"
```

```
*Tree> fmap (map toUpper) tree1
        "SEVEN"
    "FIVE"
        "SIX"
"ONE"
        "FOUR"
    "TWO"
        "THREE"
```

```
*Tree> fmap string2int tree1
        7
    5
        6
1
        4
    2
        3
```

```
*Tree> fmap length tree1
        5
    4
        3
3
        4
    3
        5
```

Fig. 1.8 Binary tree as a functor

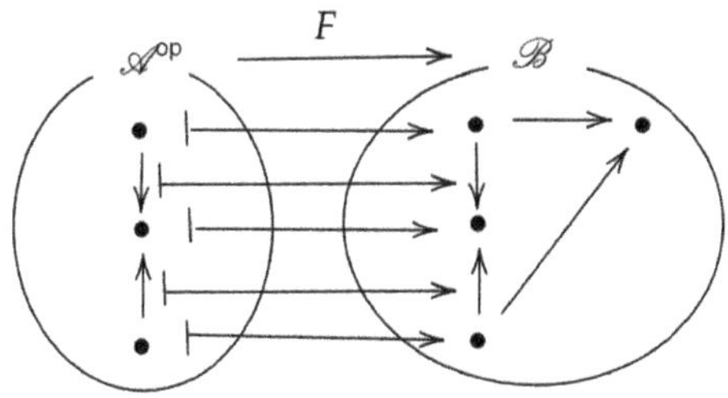

Fig. 1.9 Contravariant functor

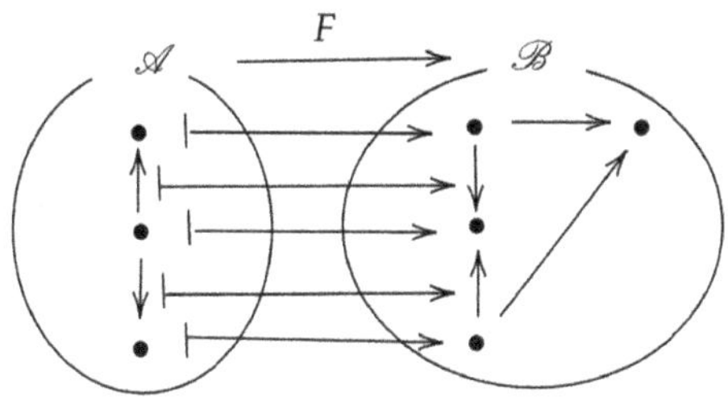

Fig. 1.10 Arrows are reversed

Example 1.23 Functor `(->)` has two variables. If we fix the first argument, we obtain a covariant functor. If we fix the second, we obtain a contravariant one. These are called Hom functors. We will discuss this in detail in Chap. 7.

In the following sample code, we do not import an extra library. We use "do it ourselves" version of `Contra` for our purpose. Note that functor `( , )` has two arguments. A functor instance for the first variable fixing the second is already defined in Prelude. We have to add an instance for the second variable by ourselves.

Listing 1.2 homfunctors.hs

```
{- defined in default startup environment
class Functor f where
  fmap :: (a -> b) -> f a -> f b

instance Functor ((->) r) where
  fmap = (.)

instance Functor ((,) a) where
  fmap f (x,y) = (x, f y)
-}

class Contra f where
  pamf ::  (a -> b) -> f b -> f a

newtype Moh b a = Moh {getHom :: a -> b}

instance Contra (Moh b) where
  pamf f (Moh g) = Moh (g . f)

newtype Riap b a = Riap {getPair :: (a,b)}

instance Functor (Riap b) where
  fmap f (Riap (x,y)) =  Riap (f x,y)
```

Lines 1–10 are excerpts from the standard library source code in the comment block. Lines 12–13 define a type class `Contra` with one class method `pamf` that corresponds to `fmap` for the `Functor` class. Line 15 introduces a wrapping `Moh` for the exchange of variables. Lines 17–18 declare an instance of `Contra` for `Moh`. Tests for `fmap` and `pamf` can be done as follows.

```
*Main> fmap (\x -> x*x) (\x -> x + 1) 10 == 121
True
*Main> getHom (pamf (\x -> x*x) (Moh (\x -> x + 1))) 10 == 101
True
```

Line 20 defines another wrapping for the exchange of variables. This way we can declare a `Functor` instance fixing the second variable. Tests can be done as follows.

```
*Main> fmap (\x -> x*x) (10,10) == (10,100)
True
*Main> getPair (fmap (\x -> x*x) (Riap (10,10))) == (100,10)
True
```

1.5 Faithful Functors and Full Functors

Like the concepts of injection and surjection for functions between sets, we introduce the concepts of faithful and full functors.

Definition 1.8 Let $F : \mathscr{A} \to \mathscr{B}$ be a functor. If its function on morphisms

$$F : \mathscr{A}(A_1, A_2) \to \mathscr{B}(F(A_1), F(A_2))$$

is injective for all pairs $A_1, A_2 \in \mathrm{Obj}(\mathscr{A})$, we say F is *faithful*. If its function on morphisms is surjective for all pairs $A_1, A_2 \in \mathrm{Obj}(\mathscr{A})$, we say F is *full*. If its function on morphisms is bijective for all pairs $A_1, A_2 \in \mathrm{Obj}(\mathscr{A})$, we say F is *full and faithful*.

Example 1.24 Let us consider a functor $F : \mathscr{A} \to \mathscr{B}$ as in Fig. 1.11. We write down functions on objects and on morphisms as follows.

$$F(A_1) = B_1, \quad F(A_2) = B_2$$
$$F(a_1) = b_1, \quad F(a_2) = b_2$$

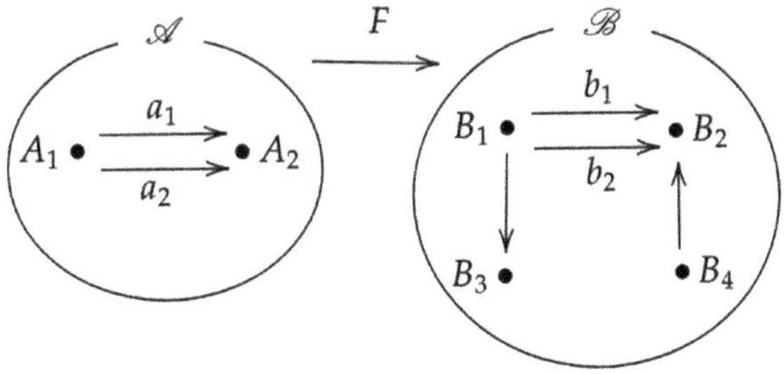

Fig. 1.11 Full and faithful functor

Notice that a functor can be full and faithful, even though its function on objects is not surjective.

Example 1.25 Let us consider a functor $F : \mathscr{A} \to \mathscr{B}$ as in Fig. 1.12. We write down functions on objects and on morphisms as follows.

$$F(A_1) = B_1, \quad F(A_2) = B_2$$
$$F(a_1) = b_1, \quad F(a_2) = b_2$$

Then, we see that F is faithful but not full since its function on morphisms is not surjective.

Example 1.26 Let us consider a functor $F : \mathscr{A} \to \mathscr{B}$ as in Fig. 1.13. We write down functions on objects and on morphisms as follows.

$$F(A_1) = B_1, \quad F(A_2) = B_2$$
$$F(a_1) = b_1, \quad F(a_2) = b_1$$

Then, we see that F is full but not faithful since its function on morphisms is not injective.

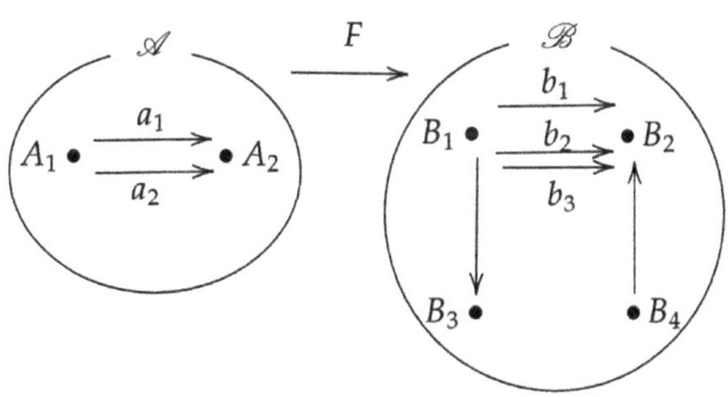

Fig. 1.12 Faithful but not full

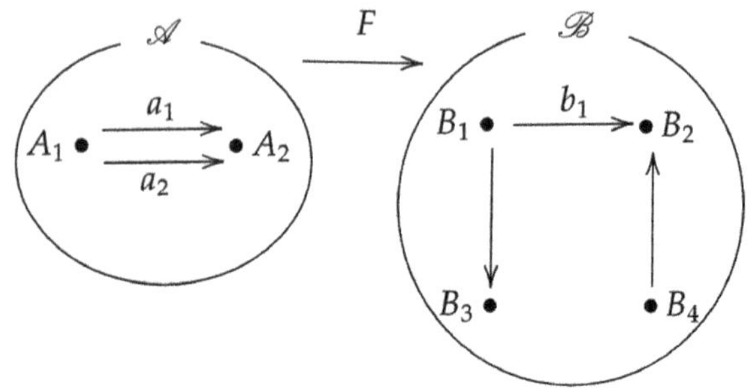

Fig. 1.13 Full but not faithful

Example 1.27 Let us consider a functor $F : \mathscr{A} \to \mathscr{B}$ as in Fig. 1.14. We write down functions on objects and on morphisms as follows.

$$F(A_1) = F(A_3) = B_1, \quad F(A_2) = F(A_4) = B_2$$
$$F(a_1) = b_1, \quad F(a_2) = b_1$$

Then, we see that F is faithful even though its function on morphisms is not injective.

We will see some examples in the wild.

Example 1.28 Let us consider an embedding functor $F : \mathbf{Ab} \to \mathbf{Grp}$. We can easily check that F is full and faithful.

Example 1.29 Let us consider a forgetful functor $F : \mathbf{Top} \to \mathbf{Set}$. Its function on objects sends a topological space to its underlying set. Its function on morphisms sends a continuous function to a function between underlying spaces. This functor is faithful but not full.

Example 1.30 A complex number consists of its real part and imaginary part. Thus, an n-dimensional complex vector space can be seen as a $2n$-dimensional real vector space.

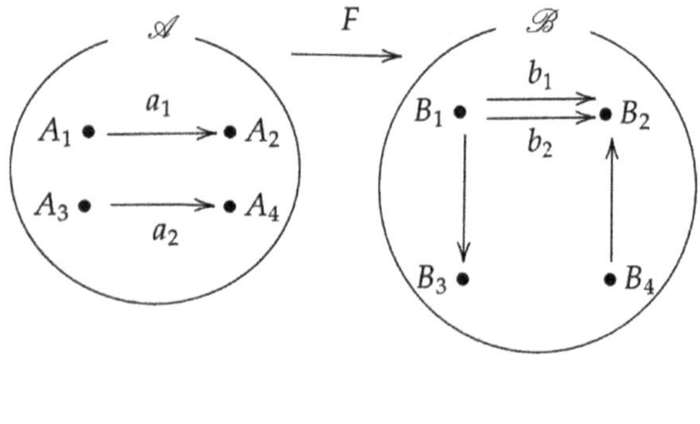

Fig. 1.14 Faithful but non-injective function on morphisms

We consider a functor $F : \text{Vect}_\mathbb{C} \to \text{Vect}_\mathbb{R}$. Let V be an object in $\text{Vect}_\mathbb{C}$, namely a complex n-dimensional vector space. Let FV be a real $2n$-dimensional vector space sharing the same underlying additive group. The difference is their scalar multiplication. So much for the function of objects.

As to the function of morphisms, a complex linear mapping can be considered as a linear mapping restricting scalar multiplication from complex to real.

Conversely, an arbitrary real linear mapping is not necessarily a complex linear mapping. So, we have

$$\text{Vect}_\mathbb{C}(V_1, V_2) \subsetneq \text{Vect}_\mathbb{R}(FV_1, FV_2).$$

Let us employ matrix representations with a fixed basis. Then, we can identify $V = \mathbb{C}^n$ and $FV = \mathbb{R}^{2n}$. Then, morphisms in $\text{Vect}_\mathbb{C}$ are represented as an $n \times n$ matrix; the corresponding morphisms in $\text{Vect}_\mathbb{R}$ are represented as an $2n \times 2n$ matrix. For example,

$$\begin{pmatrix} \sqrt{-1} & 0 \\ 0 & 10 \end{pmatrix} \begin{pmatrix} x + yi \\ u + vi \end{pmatrix}$$

is rewritten as

$$\begin{pmatrix} 0 & -1 & & \\ 1 & 0 & & \\ & & 10 & 0 \\ & & 0 & 10 \end{pmatrix} \begin{pmatrix} x \\ y \\ u \\ v \end{pmatrix}.$$

The function of morphisms acts like this:

$$\begin{pmatrix} \sqrt{-1} & 0 \\ 0 & 10 \end{pmatrix} \mapsto \begin{pmatrix} 0 & -1 & & \\ 1 & 0 & & \\ & & 10 & 0 \\ & & 0 & 10 \end{pmatrix},$$

where empty slots should be filled with 0s.

We conclude that functor F is faithful but not full.

1.6 Natural Transformations

A natural transformation is, in a sense, a morphism between functors. We will give an exact meaning for this idea in the following chapter.

Definition 1.9 Let F and G are functors from category $\mathscr{A}$ to $\mathscr{B}$. Let $\{\alpha_A : FA \to GA\}_{A \in \text{Obj}(\mathscr{A})}$ be a collection of morphisms in $\mathscr{B}$ satisfying the following condition: for any morphism $f : A_1 \to A_2$ in $\mathscr{A}$, the square to the right in Fig. 1.15 commutes. Namely,

$$\alpha_{A_2} \circ Ff = Gf \circ \alpha_{A_1}.$$

Then, we call the collection $\{\alpha_A\}$ a *natural transformation* $\alpha : F \to G$.

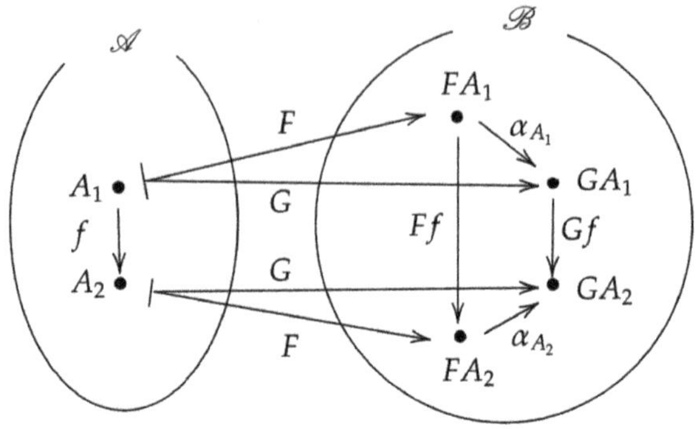

Fig. 1.15 Natural transformation $\alpha : F \to G$

A functor is a kind of photograph of $\mathscr{A}$ recorded on $\mathscr{B}$. When there are two functors, we have two photographs of $\mathscr{A}$ on $\mathscr{B}$. So, we can compare the two images on the film or CCD (that is, $\mathscr{B}$). This is just what natural transformations do.

A natural transformation $\alpha : F \to G$ is often drawn as in the diagram:

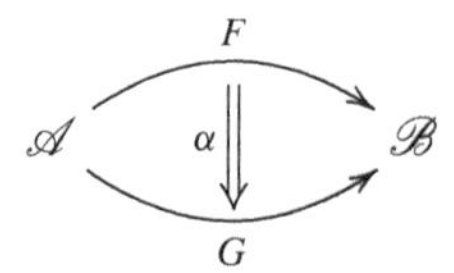

Example 1.31 Let us consider categories $\mathscr{A}$ and $\mathscr{B}$ in Fig. 1.16. $\mathscr{A}$ has three objects and two morphisms p and q other than the identities. $\mathscr{B}$ has nine objects and nine morphisms other than the identities and two diagonals obtained by transitivity, where we suppose the two squares are commutative.

F, G, H are functors from $\mathscr{A}$ to $\mathscr{B}$. There is a natural transformation $\alpha : F \to G$, but no natural transformation from $F \to H$. The reader may feel that natural transformations are kind of a homotopy. This idea is established in higher category theory, which is out of the scope of this exposition.

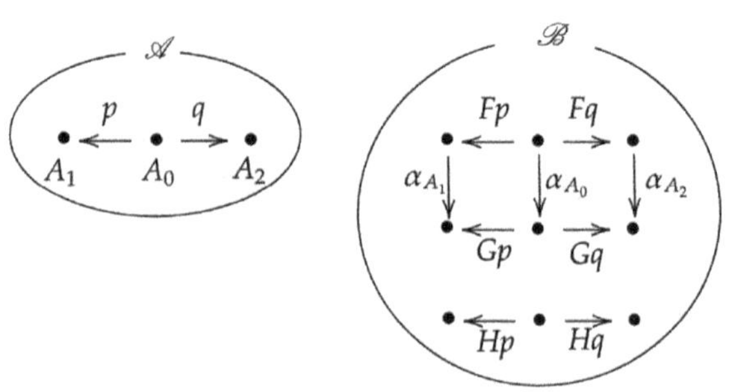

Fig. 1.16 Natural transformation and homotopy

Example 1.32 Let us consider a natural transformation `concat` from `[ ][ ]` to `[ ]`. As a Haskell function, `concat` flattens a list of lists into a single list.

```
Prelude> concat [[1,2,3],[4,5],[6,7,8]]
[1,2,3,4,5,6,7,8]
```

Figure 1.17 illustrates the naturality of `concat`. The diagram in Fig. 1.17 can be tested as follows.

```
*Main> :{
*Main| map length
*Main|     (concat [[[1,2,3]],[[4,5],[]],[[6],[7],[8]]])
*Main| :}
[3,2,0,1,1,1]
*Main> :{
*Main| concat ((map.map) length
*Main|                        [[[1,2,3]],[[4,5],[]],[[6],[7],[8]]])
*Main| :}
[3,2,0,1,1,1]
```

Example 1.33 We define a natural transformation `safehead` from the List functor to the Maybe functor as follows.

Listing 1.3 safehead.hs

```
safehead :: [] a -> Maybe a
safehead []     = Nothing
safehead (x:xs) = Just x
```

Fig. 1.17 Natural transformation `concat`

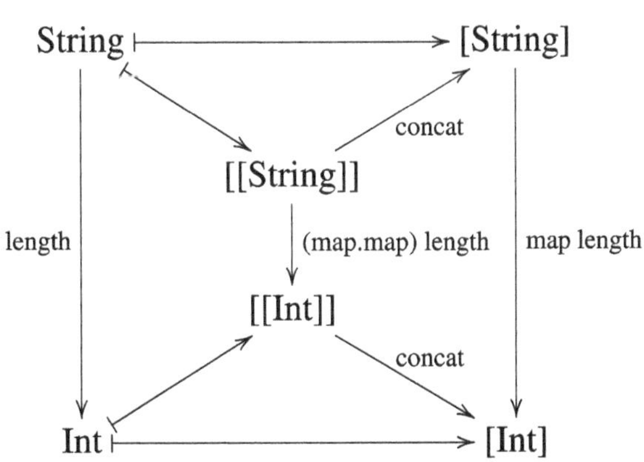

We can test like:

```
*Main> safehead []
Nothing
*Main> safehead [1..10]
Just 1
```

Figure 1.18 illustrates the naturality of `safehead`. We can test the naturality of `safehead` like this.

```
*Main> :{
*Main| fmap length
*Main|      (safehead ["hello", "Bon jour",
                        "Guten Tag", "Buenos Diaz"])
*Main| :}
Just 5
*Main> :{
*Main| safehead (map length
*Main|                 ["hello", "Bon jour",
                        "Guten Tag", "Buenos Diaz"])
*Main| :}
Just 5
*Main> fmap length (safehead [])
Nothing
*Main> safehead (map length [])
Nothing
```

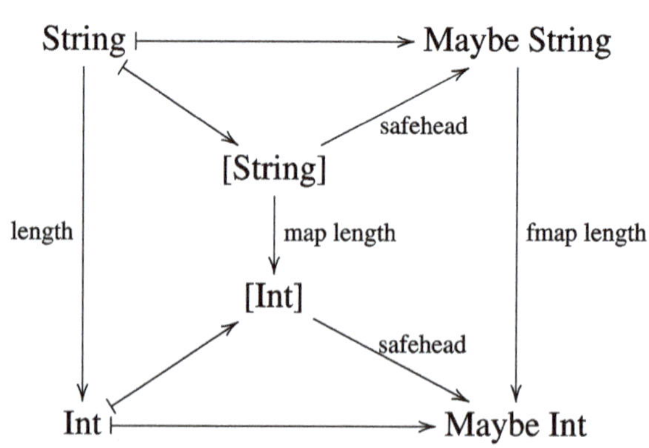

Fig. 1.18 Natural transformation from List to Maybe

Definition 1.10 A functor that maps all the objects in the source to a designated object in the target, mapping all the morphisms to the identity, is called a *constant functor*.

At first sight, such a functor is too trivial a thing. However, we will see its great roles in various situations.

Example 1.34 The length function can be seen as a constant functor. We make it clear using a home-made length, `mylength`.

Listing 1.4 mylength.hs

```
import Data.Functor.Const

mylength :: [b] -> Const Int b
mylength []     = Const 0
mylength (x:xs) = Const (1 + getConst (mylength xs))
```

We can test `mylength` like

```
*Main> mylength [1..100]
Const 100
*Main> mylength []
Const 0
*Main> :t mylength
mylength :: [b] -> Const Int b
```

The `Data.Functor.Const` module is defined like the following.

```
...
-- | The 'Const' functor.
newtype Const a b = Const { getConst :: a }
    deriving ...
...
```

In code listing 1.4, `Const Int` is a functor. It sends any Hask object, namely any type `b`, to `Int`. Note that `Const` itself is not a functor, but `Const Int` is. See Fig. 1.19 carefully. When we test `Const Int f` as in Fig. 1.20, we have to rewrite as `fmap f` in actual Haskell interaction. In Fig. 1.20, `f` stands for `\xs -> xs ++ xs`.

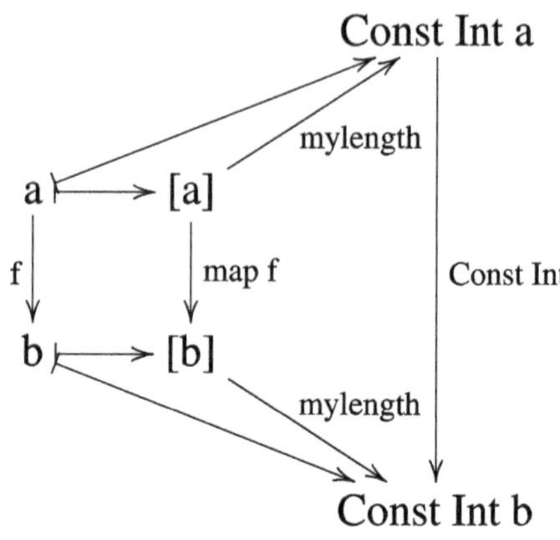

Fig. 1.19 Natural transformation from List to Const Int

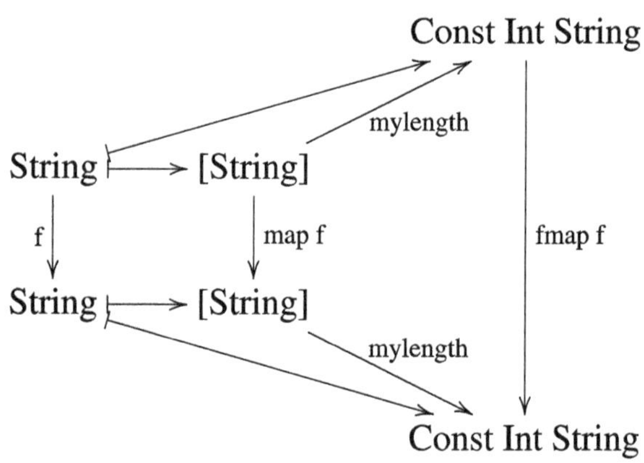

Fig. 1.20 Tests for const

```
*Main> :{
*Main| mylength . map (\xs -> xs ++ xs) $
*Main|   ["hello", "Bon jour", "Guten Tag", "Buenos Diaz"]
*Main| :}
Const 4
*Main> :{
*Main| fmap (\xs -> xs ++ xs) . mylength $
*Main|   ["hello", "Bon jour", "Guten Tag", "Buenos Diaz"]
*Main| :}
Const 4
```

Given functors $\alpha : F \to G$ and $\beta : G \to H$, we can compose these functors to obtain a functor $\gamma : F \to H$. Let us define componentwise as follows.

$$\gamma_A = \beta_A \circ \alpha_A$$

We often describe this composition with a diagram

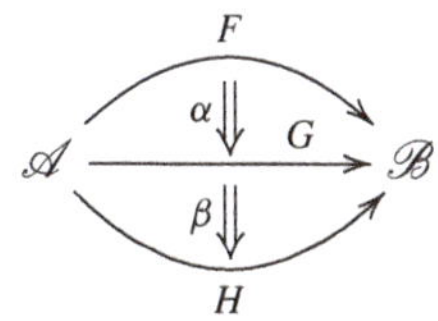

We say γ is a *vertical composition* of α and β. At this point, however, it is not obvious that γ defined componentwise as above is really a functor. To see that this is the case, we only have to check if the outer square in the following diagram commutes for arbitrary $f : A_1 \to A_2$.

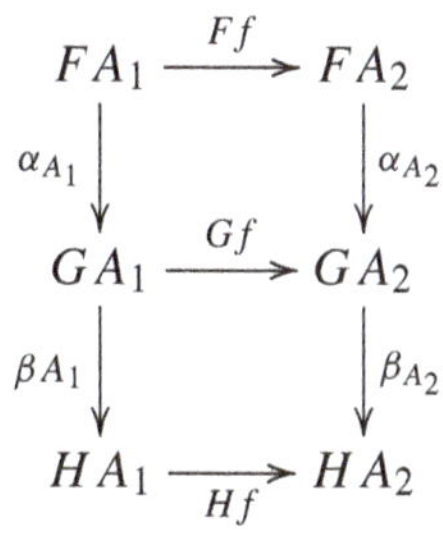

The two small rectangles commute since F and G are functors. Therefore, the outer square commutes. So much for the proof, but if you want the other way of presentation, the author suggests that you elaborate on a rewriting sequence. The goal is

$$Hf \circ \beta_{A_1} \circ \alpha_{A_1} = \beta_{A_2} \circ \alpha_{A_2} \circ Ff$$

Let us begin with LHS:

$$\begin{aligned} & Hf \circ \beta_{A_1} \circ \alpha_{A_1} \\ =& \beta_{A_2} \circ Gf \circ \alpha_{A_1} \quad (lower\ square) \\ =& \beta_{A_2} \circ \alpha_{A_2} \circ Ff \quad (upper\ square), \end{aligned}$$

thus we reached the goal.

Example 1.35 Let us try a vertical composition of natural transformations in Examples 1.32 and 1.33.

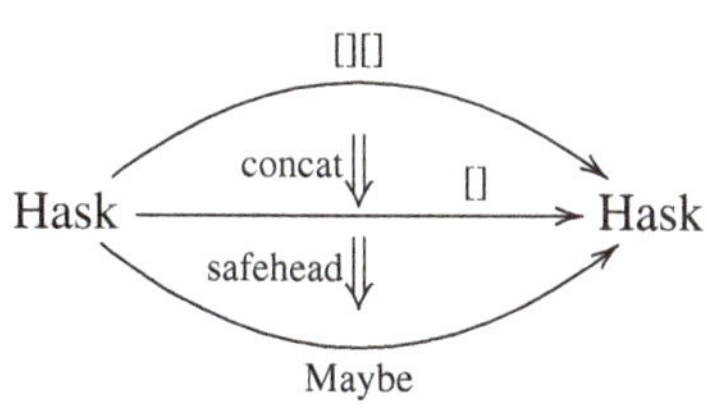

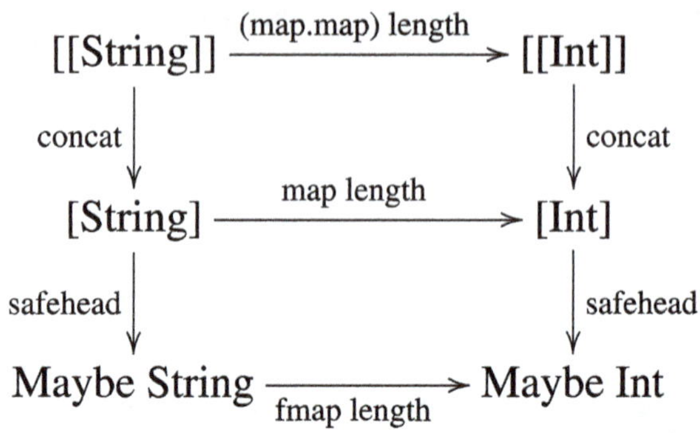

Fig. 1.21 Vertical composition of `concat` and `safehead`

To test this composition, let us take Hask objects `String` and `Int` and morphism `length`. We want to check if the outer square in Fig. 1.21 commutes. We do it as follows.

```
*Main> :{
*Main| safehead . concat . (map.map) length $
*Main| [["hello", "Bon jour"],["Guten Tag"],
         ["abc","def","gh"]]
*Main| :}
Just 5
*Main> :{
*Main| fmap length . safehead . concat $
*Main| [["hello", "Bon jour"],["Guten Tag"],
         ["abc","def","gh"]]
*Main| :}
Just 5
*Main> safehead . concat . (map.map) length $ [[]]
Nothing
*Main> fmap length . safehead . concat $ [[]]
Nothing
```

Functors are often compared to containers for some values. Natural transformations map containers to containers without seeing what's inside these containers.

Example 1.36 Let us study a program that flattens binary trees to plain lists.

Listing 1.5 Flatten.hs

```
module Flatten where
import Tree
import Data.Char
```

```

flatten :: Tree a -> [a]
flatten Empty = []
-- flatten (Node x t1 t2) = x:(flatten t1 ++ flatten t2)
flatten (Node x t1 t2) = flatten t1 ++ [x] ++ flatten t2

toupper = map toUpper

{- suggested tests; compare the results below
flatten . fmap toupper  $ tree1
map toupper . flatten $ tree1
-}
```

This program imports a module 1.1. Let us give it a try.

```
*Flatten> tree1
        "seven"
    "five"
        "six"
"one"
        "four"
    "two"
        "three"

*Flatten> flatten tree1
["three","two","four","one","six","five","seven"]
```

The flatten function is a natural transformation from the Tree functor to the List functor. The flatten function transforms the structure of container organizations, in this example, binary trees to lists. It behaves regardless of the contents in the containers. See Fig. 1.22 to get a clear understanding. The numbers in the figure are intended for tracing the container organization transform. To prove that `flatten` is really a natural transformation, we have to show that the square on the right in Fig. 1.23 commutes. Let us try a test by giving a and b concrete types.

```
*Main> flatten . fmap toupper  $ tree1
["ONE","TWO","THREE","FOUR","FIVE","SIX","SEVEN"]
*Main> map toupper . flatten $ tree1
["ONE","TWO","THREE","FOUR","FIVE","SIX","SEVEN"]
```

So much for the test for the following.

```
flatten . fmap toupper = map toupper . flatten
```

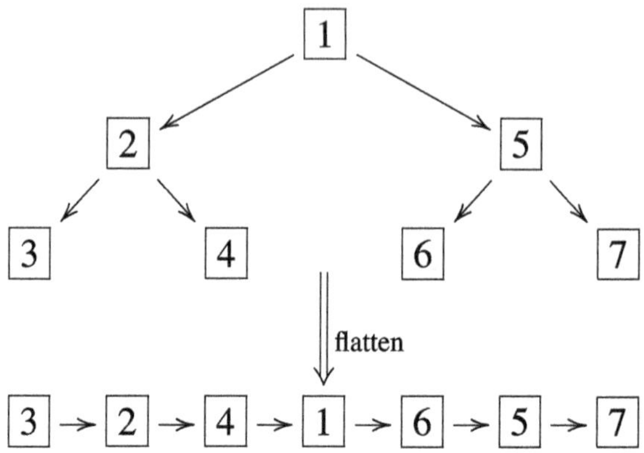

Fig. 1.22 Natural transformation from `Tree` to `List`

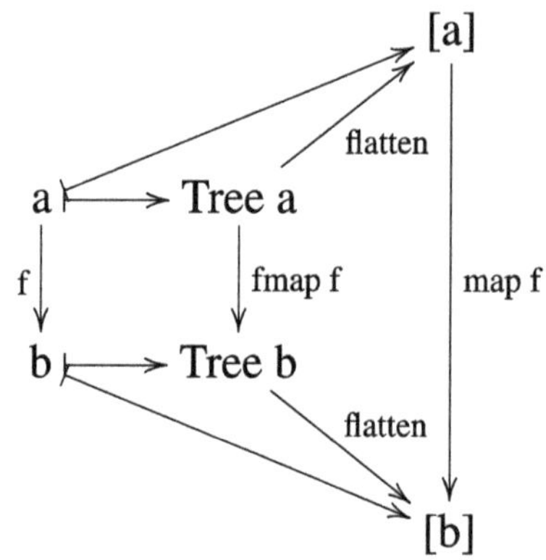

Fig. 1.23 Naturality of `flatten`

1.7 Subcategories of Hask

Polymorphic functions in Haskell that require *contexts* have signatures like

```
somefunction :: Someclass a => T1 a -> T2 a.
```

If type variable a is assigned a concrete type, then it is considered as an object of Hask. The context `Someclass a` requires that type variable a only allows the assignment of concrete types of a certain class.

Defining a type class in Haskell amounts to defining a subcategory of Hask. Both `String` and `Int` can be considered as objects in subcategory `Ord`. Function `string2int` in Example 1.1 is surely a morphism in Hask, but not a morphism in `Ord`.

Example 1.37 The sort function is a natural transformation from the List functor restricted to category Ord. The objects of Ord are Haskell types that belong to the type class Ord. The morphisms of Ord are functions that preserve order. If we can formulate a subcategory in Hask in this way, Ord is a faithful but not full subcategory of Hask.

Let us take an example of insertion sort.

Listing 1.6 sort.hs

```
import Tree
import Flatten

iSort :: Ord a => [a] -> [a]
iSort xs =
  foldr ins [] xs
    where
      ins x [] = [x]
      ins x (y:ys)
        | x <= y    = x:y:ys
        | otherwise = y: ins x ys

list2tree :: Ord a => [a] -> Tree a
list2tree xs =
  foldl sni Empty xs
    where
      sni = flip ins
      ins x Empty   = Node x Empty Empty
      ins x (Node y t1 t2)
        | x <= y    = Node y (ins x t1) t2
        | otherwise = Node y t1 (ins x t2)

iSort2 :: Ord a => [a] -> [a]
iSort2 = flatten . list2tree
```

Notice that `iSort` and `list2tree` have local functions `ins` with the same name, but they are completely different.

Look carefully at Fig. 1.24. `iSort` is a natural transformation from a functor Ord → Hask to itself.

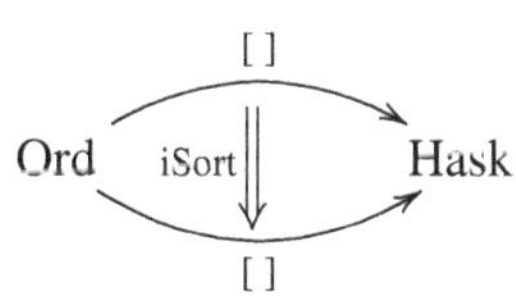

Naturality can be tested as follows.

```
*Main> map (\x -> x*2) . iSort $ [1,5,3,4,2]
[2,4,6,8,10]
*Main> iSort . map (\x -> x*2) $ [1,5,3,4,2]
[2,4,6,8,10]
```

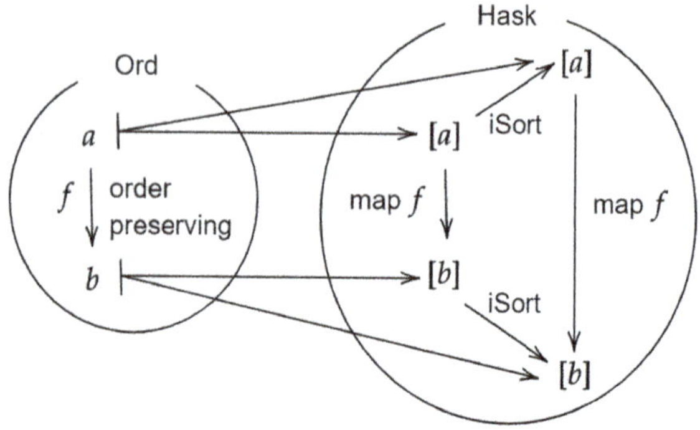

Fig. 1.24 Natural transformation iSort

We can bring any monotone functions in place of `\x -> x*2`.

Function `list2tree` is a natural transformation from a functor `[]` to Tree. See Fig. 1.25.

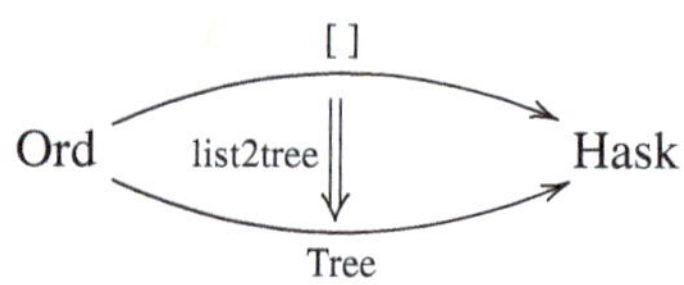

Naturality of `list2tree` can be tested in the same way.

Alternative version `iSort2` is a vertical composition of `list2tree` and `flatten`.

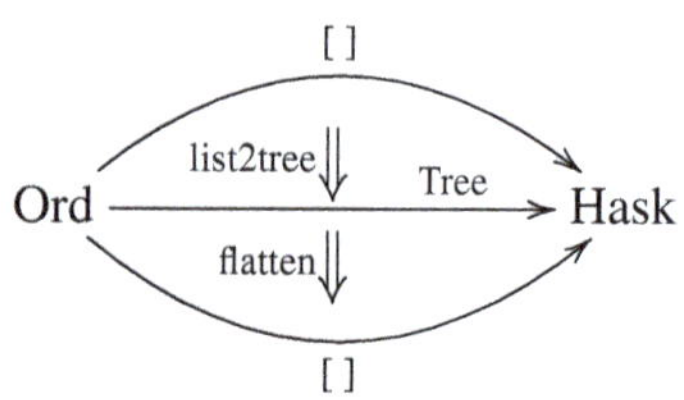

We can test this mapping monotone functions in the same way.

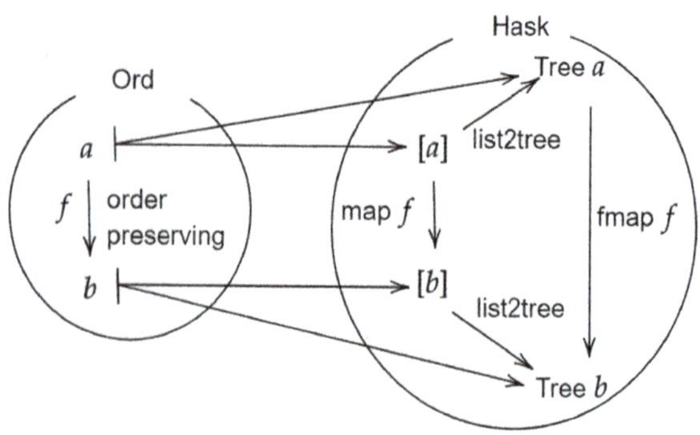

Fig. 1.25 Natural transformation list2tree

References

1. Mac Lane S (1998) Categories for the working mathematician, Graduate texts in mathematics, vol 5, 2nd edn. Springer
2. Haskell WH https://wiki.haskell.org/Hask

2 Equivalence of Categories

We introduce the concept of a functor category to better understand natural transformations. Then we study category equivalence. Most of us notice that intrinsic linear algebra, namely linear algebra independent of basis, and matrix linear algebra are essentially the same. Both can be formulated as a category of finite-dimensional linear spaces and linear transformations over a field such as $\mathbb{R}$, $\mathbb{C}$, and other variety of fields. Intrinsic linear algebras are far wider than matrix linear algebras. These two categories are clearly not isomorphic, yet we have a feeling that they are essentially the same. The concept of category equivalence is invented to tackle this dilemma.

2.1 Functor Category

We fix a small category $\mathscr{A}$ and locally small category $\mathscr{B}$. Let $[\mathscr{A}, \mathscr{B}]$ be a category with:

- the set of objects consisting of all the functors from $\mathscr{A}$ to $\mathscr{B}$ and
- the set of morphisms between two functors $F : \mathscr{A} \to \mathscr{B}$ and $G : \mathscr{A} \to \mathscr{B}$ consisting of all the natural transformations $F \to G$.

This category is called a functor category from $\mathscr{A}$ to $\mathscr{B}$.

Morphisms in a functor category, namely natural transformations, are composed componentwise, therefore associativity is automatic. The identity for a functor F is the identity natural transformation $1_F : F \to F$, where $(1_F)_A : F(A) \to F(A) = 1_{F(A)}$.

Remark 2.1 Category $\mathscr{A}$ is assumed to be small, and category $\mathscr{B}$ locally small, so that functor category $[\mathscr{A}, \mathscr{B}]$ stays locally small.

S. Yukita, *Category Theory Using Haskell*, Computer Science Foundations and Applied Logic, https://doi.org/10.1007/978-3-031-68538-5_2

Example 2.1 What we have seen in Example 1.32 can be summarized as follows. Both `List List` and `List` are objects of category [Hask, Hask]. `concat` is a morphism

$$\text{concat} \in [\text{Hask}, \text{Hask}](\text{List}^2, \text{List}).$$

What we have seen in Example 1.33 can be summarized as follows. `List` and `Maybe` are objects of category [Hask, Hask]. `safehead` is a morphism

$$\text{safehead} \in [\text{Hask}, \text{Hask}](\text{List}, \text{Maybe}).$$

What we have seen in Example 1.36 can be summarized as follows. `Tree` and `List` are objects of category [Hask, Hask]. `flatten` is a morphism

$$\text{flatten} \in [\text{Hask}, \text{Hask}](\text{Tree}, \text{List}).$$

Definition 2.1 An isomorphism in a functor category is called a *natural isomorphism.*

Example 2.2 The `mirror` function is a natural isomorphism from `Tree` to `Tree`.

Listing 2.1 mirror.hs

```haskell
import Tree

mirror :: Tree a -> Tree a
mirror Empty = Empty
mirror (Node x t1 t2) = Node x (mirror t2) (mirror t1)
```

If we run `mirror` as in Fig. 2.1, we can see `mirror` is an inverse to itself. We should check if the following two expressions evaluate the same result.

```haskell
mirror . fmap string2int $ tree1
fmap string2int . mirror $ tree1
```

```
*Main> tree1
        "seven"
    "five"
        "six"
"one"
        "four"
    "two"
        "three"
```

```
*Main> mirror tree1
        "three"
    "two"
        "four"
"one"
        "six"
    "five"
        "seven"
```

Fig. 2.1 Natural isomorphism from `Tree` to `Tree`

Let us draw a diagram that illustrates what's going on. `s2i` is short for `string2int`.

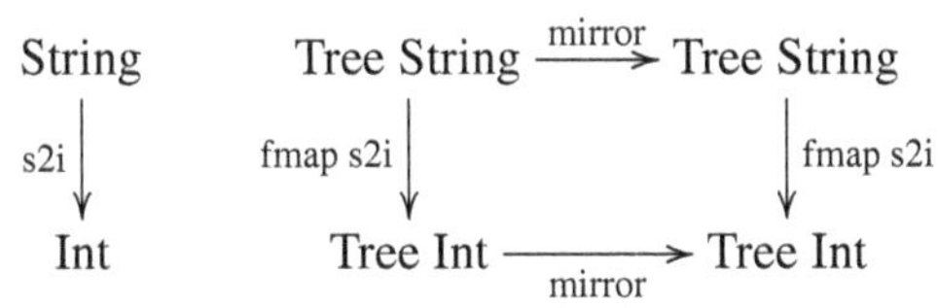

The square commutes, which means the naturality of `mirror`.

Example 2.3 We will implement a natural isomorphism from `Maybe` to `Either ()`. Functor `Either ()` is obtained by partially applying the unit type `()` to `Either`, which is in fact a functor with two arguments.

Listing 2.2 maybeEither.hs

```
1 alpha :: Maybe a -> Either () a
2 alpha Nothing  = Left ()
3 alpha (Just x) = Right x
4
5 beta :: Either () a -> Maybe a
6 beta (Left ()) = Nothing
7 beta (Right x) = Just x
8
9 {- suggested tests
10 beta . alpha $ Just 100
11 beta . alpha $ Nothing
12 alpha . beta $ Right 100
13 alpha . beta $ Left ()
14 -}
```

Performing tests suggested in lines 10-13 are left to the reader.

Lemma 2.1 *Natural transformation*

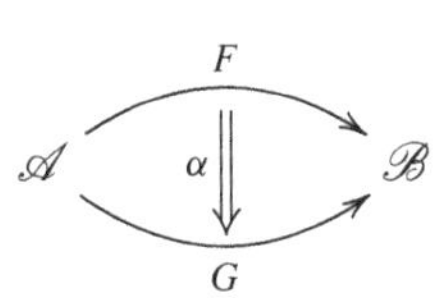

is a natural isomorphism if and only if $\alpha_A : F(A) \to G(A)$ *is an isomorphism in* $\mathscr{B}$ *for all objects* $A \in \mathrm{Obj}(\mathscr{A})$.

Remark 2.2 This lemma claims that natural isomorphism is equivalent to componentwise isomorphism. At first sight, it seems trivial, but proving it is not so trivial.

By definition, $\alpha : F \to G$ is a natural isomorphism if and only if there exists a natural transformation $\beta : G \to F$ such that

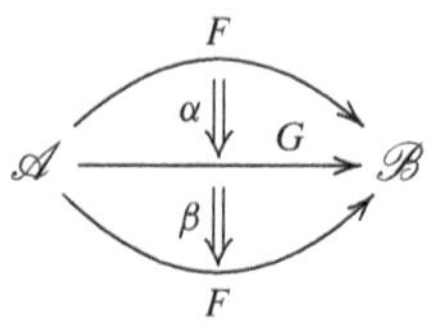

$$= 1_F$$

and

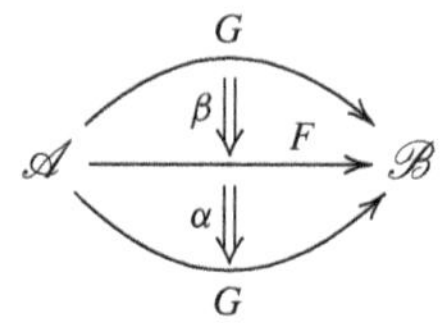

$$= 1_G .$$

Even when $\alpha_A : F(A) \to G(A)$ has an inverse $\beta_A : G(A) \to F(A)$ as morphisms in $\mathrm{Mor}(\mathscr{B})$ for all $A \in \mathrm{Obj}(\mathscr{A})$, we still do not know if the collection $\{\beta_A\}$ really forms a natural transformation.

Proof (Lemma 2.1)
Assume $\alpha : F \to G$ is a natural isomorphism. We will show that α_A is an isomorphism in $\mathscr{B}$ for all $A \in \mathrm{Obj}(\mathscr{A})$. Since $\beta \circ \alpha = 1_F$, we have $\beta_A \circ \alpha_A = 1_{F(A)}$ for all $A \in \mathrm{Obj}(\mathscr{A})$. Also, since $\alpha \circ \beta = 1_G$, we have $\alpha_A \circ \beta_A = 1_{G(A)}$ for all $A \in \mathrm{Obj}(\mathscr{A})$. Therefore, we have that α_A is an isomorphism in $\mathscr{B}$ for all $A \in \mathrm{Obj}(\mathscr{A})$.

Let us prove the other direction. Assume that a natural transformation $\alpha : F \to G$ has componentwise inverses $\beta_A : G(A) \to F(A)$, namely

$$\beta_A \circ \alpha_A = 1_{F(A)} \quad \text{and} \quad \alpha_A \circ \beta_A = 1_{G(A)}.$$

We will show that collection $\{\beta_A\}$ forms a natural transformation $\beta : G \to F$ and that $G \circ F = 1_F$ and $F \circ G = 1_G$. The goal is the naturality of β. Establishing naturality amounts to show the square in the following diagram commutes.

$$\begin{array}{ccccc} A_1 & & F(A_1) & \xleftarrow{\beta_{A_1}} & G(A_1) \\ f\downarrow & & Ff\downarrow & & \downarrow Gf \\ A_2 & & F(A_2) & \xleftarrow[\beta_{A_2}]{} & G(A_2) \end{array}$$

We add information about α to this diagram and get the following.

$$\begin{array}{ccccc} A_1 & & F(A_1) & \underset{\alpha_{A_1}}{\overset{\beta_{A_1}}{\rightleftarrows}} & G(A_1) \\ f\downarrow & & Ff\downarrow & & \downarrow Gf \\ A_2 & & F(A_2) & \underset{\beta_{A_2}}{\overset{\alpha_{A_2}}{\rightleftarrows}} & G(A_2) \end{array}$$

The assumption is the commutativity of the inner square. The assumption also says that anti-parallel pairs of morphisms are isomorphisms. The commutativity of the inner square is

$$\alpha_{A_2} \circ Ff = Gf \circ \alpha_{A_1}.$$

We compose β_{A_1} from right and β_{A_2} from left

$$\beta_{A_2} \circ \alpha_{A_2} \circ Ff \circ \beta_{A_1} = \beta_{A_2} \circ Gf \circ \alpha_{A_1} \circ \beta_{A_1}.$$

Cancelling the inverses, we have

$$Ff \circ \beta_{A_1} = \beta_{A_2} \circ Gf.$$

Now, we see that the outer square commutes. Thus, we established that β is a natural transformation. □

2.2 Equivalence of Categories

Given a pair of functors $F : \mathscr{A} \to \mathscr{A}'$ and $G : \mathscr{A}' \to \mathscr{A}$ with $GF = \mathrm{Id}_{\mathscr{A}}$ and $FG = \mathrm{Id}_{\mathscr{A}'}$, we say that $\mathscr{A}$ and $\mathscr{A}'$ are isomorphic categories. This is just what we can think of next to equality. But, unfortunately this idea yields little insight for a deeper understanding of the relations among categories. Instead, we will introduce the concept of *equivalence of categories* that is coarser than isomorphism. We will see that equivalence of categories is closely related to the concept of adjoint in Chap. 5.

Definition 2.2 We say categories $\mathscr{A}$ and $\mathscr{A}'$ are *equivalent* if there are functors $F : \mathscr{A} \to \mathscr{A}'$ and $G : \mathscr{A}' \to \mathscr{A}$, and natural isomorphisms $\eta : \mathrm{Id}_{\mathscr{A}} \to GF$ and $\varepsilon : FG \to \mathrm{Id}_{\mathscr{A}'}$.

Is so defined equivalence of categories really an equivalence relation among categories? At first sight, this relation looks non-symmetric, which is fatal to be equivalence. Further, we doubt at present that the relation is reflexive and transitive. We will show that the equivalence of categories is an equivalence relation later.

Let us first take an example of linear algebra.

Example 2.4 Let $\mathrm{Vect}_{\mathbb{C}}$ be a category of finite-dimensional complex vector spaces. Let $\mathbb{C}^n$ be an n-dimensional complex vector space consisting of column vectors. Let $\mathrm{Col}_{\mathbb{C}}$ be a category with the set of objects $\{\mathbb{C}^n\}_{n\in\mathbb{N}}$ and the set of morphisms consisting of all $m \times n$-complex matrices, where morphisms from $\mathbb{C}^n \to \mathbb{C}^m$ are matrix multiplication from the left. Note that $\mathbb{N}$ is the set of natural numbers, namely $\{n \in \mathbb{Z} \mid n \geq 0\}$.

We define a functor $F : \mathrm{Vect}_\mathbb{C} \to \mathrm{Col}_\mathbb{C}$ as follows:

- We first choose a particular basis for each object V in $\mathrm{Vect}_\mathbb{C}$. If you are familiar with the axiom of choice, you may wonder if this kind of activities are possible. Yes, we adopt the axiom.
- FV consists of column vectors that are obtained from vectors in V by taking components with respect to the designated basis of V.
- Let $T : V_1 \to V_2$ be a complex linear transformation. $FT : \mathbb{C}^n \to \mathbb{C}^m$ is a $m \times n$-complex matrix that represents T with respect to the designated bases.

F introduced in this way is clearly a functor. We skip the proof of functoriality.

Next, we consider an obvious embedding functor $G : \mathrm{Col}_\mathbb{C} \to \mathrm{Vect}_\mathbb{C}$ and proceed to the construction of natural transformations. The composition $G \circ F : \mathrm{Vect}_\mathbb{C} \to \mathrm{Vect}_\mathbb{C}$ is a full and faithful functor. The function on objects is not injective, but the function on morphisms is bijective for all pairs of objects in $\mathrm{Vect}_\mathbb{C}$. We construct a natural transformation $\eta : G \circ F \to \mathrm{Id}_{\mathrm{Vect}_\mathbb{C}}$, and show η is a natural isomorphism. We have to show that the square commutes in the following diagram.

$$\begin{array}{ccccc} V_1 & & \mathrm{Id}(V_1) & \xrightarrow{\eta_{V_1}} & G(F(V_1)) \\ {\scriptstyle f}\downarrow & & {\scriptstyle f}\downarrow & & \downarrow{\scriptstyle G(F(f))} \\ V_1 & & \mathrm{Id}(V_2) & \xrightarrow[\eta_{V_2}]{} & G(F(V_2)) \end{array}$$

Functor G is just an embedding and does nothing significant on objects and morphisms. Therefore, commutativity depends solely on F, which is a functor that promises commutativity. Since η_{V_1} and η_{V_2} are isomorphisms, $\eta : \mathrm{Id}_{\mathrm{Vect}_\mathbb{C}} \to G \circ F$ is a natural isomorphism.

Another direction is quite easy. Since $F \circ G : \mathrm{Col}_\mathbb{C} \to \mathrm{Col}_\mathbb{C}$ is already an identity functor, $\varepsilon : F \circ G \to \mathrm{Id}_{\mathrm{Col}_\mathbb{C}}$ is already there.

We got an example in which a large category is equivalent to a small category.

Theorem 2.1 *The equivalence of categories is an equivalence relation.*

Proof (1) We first show reflexivity: $\mathscr{A} \simeq \mathscr{A}$.

We set $F = G = \mathrm{Id}_\mathscr{A} : \mathscr{A} \to \mathscr{A}$, and $\eta = GF = \varepsilon = FG : \mathrm{Id}_\mathscr{A} \to \mathrm{Id}_\mathscr{A}$.

(2) We show symmetricity: $\mathscr{A} \simeq \mathscr{A}' \Rightarrow \mathscr{A}' \simeq \mathscr{A}$.

Let $\eta : \mathrm{Id}_\mathscr{A} \to GF$ and $\varepsilon : FG \to \mathrm{Id}_{\mathscr{A}'}$ be natural isomorphisms between two functors $F : \mathscr{A} \to \mathscr{A}'$ and $G : \mathscr{A}' \to \mathscr{A}$.

If we set $F' = G$, $G' = F$, $\eta' = \varepsilon$, and $\varepsilon' = \eta$, we have equivalence of categories $\mathscr{A}' \simeq \mathscr{A}$ with functors $F' : \mathscr{A}' \to \mathscr{A}$ and $G' : \mathscr{A} \to \mathscr{A}'$, and with natural isomorphisms $\eta' : \mathrm{Id}_{\mathscr{A}'} \to G'F'$ and $\varepsilon' : F'G' \to \mathrm{Id}_\mathscr{A}$.

(3) We show transitivity: $\mathscr{A} \simeq \mathscr{A}'$ and $\mathscr{A}' \simeq \mathscr{A}'' \Rightarrow \mathscr{A} \simeq \mathscr{A}''$.

Since $\mathscr{A} \simeq \mathscr{A}'$, we have natural isomorphisms η and ε:

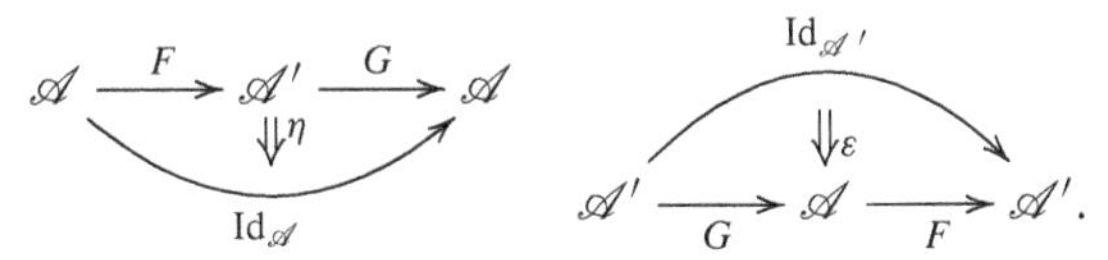

(2.1)

Since $\mathscr{A}' \simeq \mathscr{A}''$, we have natural isomorphisms η' and ε':

(2.2)

All we have to do is to construct two natural isomorphisms η'' and ε'' out of diagrams (2.1) and (2.2).

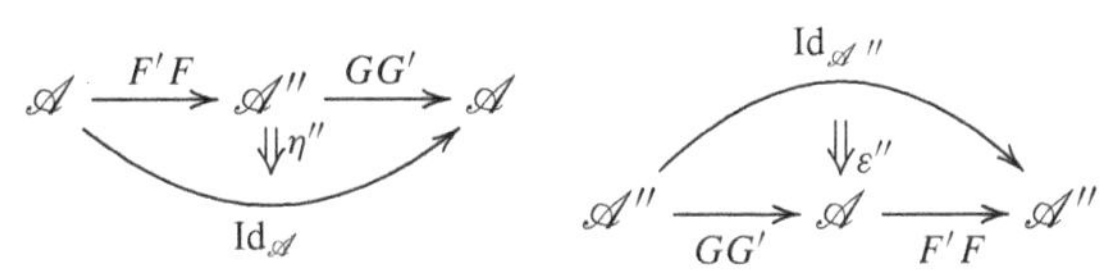

To construct η'', we place vertically two diagrams for η and η' in (2.1) and η and (2.2).

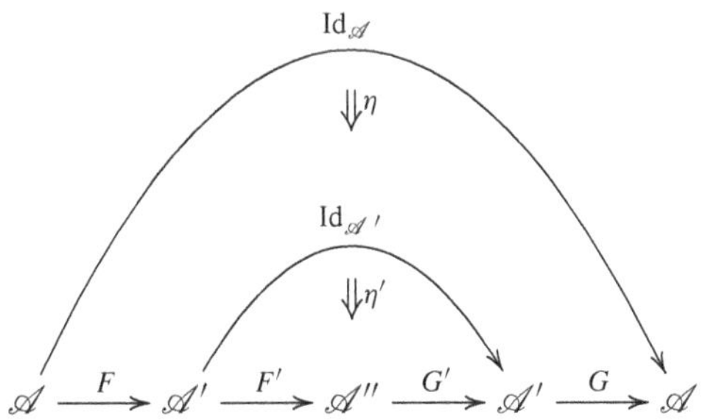

Note that we can rewrite $\eta : \mathrm{Id}_{\mathscr{A}} \to GF$ as

$$\eta : \mathrm{Id}_{\mathscr{A}} \to G\ \mathrm{Id}_{\mathscr{A}'}\ F.$$

Horizontally composing $\eta' : \mathrm{Id}_{\mathscr{A}'} \to G'F'$ with G and F, we have a natural isomorphism

$$G\eta'F : G\ \mathrm{Id}_{\mathscr{A}'}\ F \to GG'F'F.$$

Further composing vertically, we have

$$G\eta'F \circ \eta : \mathrm{Id}_{\mathscr{A}} \to GG'F'F.$$

Thus, we can define a natural isomorphism

$$\eta'' = G\eta' F \circ \eta.$$

To construct ε'', we place vertically two diagrams for ε and ε' in (2.1) and η and (2.2).

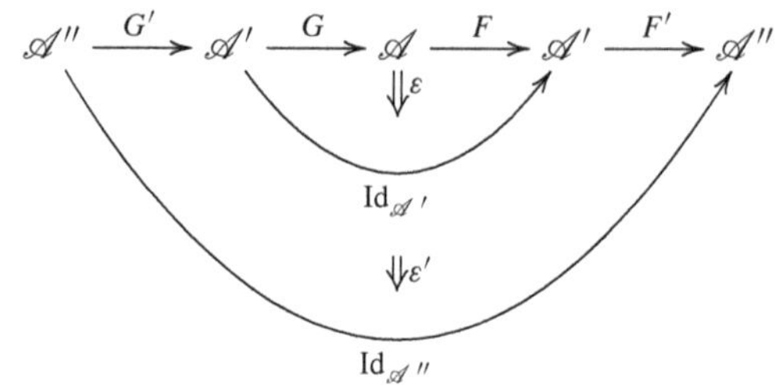

Note that we can rewrite $\varepsilon' : F'G' \to \mathrm{Id}_{\mathscr{A}''}$ as

$$\varepsilon' : F' \,\mathrm{Id}_{\mathscr{A}'}\, G' \to \mathrm{Id}_{\mathscr{A}''}.$$

Horizontally composing $\varepsilon : FG \to \mathrm{Id}_{\mathscr{A}'}$ with G' and F', we have a natural isomorphism

$$F'\varepsilon G' : F'FGG' \to F' \,\mathrm{Id}_{\mathscr{A}'}\, G'.$$

Vertically composing, we have

$$\varepsilon' \circ F'\varepsilon G' : F'FGG' \to \mathrm{Id}_{\mathscr{A}''}.$$

$$\varepsilon'' = \varepsilon' \circ F'\varepsilon G'$$

We have shown reflexivity, symmetricity, and transitivity. □

Equivalence of categories can be put in other forms with which we can decide equivalence with information about either F or G, not both.

Definition 2.3 A functor $F : \mathscr{A} \to \mathscr{B}$ is said to be *essentially surjective on objects*, if for all $B \in \mathrm{Obj}(\mathscr{B})$ there exist $A \in \mathrm{Obj}(\mathscr{A})$ such that $F(A) \cong B$.

If the function on objects is surjective, the functor, of course, is automatically essentially surjective on objects. There are cases in which the function on objects is not surjective but essentially surjective on objects. See the diagram below. An object B is out of the image of F but it has an isomorphism to B_∞ that belongs to the image of F.

$$
\begin{array}{ccc}
\mathscr{A} & \xrightarrow{F} & \mathscr{B} \\
A_1 & \longmapsto & B_1 \\
\vdots & & \vdots \\
A_\infty & \longmapsto & B_\infty \\
 & & \big\updownarrow \text{isomorphic} \\
 & & B
\end{array}
$$

Full and faithful functors have important properties concerning isomorphisms.

Lemma 2.2 *Let $F : \mathscr{A} \to \mathscr{B}$ be a full and faithful functor. For any pair $A, A' \in$ Obj$(\mathscr{A})$, we have the following:*

(a) A morphism $f : A \to A'$ is isomorphic if and only if $Ff : FA \to FA'$ is isomorphic.
(b) For all isomorphisms $g : FA \to FA'$, there exists a unique isomorphism $f : A \to A'$ with $Ff = g$.
(c) A and A' are isomorphic if and only if FA and FA' are isomorphic.

Proof Let us show (a). Suppose f is an isomorphism. Then, there is a morphism $f' : A' \to A$ such that $f' \circ f = 1_A$ and $f \circ f' = 1_{A'}$. Since F is a functor, it maps identities to identities. Function on morphisms maps like this:

$$
\begin{array}{ccc}
\mathscr{A} & \longrightarrow & \mathscr{B} \\
f' \circ f = 1_A & \longmapsto & Ff' \circ Ff = 1_{FA} \\
f \circ f' = 1_{A'} & \longmapsto & Ff \circ Ff' = 1_{FA'}.
\end{array}
$$

We see that Ff is an isomorphism.

Conversely, suppose Ff is an isomorphism. There exists a morphism $g : FA' \to FA$ such that $g \circ Ff = 1_{FA}$ and $Ff \circ g = 1_{FA'}$. Since F is full, there exists a morphism $f' : A' \to A$ such that $Ff' = g$. We see that F maps morphisms as follows:

$$
\begin{array}{ccc}
\mathscr{A} & \longrightarrow & \mathscr{B} \\
f' \circ f & \longmapsto & Ff' \circ Ff \\
 & & \| \\
1_A & \xmapsto[\text{axiom}]{} & 1_{FA} \\
f \circ f' & \longmapsto & Ff \circ Ff' \\
 & & \| \\
1_{A'} & \xmapsto[\text{axiom}]{} & 1_{FA'}.
\end{array}
$$

Since F is faithful, we have $f' \circ f = 1_A$ and $f \circ f' = 1_{A'}$. Thus, f is an isomorphism.

Let us show (b).

Assume that $g : FA \to FA'$ is an isomorphism. There exists a morphism $g' : FA' \to FA$ such that $g' \circ g = 1_{FA}$ and $g \circ g' = 1_{FA'}$. Since F is full and faithful, there exists unique morphisms such that $g = Ff$ and $g' = Ff'$. F maps morphisms as follows:

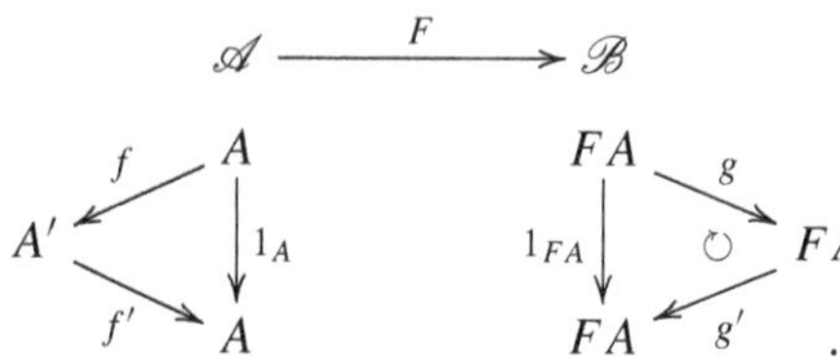

The two triangles correspond mirror symmetrically via F. By assumption, the right triangle commutes. We do not know at present about the left one. We can summarize the situation like this.

$$\begin{array}{ccc} \mathscr{A}(A, A) & \xrightarrow{F} & \mathscr{B}(FA, FA) \\ f' \circ f & \longmapsto & g' \circ g \\ & & \| \\ 1_A & \underset{\text{axiom}}{\longmapsto} & 1_{FA} \end{array}$$

Since F is faithful, we have $f' \circ f = 1_A$. This means the left triangle also commutes. Likewise, we can show that $f \circ f' = 1_{A'}$. To sum up, there exists a unique isomorphism $f : A \to A'$ such that $g = Ff$.

Let us show (c).

Suppose that $A \simeq A'$, namely that there exists an isomorphism $f : A \to A'$. By (a), we have that $Ff : FA \to FA'$ is an isomorphism. Therefore, $FA \simeq FA'$.

Conversely suppose that $FA \simeq FA'$, namely that there exists an isomorphism $g : FA \to FA'$. By (b), we see that there exists an isomorphism $f : A \to A'$, and thus $A \simeq A'$. □

Proposition 2.1 *For any functor $F : \mathscr{A} \to \mathscr{B}$, the following two conditions are equivalent.*

1. *There exists a functor $G : \mathscr{B} \to \mathscr{A}$ such that the pair (F, G) gives an equivalence of categories $\mathscr{A} \simeq \mathscr{B}$.*
2. *F is full and faithful and essentially surjective on objects.*

Remark 2.3 What is great about the second statement above is that it is described by information of F only. It is often the case that the construction of functor G is cumbersome.

Note that the roles of F and G are exchangeable, therefore both of them are full and faithful and essentially surjective on objects.

Proof (1) Let us show $1 \Rightarrow 2$.

Natural isomorphisms $\eta : \mathrm{Id}_{\mathscr{A}} \to GF$ and $\varepsilon : FG \to \mathrm{Id}_{\mathscr{B}}$ are available.

- We first prove that F is essentially surjective on objects.
  For any $B \in \mathrm{Obj}(\mathscr{B})$, we have isomorphism $\varepsilon_B : FGB \to B$. This shows that B is isomorphic to the image of $G(B)$ by F, meaning that F is essentially surjective on objects.
- Next we prove that F is faithful.
  The naturality of η claims the following diagram commutes:

$$\begin{array}{ccc} A & \xrightarrow{\eta_A} & GFA \\ {\scriptstyle f}\downarrow & & \downarrow{\scriptstyle GFf} \\ A' & \xrightarrow[\eta_{A'}]{} & GFA'. \end{array}$$

  Further, since η is an isomorphism, inverses η_A^{-1} and $\eta_{A'}^{-1}$ are available. The naturality equation is written as

$$\eta_{A'} \circ f = GFf \circ \eta_A.$$

  If we apply $\eta_{A'}^{-1} \circ -$ from the left to both sides, we get

$$f = \eta_{A'}^{-1} \circ GFf \circ \eta_A. \tag{2.3}$$

  Faithfulness means that

$$Ff_1 = Ff_2 \Rightarrow f_1 = f_2$$

  for any pair of morphisms f_1 and f_2 in $\mathscr{A}$, which is obvious by Eq. (2.3).
- Before we proceed to the fullness of F, let us examine the faithfulness of G.
  The naturality of ε claims the following diagram commutes:

$$\begin{array}{ccc} FGB & \xrightarrow{\varepsilon_B} & B \\ {\scriptstyle FGg}\downarrow & & \downarrow{\scriptstyle g} \\ FGB' & \xrightarrow{\varepsilon_{B'}} & B'. \end{array}$$

  Further, since ε is an isomorphism, inverses ε_B^{-1} and $\varepsilon_{B'}^{-1}$ are available. The naturality equation is written as

$$g \circ \varepsilon_B = \varepsilon_{B'} \circ FGg.$$

  If we apply $- \circ \varepsilon_B^{-1}$ from the right to both sides, we get

$$g = \varepsilon_{B'} \circ FGg \circ \varepsilon_B^{-1}. \tag{2.4}$$

  Faithfulness means that

$$Gg_1 = Gg_2 \Rightarrow g_1 = g_2$$

  for all pairs of morphisms g_1 and g_2 in $\mathscr{B}$, which is obvious by Eq. (2.4).

- We prove that F is full.
  We have to show that
$$\mathscr{A}(A, A') \to \mathscr{B}(FA, FA')$$
  is surjective for all pairs of $A, A' \in \mathrm{Obj}(\mathscr{A})$. We show that there exists a morphism $f : A \to A'$ such that $h = Ff$ for all morphisms $h : FA \to FA'$. Since $\eta_{A'}$ is invertible in the following diagram
$$\begin{array}{ccccc} A & \xrightarrow{\eta_A} & GFA & & FA \\ {\scriptstyle \exists f}\vdots & & \downarrow{\scriptstyle Gh} & & \downarrow{\scriptstyle \forall h} \\ A' & \xrightarrow[\eta_{A'}]{} & GFA' & & FA', \end{array}$$
  we may define morphism $f : A \to A'$ by
$$f = \eta_{A'}^{-1} \circ Gh \circ \eta_A.$$
  Recalling that $\eta : \mathrm{Id}_{\mathscr{A}} \to GF$ is a natural transformation, we draw two commuting squares as follows.
$$\begin{array}{ccc} A & \xrightarrow{\eta_A} & GFA \\ {\scriptstyle f}\downarrow & & \downarrow{\scriptstyle \underline{Gh}} \\ A' & \xrightarrow[\eta_{A'}]{} & GFA' \end{array} \qquad \begin{array}{ccc} A & \xrightarrow{\eta_A} & GFA \\ {\scriptstyle f}\downarrow & & \downarrow{\scriptstyle \underline{GFf}} \\ A' & \xrightarrow[\eta_{A'}]{} & GFA' \end{array}$$
  Since underlined morphisms Gh and GFf both equal to
$$\eta_{A'} \circ f \circ \eta_A^{-1},$$
  we have $Gh = GFf$. We have already shown that G is faithful, therefore $h = Ff$. This shows the fullness of F.

(2) Let us show $2 \Rightarrow 1$.
We will construct a functor $G : \mathscr{B} \to \mathscr{A}$ and two natural isomorphisms $\eta : \mathrm{Id}_{\mathscr{A}} \to GF$ and $\varepsilon : FG \to \mathrm{Id}_{\mathscr{B}}$ out of a given functor $F : \mathscr{A} \to \mathscr{B}$.
Since F is essentially surjective on objects, for all $B \in \mathrm{Obj}(\mathscr{B})$ we can find $A \in \mathrm{Obj}(\mathscr{A})$ and a morphism $\varepsilon_B : FA \to B$ as in
$$\begin{array}{ccc} \mathscr{A} & \xrightarrow{F} & \mathscr{B} \\ A & \longmapsto & FA \\ & & \downarrow{\scriptstyle \varepsilon_B} \\ & & B. \end{array}$$
Neither A nor ε_B is unique. We choose them arbitrarily and fix for each B. We define a function on objects G as $A = GB$. At this point, we do not know either if the collection $\{\varepsilon_B\}_{B \in \mathrm{Obj}(\mathscr{B})}$ forms a natural isomorphism, or if G is a functor.

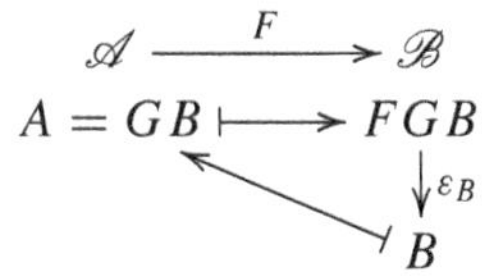

Now, we want to extend G to morphisms. Given a morphism $g : B \to B'$, we study a diagram

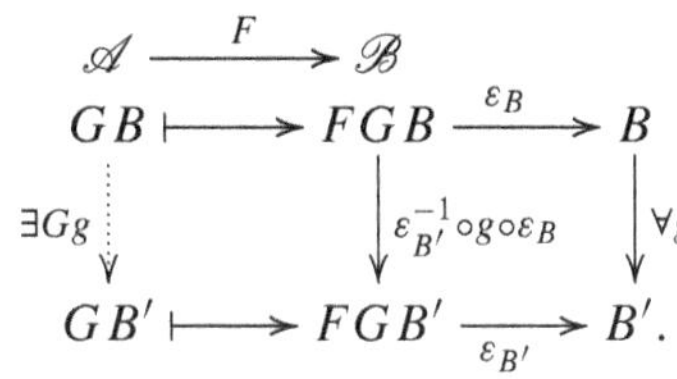

Since $\varepsilon_{B'}$ is invertible,

$$\varepsilon_{B'}^{-1} \circ g \circ \varepsilon_B : FGB \to FGB'$$

makes the right square commute. Since F is full and faithful, there is a unique morphism Gg that makes the left square commute. Thus, we obtain a function on morphisms.

- Let us show that G is a functor.
  Given two morphisms $g : B \to B'$ and $g' : B' \to B''$, we must show

$$G(g' \circ g) = Gg' \circ Gg. \tag{2.5}$$

Let us consider the following diagram.

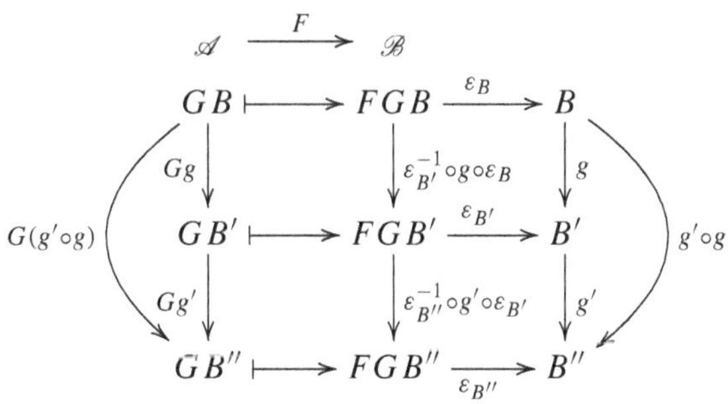

If we compose two morphisms in the center column, we obtain

$$\varepsilon_{B''}^{-1} \circ g' \circ g \circ \varepsilon_B = \varepsilon_{B''}^{-1} \circ (g' \circ g) \circ \varepsilon_B.$$

We now see that $Gg' \circ Gg$ and $G(g' \circ g)$ are mapped by F to the same morphism $\varepsilon_{B''}^{-1} \circ g' \circ g \circ \varepsilon_B$. For better illustration, we draw morphism mapping as follows.

$$\begin{array}{ccc} \mathscr{A}(GB, GB'') & \xrightarrow{F} & \mathscr{B}(FGB, FGB'') \\ Gg' \circ Gg & \longmapsto & \varepsilon_{B''}^{-1} \circ g' \circ g \circ \varepsilon_B \\ & & \parallel \\ G(g' \circ g) & \longmapsto & \varepsilon_{B''}^{-1} \circ (g' \circ g) \circ \varepsilon_B \end{array}$$

By assumption F is faithful, so the two morphisms on the left must coincide, which proves Eq. (2.5). Now, G is a functor.

- Let us show that ε is a natural transformation.
  We know that ε_B is an isomorphism for each $B \in \mathrm{Obj}(\mathscr{B})$ by definition. All we have to do is to show the collection $\{\varepsilon_B\}_{B \in \mathrm{Obj}(\mathscr{B})}$ forms a natural transformation. Let us review the definition of G.

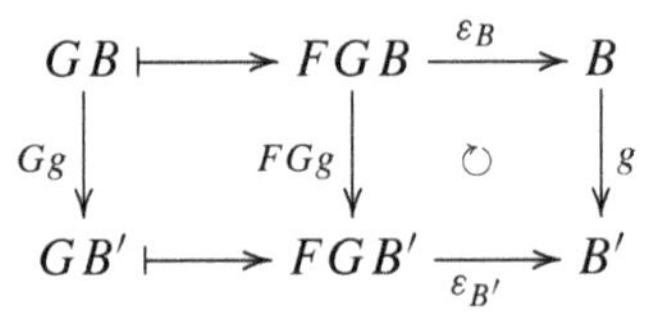

  We constructed $\varepsilon_{B'}^{-1} \circ g \circ \varepsilon_B$ in the place of FGg so that the right square commutes. Since F is full and faithful, we can equate $FGg = \varepsilon_{B'}^{-1} \circ g \circ \varepsilon_B$. We proved that G is a functor. The commutative square means that ε is a natural transformation.
- Let us construct a natural isomorphism $\eta : \mathrm{Id}_{\mathscr{A}} \to GF$.
  Since ε_{FA} is an isomorphism, we have the inverse $\varepsilon_{FA}^{-1} : FA \to FGFA$. Since F is full and faithful, there exists a unique η_A in the diagram:

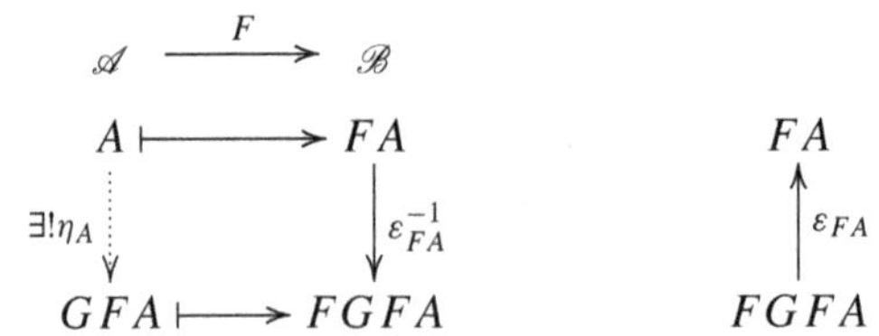

  η_A is an isomorphism by lemma 2.2.

Next, we show the naturality of η.

$$\mathscr{A} \xrightarrow{\quad F \quad} \mathscr{B}$$

$$\begin{array}{ccc} A & \xrightarrow{\eta_A} & GFA \\ {\scriptstyle f}\big\downarrow & & \big\downarrow{\scriptstyle GFf} \\ A' & \xrightarrow[\eta_{A'}]{} & GFA' \end{array} \qquad\qquad \begin{array}{ccc} FA & \xrightarrow{\varepsilon_{FA}^{-1}} & FGFA \\ {\scriptstyle Ff}\big\downarrow & \circlearrowleft & \big\downarrow{\scriptstyle FGFf} \\ FA' & \xrightarrow[\varepsilon_{FA}^{-1}]{} & FGFA' \end{array}$$

The square on the right commutes for ε is a natural isomorphism. The function F on morphisms show

$$\begin{array}{ccc} \mathscr{A}(A, GFA') & \xrightarrow{F} & \mathscr{B}(FA, FGFA') \\ GFf \circ \eta_A & \longmapsto & FGFf \circ \varepsilon_{FA}^{-1} \\ & & \| \\ \eta_{A'} \circ f & \longmapsto & \varepsilon_{FA}^{-1} \circ Ff. \end{array}$$

By the faithfulness of F, we have

$$GFf \circ \eta_A = \eta_{A'} \circ f.$$

Thus, we have that the square on the left commutes, which means that η is a natural isomorphism.

□

Universality and Limits

3

We will see universality and limits. We do not define the term *universal property*, but the whole discussion is about universality.

Emphasis is placed on the actual calculation of limit cones, and many examples of calculations with pencil and paper are given accompanied by sample code. In the colimit examples, we often calculate quotient structures, that is, division by equivalence relations, which requires iterative and recursive thinking. Sample code uses the `fold` function, which embodies a design pattern of functional programming. That gives a consistent style, but the algorithm is not so obvious at a glance. So, we include demonstrations of tracing recursive calls.

3.1 Initial and Terminal Objects

Definition 3.1 An *initial object* of $\mathscr{A}$ is an object A such that for all $X \in \mathrm{Obj}(\mathscr{A})$ there exists a unique morphism $A \to X$.

A *terminal object* of $\mathscr{A}$ is an object A such that for all $X \in \mathrm{Obj}(\mathscr{A})$ there exists a unique morphism $X \to A$.

Example 3.1 The empty set $\emptyset$ is an initial object of **Set**. A singleton set, namely a set with only one element, is a terminal object of **Set**. In set theory, a function $f : X \to Y$ is defined as a subset of $X \times Y$, a graph of f, $\{(x, f(x)) \mid x \in X\}$. See Fig. 3.1.

Not all subsets of $X \times Y$ are functions (graphs of a function). A necessary and sufficient condition for a subset $G \subset X \times Y$ to be a function (graph of a function) is

$$\forall x \;.\; (x \in X \Rightarrow \exists! y \;.\; y \in Y \text{ and } (x, y) \in G).$$

S. Yukita, *Category Theory Using Haskell*, Computer Science Foundations and Applied Logic, https://doi.org/10.1007/978-3-031-68538-5_3

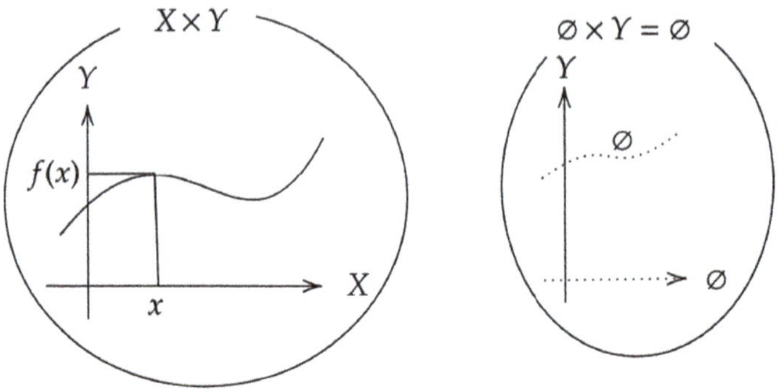

Fig. 3.1 A function is a subset of $X \times Y$

$\exists!$ means there exists a unique so-and-so.

Think of a function $f : \emptyset \to Y$. We can identify f with a subset of $\emptyset \times Y$, which must be $\emptyset$. The empty set qualifies to be a graph of a function since $p \Rightarrow q$ is true when p is false. In **Set**, there is a unique initial object $\emptyset$ and infinitely many terminal objects.

Proposition 3.1 *If A_1 and A_2 are two initial objects of $\mathscr{A}$, then they are isomorphic.*

Proof Since A_1 is initial, there exists a unique morphism $f : A_1 \to A_2$. Since A_2 is also initial, there exists a unique morphism $g : A_2 \to A_1$. The composite $g \circ f : A_1 \to A_1$ is a morphism from A_1 to A_1, which must be unique due to the initiality of A_1, therefore the composite equals to 1_{A_1}. Likewise, we have $f \circ g = 1_{A_2}$. This completes the proof. □

Proposition 3.2 *If A_1 and A_2 are two terminal objects of $\mathscr{A}$, then they are isomorphic.*

Proof Since A_2 is terminal, there exists a unique morphism $f : A_1 \to A_2$. Since A_1 is also terminal, there exists a unique morphism $g : A_2 \to A_1$. The composite $g \circ f : A_1 \to A_1$ is a morphism from A_1 to A_1, which must be unique due to the terminality of A_1, therefore the composite equals to 1_{A_1}. Likewise, we have $f \circ g = 1_{A_2}$. This completes the proof. □

Example 3.2 Given a set $X = \{1, 2, 3\}$, its power set is:

$$2^X = \{\emptyset, \{1\}, \{2\}, \{3\}, \{1, 2\}, \{2, 3\}, \{1, 3\}, X\}.$$

We define a category $\mathscr{P}(X)$, the category of power sets of X, as follows.

- $\text{Obj}(\mathscr{P}(X)) = 2^X$,

- for any objects A and B, if $A \subset B$ there exists a unique morphism from A to B. Otherwise, there is no morphism from A to B.

If we draw the set of generators omitting the identity morphisms, we have the following:

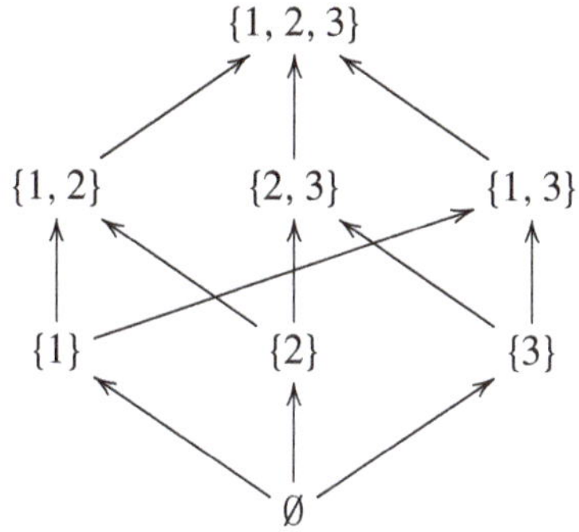

$\emptyset$ is a unique initial object, and X is a unique terminal object.

We can construct a faithful functor from $\mathscr{P}(X)$ to **Set** as follows:

- function on objects is obvious.
- function on morphisms maps $f : A \to B$ to the inclusion map. For example, a morphism $\{2, 3\} \to \{1, 2, 3\}$ of $\mathscr{P}(X)$ is mapped as

$$\begin{aligned} f :&\{2, 3\} \to \{1, 2, 3\} \\ &2 \mapsto 2, \quad 3 \mapsto 3 \end{aligned}$$

Notice that this functor is not full.

3.2 Products

We first observe how Cartesian products interact with functions.

Example 3.3 Let $A = \{7, 8, 9\}$, $B = \{a, b, c\}$, $X = \{1, 2, 3, 4\}$, and consider functions $x_1 : X \to A$, $x_2 : X \to B$:

$$\begin{aligned} x_1 :&1 \mapsto 7, \quad 2 \mapsto 8, \quad 3 \mapsto 9, \quad 4 \mapsto 7, \\ x_2 :&1 \mapsto a, \quad 2 \mapsto b, \quad 3 \mapsto c, \quad 4 \mapsto c. \end{aligned}$$

The pair of these functions can uniquely be combined as

$$f : 1 \mapsto (7, a), \quad 2 \mapsto (8, b), \quad 3 \mapsto (9, c), \quad 4 \mapsto (7, c)$$

that makes the following diagram commute:

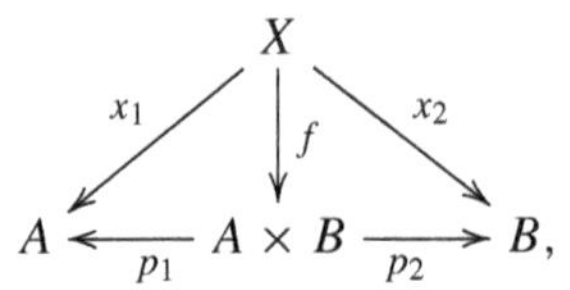

where p_1 and p_2 are projections. We fix A and B. For each triple (X, x_1, x_2) we get a unique $f : X \to A \times B$ that makes the diagram above commute.

Based on this observation, we try to define products without referring to elements of sets involved.

We fix two objects A_1 and A_2 of $\mathscr{A}$. Take any object $X \in \mathrm{Obj}(\mathscr{A})$ and consider a pair of morphisms $x_1 : X \to A_1$ and $x_2 : X \to A_2$.

$$A_1 \xleftarrow{x_1} X \xrightarrow{x_2} A_2$$

This triple (X, x_1, x_2) is call a *span* from *apex* X to $\{A_1, A_2\}$.

We consider a category $\mathrm{Span}(A_1, A_2)$ of spans to $\{A_1, A_2\}$:

- $\mathrm{Obj}(\mathrm{Span}(A_1, A_2))$ consists of all the spans $A_1 \xleftarrow{x_1} X \xrightarrow{x_2} A_2$.
- a morphism from $A_1 \xleftarrow{x_1} X \xrightarrow{x_2} A_2$ to $A_1 \xleftarrow{y_1} Y \xrightarrow{y_2} A_2$ is a morphism $f : X \to Y$ in $\mathscr{A}$ that makes the two triangles in the following diagram commute:

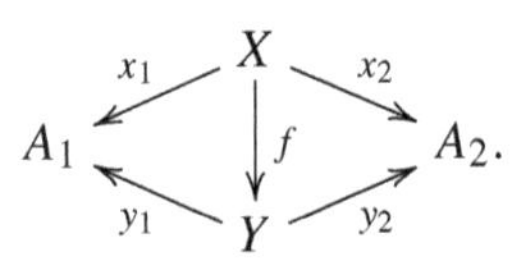

We can easily check associativity and other properties required.

We will study initial objects and terminal objects in $\mathrm{Span}(A_1, A_2)$. But before stepping into the main discussion, we consider the influence of the initial and terminal objects in $\mathscr{A}$ on those in $\mathrm{Span}(A_1, A_2)$.

Suppose there is an initial object 0 in $\mathscr{A}$. Then, the two triangles in a diagram

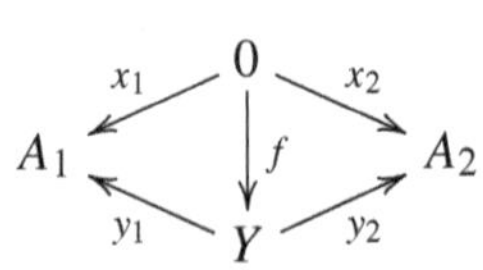

automatically commute with a unique morphism f, and thus we get a unique morphism in $\mathrm{Span}(A_1, A_2)$. In fact, both x_1 and $y_1 \circ f$ are morphisms from 0 to A_1, and we know such a morphism exists at most one. Thus, we have $x_1 = y_1 \circ f$. Likewise, we have $x_2 = y_2 \circ f$. If this is the case, we get no meaningful results in seeking for initial objects in $\mathrm{Span}(A_1, A_2)$. And this is often the case with examples we encounter. So, it is not a good idea to think about initial objects in $\mathrm{Span}(A_1, A_2)$.

Then, how about thinking of terminals? Suppose there is a terminal object 1 in $\mathscr{A}$. Then, the two triangles in a diagram

$$\begin{array}{ccccc} & & X & & \\ & \swarrow^{x_1} & \downarrow f & \searrow^{x_2} & \\ A_1 & & & & A_2 \\ & \nwarrow_{y_1} & & \nearrow_{y_2} & \\ & & 1 & & \end{array}$$

do not automatically commute, even though we have a unique morphism f. We have no hope to get a unique morphism in $\mathrm{Span}(A_1, A_2)$. Therefore, we expect meaningful results in seeking terminal objects in $\mathrm{Span}(A_1, A_2)$, despite the possible existence of terminals in $\mathscr{A}$. So, it is a good idea to study terminal objects in $\mathrm{Span}(A_1, A_2)$.

A terminal span $A_1 \xleftarrow{z_1} Z \xrightarrow{z_2} A_2$ ensures that for any span $A_1 \xleftarrow{x_1} X \xrightarrow{x_2} A_2$ there exists a unique morphism $f : X \to Z$ that makes the two triangles in the following diagram commute:

$$\begin{array}{ccccc} & & X & & \\ & \swarrow^{x_1} & \downarrow f & \searrow^{x_2} & \\ A_1 & & & & A_2. \\ & \nwarrow_{z_1} & & \nearrow_{z_2} & \\ & & Z & & \end{array}$$

Definition 3.2 A terminal object $A_1 \xleftarrow{z_1} Z \xrightarrow{z_2} A_2$ in $\mathrm{Span}(A_1, A_2)$ is called a product of A_1 and A_2 and written as $A_1 \xleftarrow{p_1} A_1 \times A_2 \xrightarrow{p_2} A_2$ or simply $A_1 \times A_2$ with implicit projections.

Not all pairs have a product in general. If all the pairs in $\mathscr{A}$ have products, we say that $\mathscr{A}$ has products, or $\mathscr{A}$ is a *category with products.*

Example 3.4 Let us consider Example 3.2 again. Let $X = \{1, 2, 3\}$. We study products in category $\mathscr{P}(X)$.

If a product of $\{1, 2\}$ and $\{2, 3\}$ exists in $\mathscr{P}(X)$, we put $z = \{1, 2\} \times \{2, 3\}$, and write as

$$\{1, 2\} \leftarrow z \rightarrow \{2, 3\}.$$

Candidates for z are only $\{2\}$ and $\emptyset$ since morphisms are very limited in this category. In other words, we can write as

$$\begin{aligned} &\mathrm{Obj}(\mathrm{Span}(\{1, 2\}, \{2, 3\})) \\ =&\{\, \{1, 2\} \leftarrow \{2\} \rightarrow \{2, 3\}, \quad \{1, 2\} \leftarrow \emptyset \rightarrow \{2, 3\} \,\}. \end{aligned}$$

$\{1, 2\} \leftarrow \{2\} \rightarrow \{2, 3\}$ is clearly a terminal object in $\mathrm{Span}(\{1, 2\}, \{2, 3\})$. So, we may write as $\{2\} = \{1, 2\} \times \{2, 3\}$. Notice that $\{2\}$ is the greatest lower bound of $\{1, 2\}$ and $\{2, 3\}$.

Proposition 3.3 *A product, if it exists, is* unique up to isomorphisms.

This comes directly from the uniqueness of terminal objects in general. But, let us repeat the proof in a new context.

Proof Suppose there are two products (X, x_1, x_2) and (Y, y_1, y_2) of A_1 and A_2. All the triangles, including large ones, in the left diagram

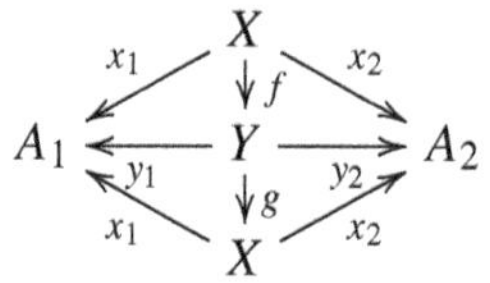

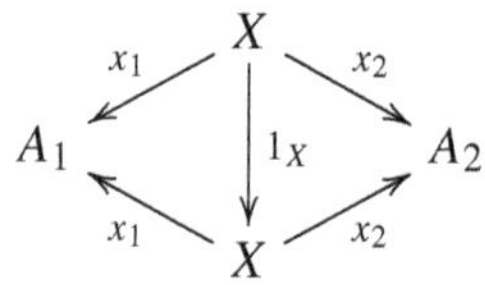

commute with uniquely determined f and g. The right one is a trivial commuting diagram. Morphism from (X, x_1, x_2) to itself must be unique, namely, 1_X. Thus, we have $g \circ f = 1_X$.

The same argument goes with X and Y exchanged. So, we have $f \circ g = 1_Y$. □

Example 3.5 We visit Example 3.3 again. In the diagram

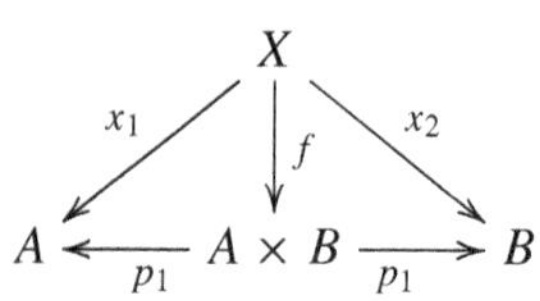

we fix A and B. We consider various spans (X, x_1, x_2). For each span (X, x_1, x_2), there exists a unique $f : X \to A \times B$ that makes the two triangles commute. We simulate this situation with a program, where products are implemented with the type constructor `(,)`.

Listing 3.1 product.hs

```
data SetA = A7 | A8 | A9 deriving (Enum,Show)

data SetB = Ba | Bb | Bc deriving (Enum,Show)

data SetX = X1 | X2 | X3 | X4 deriving (Enum,Show)

x1 :: SetX -> SetA
x1 X1 = A7
x1 X2 = A8
x1 X3 = A9
x1 X4 = A7

x2 :: SetX -> SetB
x2 X1 = Ba
x2 X2 = Bb
x2 X3 = Bc
x2 X4 = Bc
```

```
factor :: (x -> a) -> (x -> b) -> (x -> (a,b))
factor p1 p2 x = (p1 x, p2 x)
```

We can check the signature of the `factor` function as follows:

```
*Main> :t factor x1 x2
factor x1 x2 :: SetX -> (SetA, SetB)
```

We can see that the function `factor x1 x2` returns a value of type $X \to A \times B$. Tests can be done as follows:

```
*Main> [(x, factor x1 x2 x) | x <- [X1 .. X4] ]
[(X1,(A7,Ba)),(X2,(A8,Bb)),(X3,(A9,Bc)),(X4,(A7,Bc))]
```

3.3 Coproducts

We first observe how disjoint unions or direct sums interact with functions. The disjoint union of A and B is denoted by $A \sqcup B$ or $A + B$. When we want a disjoint union of A and B with $A \cup B \neq \emptyset$, we adopt a brute force method: we attach different tags to elements of both sets. For example,

$$\{1, 2, 3\} \sqcup \{3, 4\} = \{(L, 1), (L, 2), (L, 3), (R, 3), (R, 4)\}.$$

This is just Haskell type constructor `Either` does, which we will see sample code later.

Example 3.6 Let $A = \{7, 8, 9\}$, $B = \{a, b\}$, $X = \{1, 2, 3, 4, 5, 6\}$, and consider functions $x_1 : A \to X, x_2 : B \to X$:

$$\begin{aligned} x_1 &: 7 \mapsto 1, \quad 8 \mapsto 2, \quad 9 \mapsto 3 \\ x_2 &: a \mapsto 3, \quad b \mapsto 4. \end{aligned}$$

The pair of these functions can uniquely be combined as

$$f : 7 \mapsto 1, \quad 8 \mapsto 2, \quad 9 \mapsto 3, \quad a \mapsto 3, \quad b \mapsto 4$$

that makes the following diagram commute:

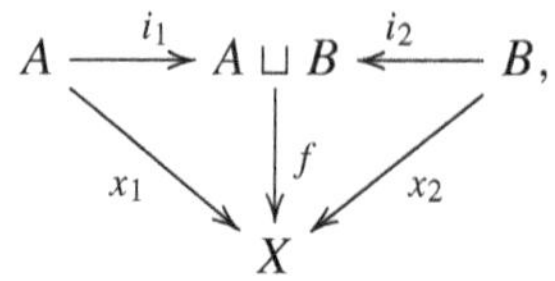

where i_1 and i_2 are *coprojections*. We fix A and B. For each triple (x_1, x_2, X) we get a unique $f : A \sqcup B \to X$ that makes the diagram above commute.

Based on this observation, we try to define coproducts without referring to elements of sets involved.

We fix two objects A_1 and A_2 of $\mathscr{A}$. Take any object $X \in \mathrm{Obj}(\mathscr{A})$ and consider a pair of morphisms $x_1 : A_1 \to X$ and $x_2 : A_2 \to X$.

$$A_1 \xrightarrow{x_1} X \xleftarrow{x_2} A_2$$

This triple (x_1, x_2, X) is call a *cospan* from $\{A_1, A_2\}$ to X. We consider a category $\mathrm{Cospan}(A_1, A_2)$ of cospans from $\{A_1, A_2\}$:

- $\mathrm{Obj}(\mathrm{Cospan}(A_1, A_2))$ consists of all the cospans $A_1 \xrightarrow{x_1} X \xleftarrow{x_2} A_2$.
- a morphism from $A_1 \xrightarrow{x_1} X \xleftarrow{x_2} A_2$ to $A_1 \xrightarrow{y_1} Y \xleftarrow{y_2} A_2$ is a morphism $f : X \to Y$ in $\mathscr{A}$ that makes the two triangles in the following diagram commute:

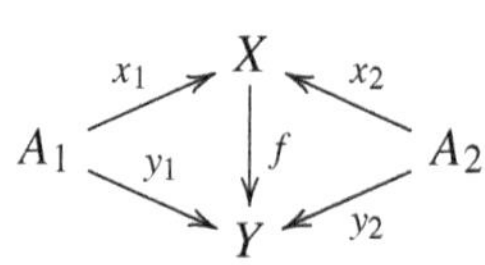

We can easily check associativity and other properties required.

In this way, the category of cospans is obtained by fixing A_1 and A_2. In this category of cospans, we consider initial and terminal objects.

Suppose the category $\mathscr{A}$ itself has a terminal object 1. In diagram

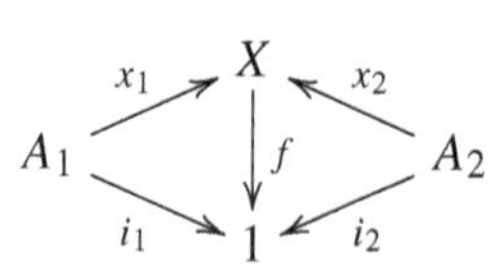

i_1, i_2, and f are all uniquely determined. $f \circ x_1$ is a morphism from A_1 to 1. Since only one such morphism is allowed, it matches i_1. So the left triangle is automatically commutative. Similarly, the triangle on the right is automatically commutative. From this observation, we can see that thinking about terminal objects in the category of cospans is too trivial to give us any new information.

When category $\mathscr{A}$ itself has initial object 0, only f is known to be unique in the following diagram.

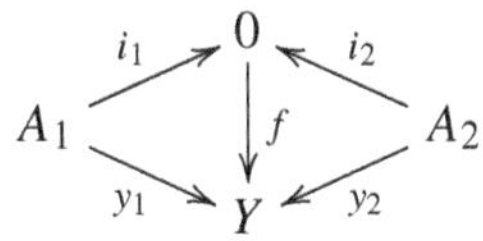

The commutativity of the two triangles is not guaranteed. Therefore, it makes sense to consider the initial object in the category of cospans.

Definition 3.3 Let A_1 and A_2 be objects in $\mathscr{A}$. The *coproduct* of A_1 and A_2 is an initial object in the category $\mathrm{Cospan}(A_1, A_2)$, written as

$$A_1 \sqcup A_2 = A_1 \xrightarrow{i_1} Z \xleftarrow{i_2} A_2,$$

which is unique up to isomorphisms if it exists. We often (almost always) identify $A_1 \sqcup A_2$ with its coapex Z by abuse of terminology. If a coproduct exists for any pair of objects in $\mathscr{A}$, the category $\mathscr{A}$ is said to *have coproducts* or said to be a *category with coproducts* .

Example 3.7 Let us search for coproducts in the category $\mathscr{P}(X)$ of the power set of Example 3.2. Consider $\{1\} \sqcup \{1, 2\}$. Does it exist in the first place? Let z be a coproduct if it exists. For a cospan

$$\{1\} \to z \leftarrow \{1, 2\}$$

candidates for z are $\{1, 2\}$ and $\{1, 2, 3\}$. The category of cospans consists of only two objects and obvious morphisms. The initial object is $\{1, 2\}$, which is the least upper bound of $\{1\}$ and $\{1, 2\}$. We have

$$\{1\} \sqcup \{1, 2\} = \{1, 2\}.$$

Proposition 3.4 *Given two objects in a category, their coproduct is unique up to isomorphisms, if it exists.*

This comes directly from the uniqueness of initial objects in general. But, let us repeat the proof in a new context.

Proof Let (x_1, x_2, X) and (y_1, y_2, Y) be two coproducts of A_1 and A_2. Observe the following two diagrams.

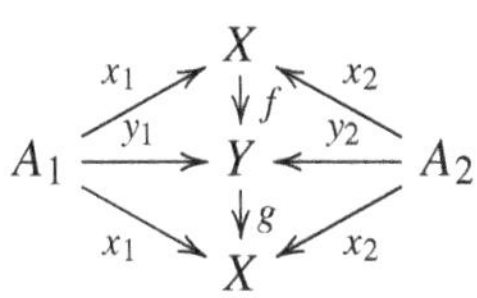

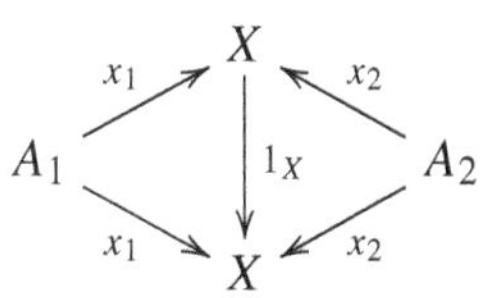

All triangles, small or large, in the left diagram commute where f and g are uniquely determined. The diagram on the right is a trivial commuting diagram. Since (x_1, x_2, X) is initial in the category of cospans, we have $g \circ f = 1_X$. Exchanging X and Y, we also have $f \circ g = 1_Y$. This means that X and Y are isomorphic, and that (x_1, x_2, X) and (y_1, y_2, Y) are isomorphic. □

Example 3.8 We reexamine Example 3.6. Given a diagram

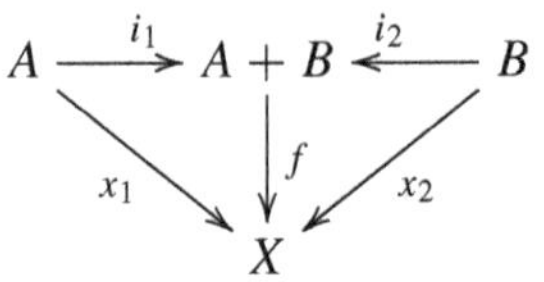

we fix A and B and let data (x_1, x_2, X) vary arbitrarily. For any (x_1, x_2, X), we get a unique $f : A + B \to X$ such that all the triangles commute in the above diagram. Let us simulate the situation with Haskell. Coproduct is realized by type constructor `Either` with two type parameters.

Listing 3.2 coproduct.hs

```
data SetA = A7 | A8 | A9 deriving (Enum,Show)

data SetB = Ba | Bb deriving (Enum,Show)

data SetX = X1 | X2 | X3 | X4 | X5 | X6 deriving (Enum,Show)

x1 :: SetA -> SetX
x1 A7 = X1
x1 A8 = X2
x1 A9 = X3

x2 :: SetB -> SetX
x2 Ba = X3
x2 Bb = X4

factor :: (a -> x) -> (b -> x) -> (Either a b -> x)
factor i1 i2 (Left a)  = i1 a
factor i1 i2 (Right b) = i2 b
```

The signature of `factor` is confirmed as follows.

```
*Main> :t factor x1 x2
factor x1 x2 :: Either SetA SetB -> SetX
```

`factor x1 x2` is a function that takes a parameter from `Either SetA SetB` and returns a value in `SetX`.

```
*Main> [(Left x, factor x1 x2 (Left x)) | x <- [A7 .. A9] ]
[(Left A7,X1),(Left A8,X2),(Left A9,X3)]
*Main> [(Right x, factor x1 x2 (Right x)) | x <- [Ba,Bb] ]
[(Right Ba,X3),(Right Bb,X4)]
```

3.4 Limits

What we did with products and coproducts can be extended to the notions of limit and colimit. In this section, we generalize the concept of product to that of limit.

Definition 3.4 Let J be a small category and $\mathscr{A}$ a category. A *diagram in* $\mathscr{A}$ *of shape* J is a functor $F : J \to \mathscr{A}$. J is called an *index category*. F is often called a J*-shaped diagram* in $\mathscr{A}$.

Let A be an object of $\mathscr{A}$. A *cone* from A to F is a set of data:

- For each object j of J, there is a morphism $p_j : A \to F(j)$ called *projection*. A is called the *apex* of a cone.
- For all morphisms $f : j_1 \to j_2$ of J, the diagram

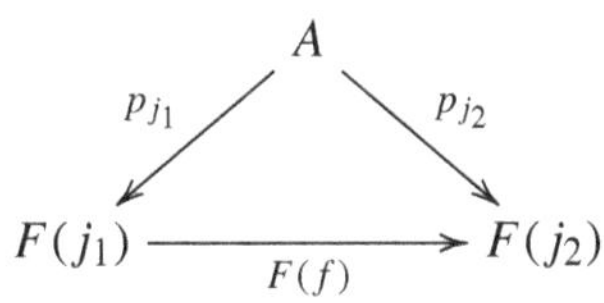

  commutes.

A cone is presented as a triple $(A, \{p_j\}, F)$, or a pair $(A, \{p_j\})$, or simply $\{p_j\}$, all of which contain the same amount of information.

Definition 3.5 Let J be a small category and $\mathscr{A}$ a category. We fix a J-shaped diagram $F : J \to \mathscr{A}$. We consider a category $(\mathscr{A}, F)$ defined as follows.

- Objects are cones $(A, \{p_j\})$.
- A morphism from $(A, \{p_j\})$ to $(A', \{q_j\})$ is a morphism $u : A \to A'$ of $\mathscr{A}$ such that all the triangles below commute for all morphisms f of J.

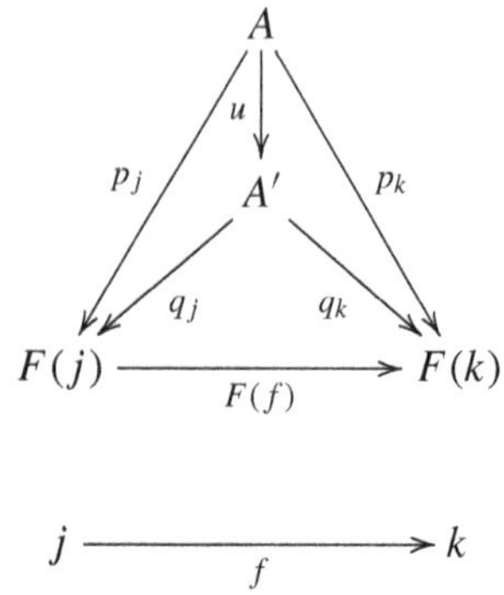

If category $(\mathscr{A}, J, F)$ has a terminal object $(L, \{r_j\})$, we call it a *limit* of the diagram F. We often call its apex L a limit of F with implicit $\{r_j\}$. Cone $(L, \{r_j\})$ is also called a *universal cone* or *limiting cone*. From the general argument over terminal objects, L is unique up to isomorphism. It is often written as $L = \varprojlim F$. Given an arbitrary cone, a unique morphism from it to the limiting cone is called a *mediating morphism*.

Remark 3.1 The set of all cones from A to F is written as $\mathrm{Cones}(A, F)$. We have the following isomorphism.

$$\mathscr{A}(A, \varprojlim F) \simeq \mathrm{Cones}(A, F)$$

When $\mathscr{A}$ is locally small, using the language of Chap. 7, the above isomorphism means that $\varprojlim F$ is an object that represents the contravariant functor $\mathrm{Cones}(-, F) : \mathscr{A}^{\mathrm{op}} \to \mathbf{Set}$. See also Chap. 5.

Example 3.9 Let the index category J be the category with only two objects 1 and 2 and the identity morphisms. Determining a functor F from J to the category $\mathscr{A}$ is the same as choosing two objects in the category $\mathscr{A}$. Given $F(1) = A_1$ and $F(2) = A_2$, the span from any object X in the category $\mathscr{A}$

$$A_1 \xleftarrow{x_1} X \xrightarrow{x_2} A_2$$

is the same as considering the cone from X to a J-shaped diagram F. Considering the category of spans $\mathrm{Span}(A_1, A_2)$, and the category of cones $(\mathscr{A}, F)$ is the same. So it turns out that the limit is a generalization of the product.

Example 3.10 We consider an index category J with natural numbers $1, 2, 3, \cdots$ as objects and

$$1 \leftarrow 2 \leftarrow 3 \leftarrow \cdots$$

as generators of morphisms. Let $F : J \to \mathbf{Set}$ be a functor where

- function on objects is given by $F(n) = \mathbb{R}^n$,
- function on morphisms is given by $F(n \leftarrow n+1) = p_n$, where p_n is a projection

$$\begin{aligned} p_n : & \mathbb{R}^{n+1} \to \mathbb{R}^n \\ & (x_1, x_2, \cdots, x_n, x_{n+1}) \mapsto (x_1, x_2, \cdots, x_n). \end{aligned}$$

Let Y be an arbitrary set. A cone from Y to a J-shaped diagram F is given by a family of morphisms $\{y_n\}$ that makes all the triangles with edges y_n and p_n commute:

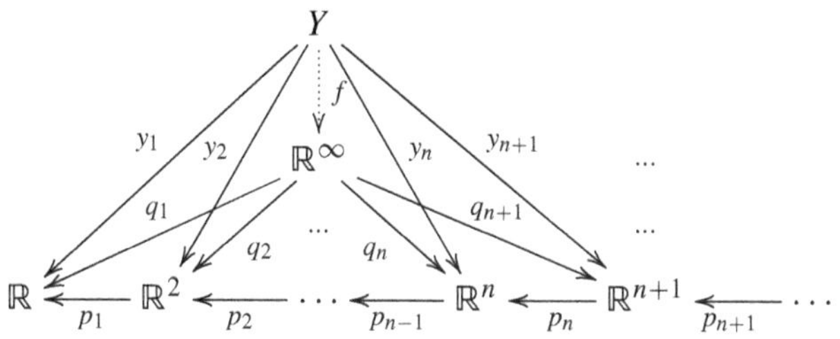

We have $\varprojlim F = \mathbb{R}^\infty$. In fact, we use the data $\{y_n\}$ to uniquely define $f : Y \to \mathbb{R}^\infty$ as

$$f(z) = (f_1(z), f_2(z), \cdots, f_n(z), \cdots),$$

where $y_n(z) \in \mathbb{R}^n$ and $f_n(z)$ is its n-th component.

Definition 3.6 We consider an index category J with objects 0, 1, and 2. The following two morphisms

$$1 \to 0 \leftarrow 2$$

generate the whole morphisms. A limit of a J-shaped diagram is call a *pullback* or *fiber product*. Let $F : J \to \mathscr{A}$ be a functor given by:

- $F(0) = A_0$, $F(1) = A_1$, $F(2) = A_2$, and
- $F(1 \to 0) = a_1$, $F(2 \to 0) = a_2$.

We consider a cone $(X, \{x_0, x_1, x_2\}, F)$, which is depicted as the left diagram in Eq. (3.1).

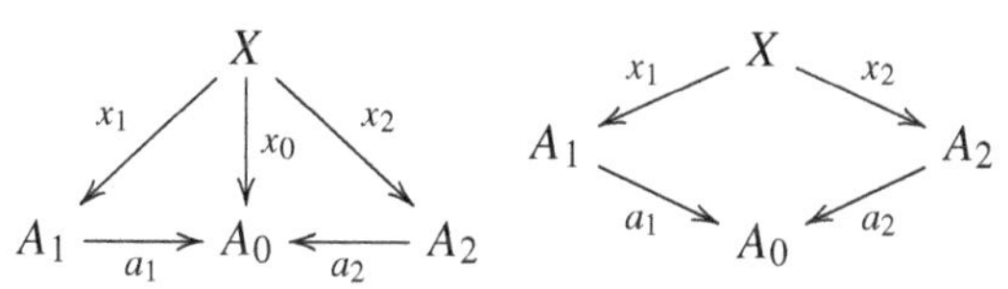

(3.1)

The qualification test for being a cone is

$$x_0 = a_1 \circ x_1 \quad \text{and} \quad x_0 = a_2 \circ x_2,$$

which is reduced to a single equation $a_1 \circ x_1 = a_2 \circ x_2$. We can confirm this fact as follows. Omitting x_0 gives the diagram on the right. If this diamond is commutative, we can regain morphism x_0 by $x_0 = a_1 \circ x_1 = a_2 \circ x_2$. Thus, both commutative diagrams contain the same information.

Let us suppose there is a universal cone $(L, \{p_0, p_1, p_2\}, F)$. Taking the above argument into account, all we have to do is to check the commutativity of all the triangles and squares in the following diagram.

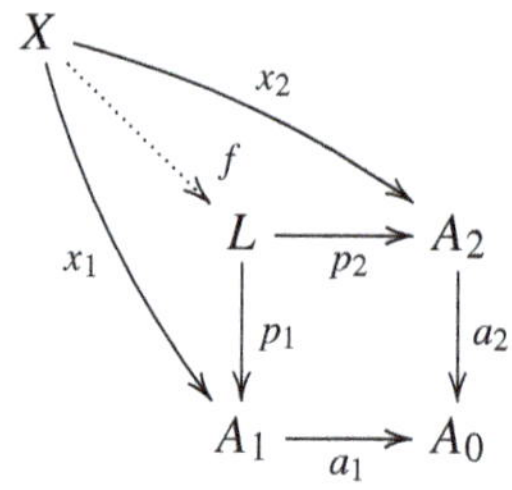

By universality, morphism f is uniquely determined. Note that morphism p_0 is omitted but easily regained by setting $p_0 = a_1 \circ p_1 = a_2 \circ p_2$.

A pullback $\varprojlim F$ is often written as

$$L = A_1 \times_{A_0} A_2.$$

Example 3.11 Let $A_0 = \{a, b\}$, $A_1 = \{1, 2, 3, 4\}$, and $A_2 = \{5, 6, 7, 8\}$. We consider functions $a_1 : A_1 \to A_0$ and $a_2 : A_2 \to A_0$ as follows.

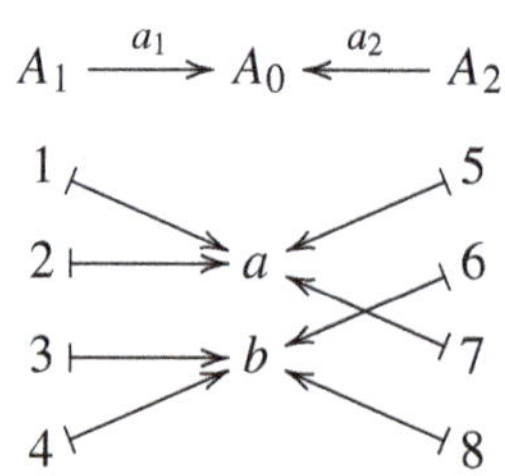

A pullback or fiber product is given by

$$A_1 \times_{A_0} A_2 = (\{1, 2\} \times \{5, 7\}) \sqcup (\{3, 4\} \times \{6, 8\}). \tag{3.2}$$

$\{1, 2\}$ and $\{5, 7\}$ are fibers over a of a_1 and a_2, respectively. $\{3, 4\}$ and $\{8, 9\}$ are fibers over b of a_1 and a_2, respectively. The RHS of (3.2) is a disjoint union of the pointwise direct products of fibers over a single point.

Haskell code is given below.

Listing 3.3 fiberproduct.hs

```
data A0 = Va | Vb  deriving (Eq, Show)

data A1 = V1 | V2 | V3 | V4 deriving (Eq, Enum, Show)

data A2 = V5 | V6 | V7 | V8 deriving (Eq, Enum, Show)
```

```
6
7  fiberProduct :: (Eq a, Eq b, Eq c) =>
8    (b -> a) -> (c -> a) -> [b] -> [c] -> [(a,b,c)]
9  fiberProduct p q xs ys =
10   [(p x,x,y) | x <- xs, y <- ys, p x == q y]
11
12 getFiber :: Eq a => a -> [(a,b,c)] -> [(b,c)]
13 getFiber bp zxys = [(x,y) | (z,x,y)<- zxys, z == bp]
14
15 -- testdata
16 a1s = [V1 .. V4]
17 a2s = [V5 .. V8]
18
19 a1 :: A1 -> A0
20 a1 V1 = Va
21 a1 V2 = Va
22 a1 V3 = Vb
23 a1 V4 = Vb
24
25 a2 :: A2 -> A0
26 a2 V5 = Va
27 a2 V6 = Vb
28 a2 V7 = Va
29 a2 V8 = Vb
30
31 {-- suggested tests
32 fiberProduct a1 a2 a1s a2s
33 getFiber Va $ fiberProduct a1 a2 a1s a2s
34 getFiber Vb $ fiberProduct a1 a2 a1s a2s
35 -}
```

Lines 1–5 define data types for A_0, A_1, and A_2 in Example 3.11.
Lines 7–10 define a function that realizes the fiber product.
Lines 12–13 define a getter function for fibers.
Lines 15–29 give data structures for all players, namely objects and morphisms.
A test goes like this.

```
*Main> getFiber Va $ fiberProduct a1 a2 a1s a2s
[(V1,V5),(V1,V7),(V2,V5),(V2,V7)]
*Main> getFiber Vb $ fiberProduct a1 a2 a1s a2s
[(V3,V6),(V3,V8),(V4,V6),(V4,V8)]
```

The results are fibers over `a` and `b`.

Definition 3.7 We consider a category J with objects 0 and 1. Generators of morphisms are

$$0 \underset{b}{\overset{a}{\rightrightarrows}} 1.$$

A limit of a J-shaped diagram in category $\mathscr{A}$ is called an *equalizer*. Let us look closer into the situation. Let a functor $F : J \to \mathscr{A}$ be defined by the following:

- function on objects is given by $F(0) = A_1$ and $F(1) = A_2$.
- function on morphisms is given by $F(a) = f$ and $F(b) = g$.

We consider a cone from X to F.

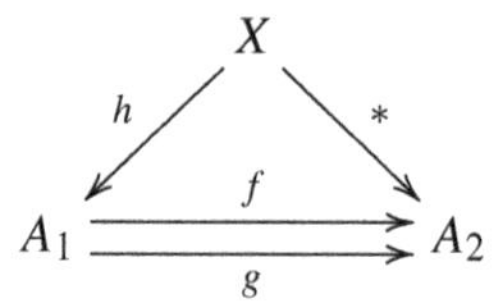

Any cone must pass a qualification test for being really a cone:

$$* = f \circ h \quad \text{and} \quad * = g \circ h$$

Thus, we have

$$f \circ h = g \circ h.$$

To sum up, getting a limit of a J-shaped diagram is completely the same as finding a universal object L giving a unique h' that makes the following diagram commute.

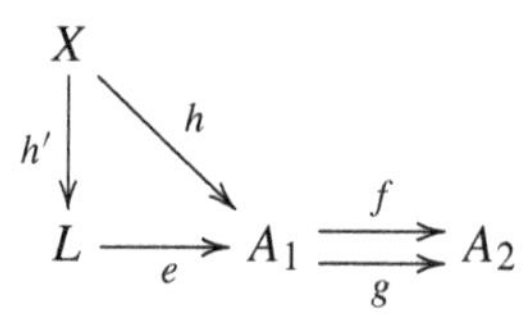

Example 3.12 Let $A_1 = \{1, 2, 3, 4\}$ and $A_2 = \{a, b, c\}$. We consider the following two functions $f, g : A_1 \to A_2$:

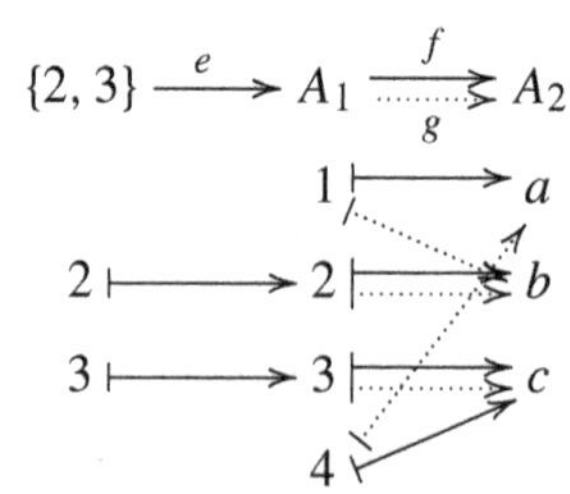

g is depicted with dotted lines. Function $e : \{2, 3\} \to A_1$ is an equalizer.

The Haskell code for this example is given below.

Listing 3.4 equalizer.hs

```
1 data A1 = V1 | V2 | V3 | V4
2   deriving (Eq, Enum, Show)
3
4 data A2 = Va | Vb | Vc  deriving (Eq, Enum, Show)
5
6 -- We will be content with
7 -- finding a subobject(subset) of A1
8 equalizer :: (Eq a, Enum a, Eq b) =>
9    (a -> b) -> (a -> b) -> [a] -> [a]
10 equalizer f g xs
11   = [x | x <- xs, f x == g x]
12
13 f :: A1 -> A2
14 f V1 = Va
15 f V2 = Vb
16 f V3 = Vc
17 f V4 = Vc
18
19 g :: A1 -> A2
20 g V1 = Vb
21 g V2 = Vb
22 g V3 = Vc
23 g V4 = Va
24
25 {-- suggested test
26 equalizer f g [V1 .. V4]
27 -}
```

Lines 1–4 define A_1 and A_2. Lines 10–11 define `equalizer`. It would be better to give this function a signure like `(a -> b) -> (a -> b) -> [a]`, but we compromise for ease of implementation.

Let us give it a try.

```
*Main> equalizer f g [V1 .. V4]
[V2,V3]
```

Definition 3.8 A category with no morphisms other than identities is called a *discrete category*. Let J be a small discrete category, namely a set. If a J-shaped diagram $F : J \to \mathscr{A}$ has a limit, we call this limit in particular a *product* or *direct product*. The apex of a limit cone is often written as $\prod_J F(j)$. Note that J is not limited to a finite set.

Given a cone from X to F, namely a set of morphisms $\{f_j : X \to F(j)\}_{j \in \mathrm{Obj}(J)}$, the mediating morphism u in the following diagram

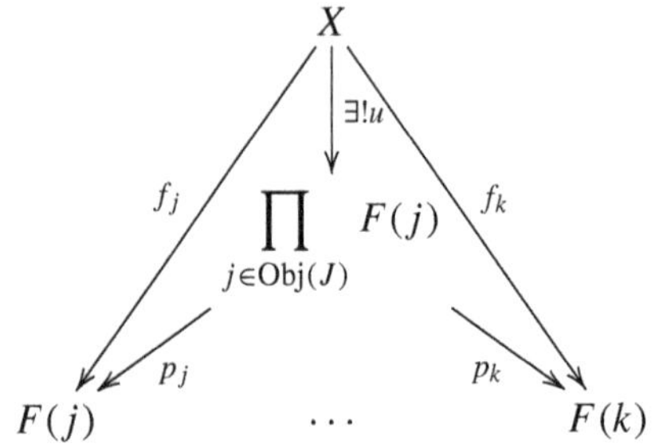

is often written as

$$u = \prod_{j \in \mathrm{Obj}(J)} f_j.$$

A *product of morphisms* accompanies with the universality of product.

Proposition 3.5 *We use the same notation in Definition 3.8. Consider an arbitrary morphism $g : Y \to X$. We have the following.*

$$\left(\prod_{j \in \mathrm{Obj}(J)} f_j \right) \circ g = \prod_{j \in \mathrm{Obj}(J)} (f_j \circ g).$$

The operation of composing g from right and the operation of taking a limit commute.

Proof First, observe that all the triangles in the diagram

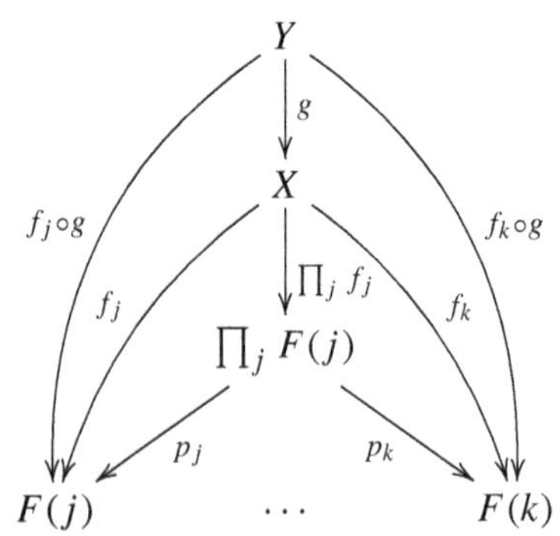

commute. Large triangles obtained by taking the union of the adjacent small triangles also commute. Consider a cone $(Y, \{f_j \circ g\})$. The mediating morphism from Y to the universal cone is given by

$$\prod_{j \in \mathrm{Obj}(J)} (f_j \circ g). \tag{3.3}$$

While morphism

$$\left(\prod_{j \in \mathrm{Obj}(J)} f_j\right) \circ g \tag{3.4}$$

makes all the large triangles

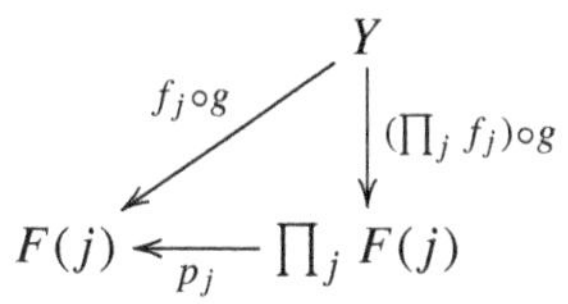

commute. Therefore, it is a mediating morphism from Y to the universal cone. Since the mediating morphism is uniquely determined, we see that (3.3) and (3.4) coincide. □

3.5 Colimits

In this section, we generalize the concept of coproduct to that of colimit. The arguments are almost parallel to those in the previous section.

We introduce the dual concept of cone.

Definition 3.9 Let J be a small category and $\mathscr{A}$ a category. Let $F : J \to \mathscr{A}$ be a J-shaped diagram and A an object of $\mathscr{A}$. A *cocone* from F to A is a set of data:

- For each object j of J, there is a morphism $p_j : F(j) \to A$ called *coprojection*. A is called the *apex* of a cocone.
- For all morphisms $f : j_1 \to j_2$ of J, the diagram

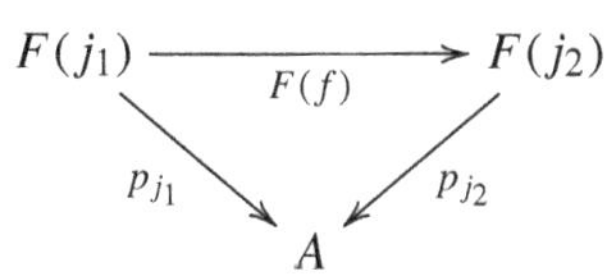

  commutes.

A cone is presented as a triple $(F, \{p_j\}, A)$, or a pair $(\{p_j\}, A)$, or simply $\{p_j\}$, all of which contain the same amount of information.

Definition 3.10 Let J be a small category and $\mathscr{A}$ a category. We fix a J-shaped diagram $F : J \to \mathscr{A}$. We consider a category $(F, \mathscr{A})$ defined as follows.

- Objects are cocones $(\{p_j\}, A)$.

- A morphism from $(\{p_j\}, A)$ to $(\{q_j\}, A')$ is a morphism $u : A \to A'$ of $\mathscr{A}$ such that all the triangles below commute for all morphisms f of J.

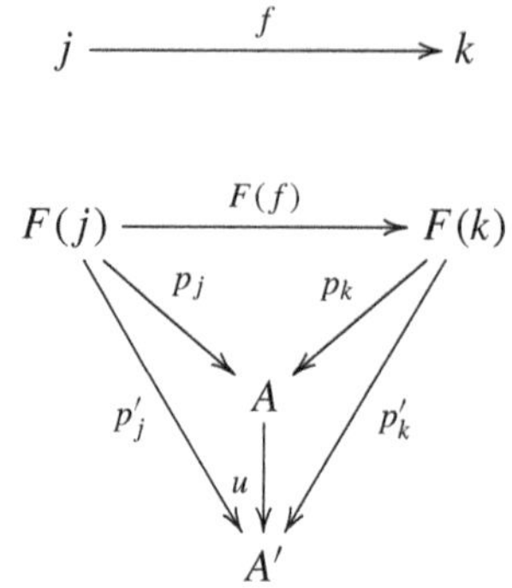

If category $(F, \mathscr{A})$ has an initial object $(\{r_j\}, L)$, we call it a *colimit* of the diagram F. We often call its apex L a colimit of F with implicit $\{r_j\}$. Cone $(L, \{r_j\})$ is also called a *universal cocone* or *limiting cocone*. From the general argument over initial objects, L is unique up to isomorphism. It is often written as $L = \varinjlim F$. Given an arbitrary cocone, a unique morphism to it from the limiting cone is called a *mediating morphism*.

Remark 3.2 The set of all cocones from F to A is written as $\mathrm{Cones}(F, A)$. We have the following isomorphism.

$$\mathscr{A}(\varinjlim F, A) \simeq \mathrm{Cones}(F, A)$$

When $\mathscr{A}$ is locally small, using the language of Chap. 7, the above isomorphism mean that $\varinjlim F$ is an object that represents the covariant functor $\mathrm{Cones}(F, -) : \mathscr{A} \to \mathbf{Set}$. See also Chap. 5.

Example 3.13 Let the index category J be the category with only two objects 1 and 2 and the identity morphisms. Determining a functor F from J to the category $\mathscr{A}$ is the same as choosing two objects in the category $\mathscr{A}$. Given $F(1) = A_1$ and $F(2) = A_2$, the cospan to any object X in the category $\mathscr{A}$

$$A_1 \xrightarrow{x_1} X \xleftarrow{x_2} A_2$$

is the same as considering the cocone from J-shaped diagram F to an object of $\mathscr{A}$. Considering the category of cospans $\mathrm{Cospan}(A_1, A_2)$, and the category of cocones $(F, \mathscr{A})$ is the same. So it turns out that colimit is a generalization of coproduct.

Example 3.14 We consider an index category J with natural numbers $1, 2, 3, \cdots$ as objects and

$$1 \to 2 \to 3 \to \cdots$$

as generators of morphisms. Let $F : J \to \mathbf{Set}$ be a functor where

- function on objects is given by $F(n) = \mathbb{R}^n$,
- function on morphisms is given by $F(n \to n+1) = i_n$, where i_n is an embedding

$$\begin{aligned} i_n : &\mathbb{R}^n \to \mathbb{R}^{n+1} \\ &(x_1, x_2, \cdots, x_n) \mapsto (x_1, x_2, \cdots, x_n, 0). \end{aligned}$$

Let Y be an arbitrary set. A cocone from a J-shaped diagram F to Y is given by a family of morphisms $\{y_n\}$ that makes all the triangles with edges y_n and i_n commute:

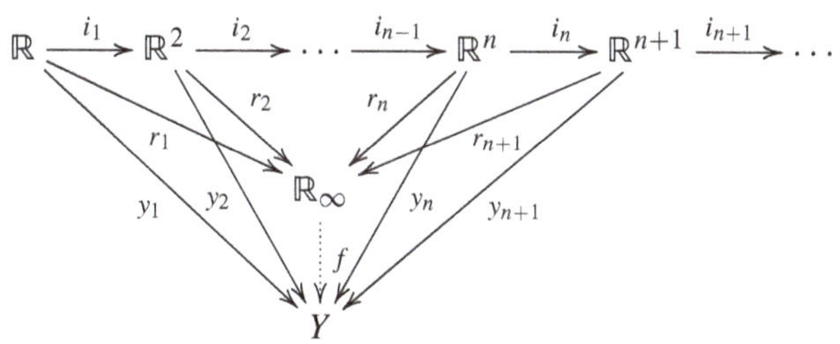

A colimit of the J-shaped diagram is $\mathbb{R}_\infty$, namely an infinite dimensional space consisting of vectors with a finite number of nonzero components. Note that $\mathbb{R}_\infty \neq \mathbb{R}^\infty$. We write $\varinjlim F = \mathbb{R}_\infty$ or

$$\bigcup_n \mathbb{R}^n.$$

Using cocone data $\{y_n\}$, we can construct the unique mediating morphism $f : \mathbb{R}_\infty \to Y$. For any $z \in \mathbb{R}_\infty$ there is a natural number n such that we can regard $z \in \mathbb{R}^n$. With this n we define as

$$f(z) = y_n(z).$$

We may choose a greater n, which gives the same result due to the commutativity of the triangles.

Definition 3.11 We consider an index category J with objects 0, 1, and 2. The following two morphisms

$$1 \leftarrow 0 \rightarrow 2$$

generate the whole morphisms. A colimit of a J-shaped diagram is call a *pushout* or *fibered coproduct*, the dual of pullback. Let $F : J \to \mathscr{A}$ be a functor given by:

- $F(0) = A_0$, $F(1) = A_1$, $F(2) = A_2$, and
- $F(0 \to 1) = a_1$, $F(0 \to 2) = a_2$.

We consider a cocone $(F, \{x_0, x_1, x_2\}, X)$, which is depicted as the left diagram in Eq. (3.5).

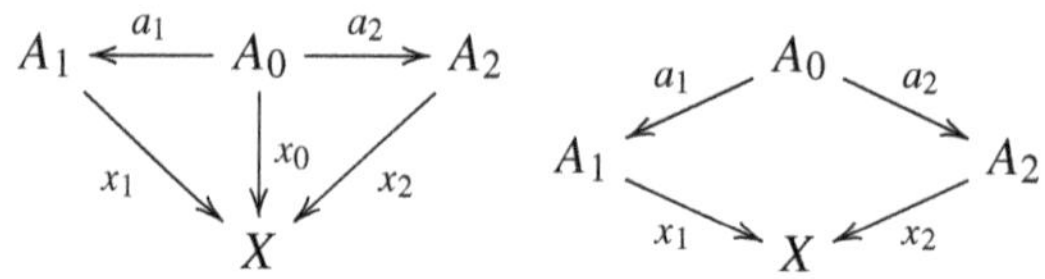

(3.5)

The qualification test for being a cocone is

$$x_0 = x_1 \circ a_1 \quad \text{and} \quad x_0 = x_2 \circ a_2,$$

which is reduced to a single equation $x_1 \circ a_1 = x_2 \circ a_2$. We can confirm this fact as follows. Omitting x_0 gives the diagram on the right. If this diamond is commutative, we can regain morphism x_0 by $x_0 = x_1 \circ a_1$. Thus, both commutative diagrams contain the same information.

Let us suppose there is a universal cocone $(F, \{i_0, i_1, i_2\}, L)$. Taking the above argument into account, all we have to do is to check the commutativity of all the triangles and squares in the following diagram.

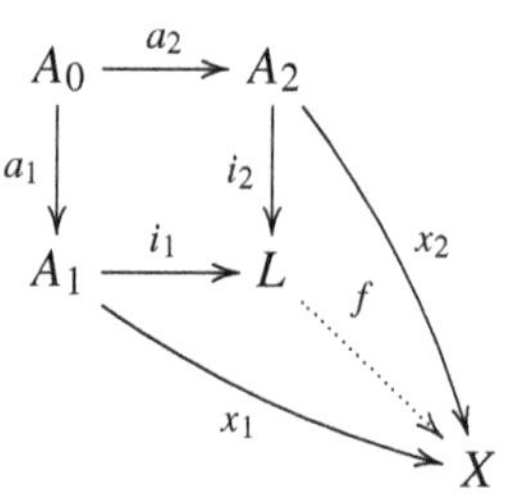

By universality, morphism f is uniquely determined. Note that morphism i_0 is omitted but easily regained by setting $i_0 = i_1 \circ a_1 = i_2 \circ a_2$.

Remark 3.3 Pushouts can be given an alternative definition as follows. We fixed a diagram at the left. Consider a category such that objects are triples (x_1, x_2, X) that makes the diagram at right commute.

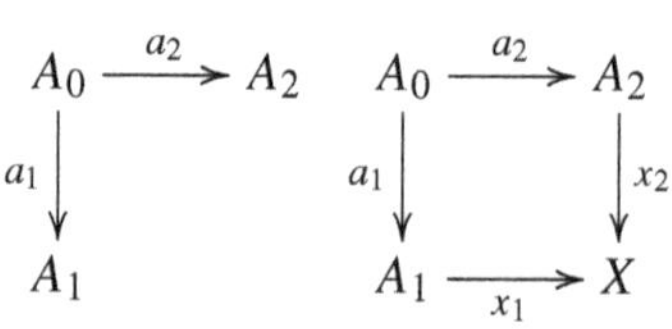

A morphism from (x_1, x_2, X) to (y_1, y_2, Y) must make all the triangles and squares in the following diagram commute.

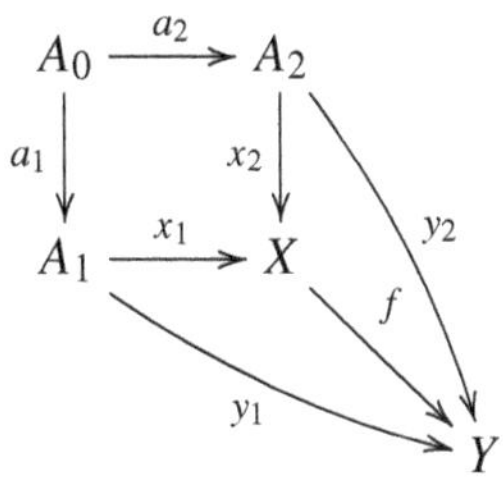

The pushout (i_1, i_2, L) is an initial object of the category so obtained.

Example 3.15 Let $A_0 = \{a, b, c, d\}$, $A_1 = \{1, 2, 3, 4\}$, and $A_2 = \{5, 6, 7, 8\}$. We consider functions $a_1 : A_0 \to A_1$ and $a_2 : A_0 \to A_2$ as follows.

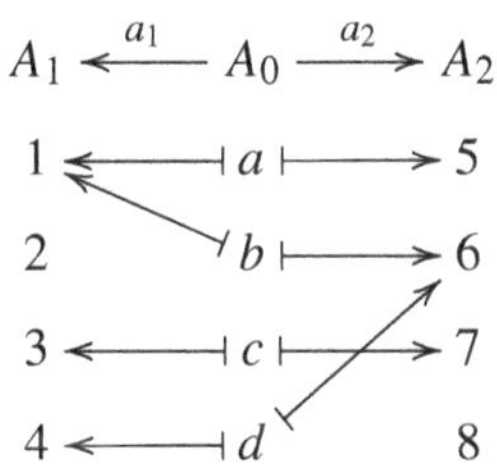

A pushout or fibered coproduct is given by

$$A_1 \bigcup_{A_0} A_2 = \{L2, \quad G[1, 4][5, 6], \quad G[3][7], \quad R8\},$$

where notation needs explanation. $L2$ and $R8$ are points that escape from being glued. $L2$ is an element that comes from the set at left. $R8$ comes from the set at right. $G[1, 4][5, 6]$ is an element formed by gluing 1 and 4 from the set at left, 5 and 6 from the set at right. $G[3][7]$ is an element obtained by gluing 3 and 7.

The above construction went as follows. We first simply take the disjoint union of A_1 and A_2. Then, the gluing is carried out by identifying points $x_1 \in A_1$ and $x_2 \in A_2$ when there is x_0 such that $x_1 = a_1(x_0)$ and $x_2 = a_2(x_0)$. Such identification must be extended transitively. This construction is called a *disjoint union with gluing*.

Haskell code may help you understand what is going on.

Listing 3.5 fibercoproduct.hs

```
data A0 = Va | Vb | Vc | Vd deriving (Ord, Eq, Enum, Show)
data A1 = V1 | V2 | V3 | V4 deriving (Ord, Eq, Enum, Show)
data A2 = V5 | V6 | V7 | V8 deriving (Ord, Eq, Enum, Show)

-- Disjoint Union with Gluing
data DUwG a b c =
```

```
  L b | R c | G [b] [c]  deriving (Ord, Eq, Show)

fibercoproduct :: (Ord a, Ord b, Ord c) =>
  (a -> b) -> (a -> c) -> [a] -> [b] -> [c] -> [DUwG a b c]
fibercoproduct p q as xs ys =
  [L x | x <- xs, not (x `elem` (fmap p as)) ] ++
  glue p q as xs ys ++
  [R y | y <- ys, not (y `elem` (fmap q as)) ]

glue :: (Ord a, Ord b, Ord c) =>
  (a -> b) -> (a -> c) -> [a] -> [b] -> [c] -> [DUwG a b c]
glue p q as xs ys =
  let
      oneStepGlues = [([p a],[q a]) | a<-as]
      classes = collect oneStepGlues
  in
      map (\xys -> G (fst xys) (snd xys)) classes

collect :: (Ord b, Ord c) => [([b],[c])] -> [([b],[c])]
collect []     = []
collect [x]    = [x]
collect (x:xs) =
    fst folded : snd folded
  where
    folded = foldl f (x,[]) (collect xs)
    f (z,zs) y
      | equivQ z y  = (merge z y, zs)
      | otherwise   = (z, y:zs)

equivQ :: (Ord b, Ord c) => ([b],[c]) -> ([b],[c]) -> Bool
equivQ (bs1,cs1) (bs2,cs2) =
    [ x | x <- bs1, y <- bs2, x == y] /= [] ||
    [u | u <- cs1, v <- cs2, u == v] /= []

merge :: (Ord b, Ord c) => ([b],[c]) -> ([b],[c]) -> ([b],[c])
merge (bs1,cs1) (bs2,cs2) =
    (uniq (bs1++bs2), uniq (cs1++cs2))

-- unique sort
uniq :: Ord a => [a] -> [a]
uniq xs =
  foldr ins [] xs
    where
      ins y [] = [y]
      ins y (z:zs)
        | y<z   = y:z:zs
        | y==z  = z:zs
        | otherwise = z:ins y zs

---------------------------------
-- testdata
-- Both a1 and a2 are injective.

a0s = [Va .. Vd]
a1s = [V1 .. V4]
```

```
a2s = [V5 .. V8]

a1 :: A0 -> A1
a1 Va = V1
a1 Vb = V2
a1 Vc = V3
a1 Vd = V4

a2 :: A0 -> A2
a2 Va = V5
a2 Vb = V6
a2 Vc = V7
a2 Vd = V8

{-- suggested tests
fibercoproduct a1 a2 a0s a1s a2s
==> [G [V1] [V5],G [V3] [V7],G [V4] [V8],G [V2] [V6]]
-}

----------------------------------
-- b2 is injective while b1 is not.

b1 :: A0 -> A1
b1 Va = V1
b1 Vb = V1
b1 Vc = V3
b1 Vd = V3

b2 :: A0 -> A2
b2 Va = V5
b2 Vb = V6
b2 Vc = V7
b2 Vd = V8

{-- suggested test
fibercoproduct b1 b2 a0s a1s a2s
==> [L V2,L V4,G [V1] [V5,V6],G [V3] [V7,V8]]
-}

--------------------------------------------------
-- c1 and c2 are non-injective.

c1 :: A0 -> A1
c1 Va = V1
c1 Vb = V1
c1 Vc = V3
c1 Vd = V3

c2 :: A0 -> A2
c2 Va = V6
c2 Vb = V6
c2 Vc = V8
c2 Vd = V8

{-- suggested test
```

```
117 fibercoproduct c1 c2 a0s a1s a2s
118 ==> [L V2,L V4,G [V1] [V6],G [V3] [V8],R V5,R V7]
119 -}
120
121 -----------------------------------------------
122 -- non trivial transition
123 d1 :: A0 -> A1
124 d1 Va = V1
125 d1 Vb = V1
126 d1 Vc = V3
127 d1 Vd = V4
128
129 d2 :: A0 -> A2
130 d2 Va = V5
131 d2 Vb = V6
132 d2 Vc = V7
133 d2 Vd = V6
134
135 {- suggested test
136 fibercoproduct d1 d2 a0s a1s a2s
137 ==> [L V2,G [V1,V4] [V5,V6],G [V3] [V7],R V8]
138 -}
```

Lines 1–3 define data types for A_0, A_1, and A_2 in Example 3.15.

Lines 6–7 define a data type for disjoint union with gluing. Value constructors L and R are for points that escape from being glued. Value constructor G takes two parameters. The first parameter is the list of elements in A_1, and the second is the list of elements in A_2, which are glued into a single point.

Lines 9–14 define a function that constructs a fibered coproduct. The return value is the concatenation of three lists. The first one is the list of points coming from the left (object) that escape from the gluing. The third one is the list of points coming from the right that escape from the gluing. The second one is the list of points that are glued in some way.

The gluing strategy, implemented by the function `glue` in lines 16–23, is as follows:

- the list `oneStepGlues` consists of pairs. We first seek for a pair `(p a, q a)` for each $a \in A_0$. Then, we box `p a` and `p q` in the singleton lists `[p a]` and `[q a]`, which will be used as seeds for the following transitive closure operation.
- the equivalence classes are recursively calculated by the `collect` function in lines 25–34, which is equivalent to taking the transitive closure.

The suggested test is commented on at the bottom of the sample code runs like this, which is derived from Example 3.15.

```
*Main> fibercoproduct d1 d2 a0s a1s a2s
[L V2,G [V1,V4] [V5,V6],G [V3] [V7],R V8]
```

Definition 3.12 We consider a category J with objects 0 and 1. Generators of morphisms are

$$0 \underset{b}{\overset{a}{\rightrightarrows}} 1.$$

A colimit of a J-shaped diagram in category $\mathscr{A}$ is called a *coequalizer*. Let us have a closer look at the situation. Let a functor $F : J \to \mathscr{A}$ be defined by the following:

- function on objects is given by $F(0) = A_1$ and $F(1) = A_2$.
- function on morphisms is given by $F(a) = f$ and $F(b) = g$.

We consider a cocone from F to X.

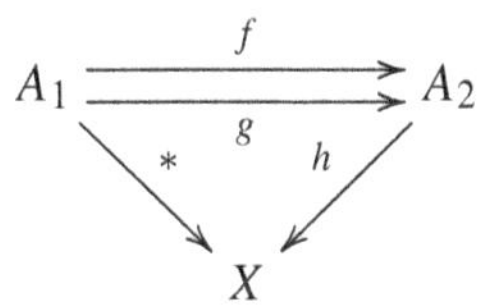

Any cocone must pass a qualification test for being really a cocone:

$$h \circ f = * \quad \text{and} \quad h \circ g = *$$

Notice that we can ignore morphism $*$ without losing any information, considering the following single equation.

$$h \circ f = h \circ g$$

To sum up, getting a colimit of a J-shaped diagram is completely the same as finding a universal object L giving a unique h' that makes the following diagram commute.

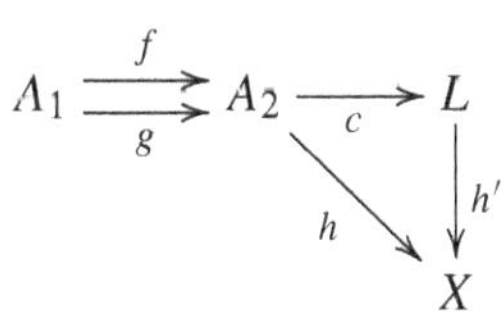

Example 3.16 Let $A_1 = \{1, 2, 3, 4, 5\}$ and $A_2 = \{a, b, c, d, e\}$. We consider the following two functions $f, g : A_1 \to A_2$:

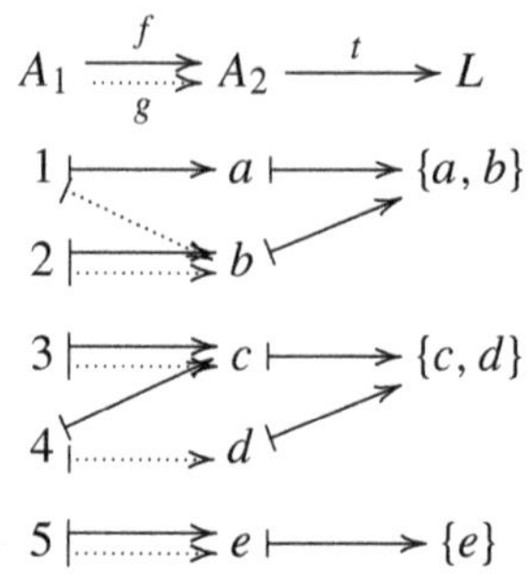

g is depicted with dotted lines. Function $t : A_2 \to \{\{a, b\}, \{c, d\}, \{e\}\}$ is a coequalizer. If the images of f and g of the same point differ, we bundle these points in L. This eliminates the differences

$$f(1) \neq g(1), \quad f(4) \neq g(4).$$

The Haskell code for this example is given below.

Listing 3.6 product.hs

```
1  data SetA = A7 | A8 | A9 deriving (Enum,Show)
2
3  data SetB = Ba | Bb | Bc deriving (Enum,Show)
4
5  data SetX = X1 | X2 | X3 | X4 deriving (Enum,Show)
6
7  x1 :: SetX -> SetA
8  x1 X1 = A7
9  x1 X2 = A8
10 x1 X3 = A9
11 x1 X4 = A7
12
13 x2 :: SetX -> SetB
14 x2 X1 = Ba
15 x2 X2 = Bb
16 x2 X3 = Bc
17 x2 X4 = Bc
18
19 factor :: (x -> a) -> (x -> b) -> (x -> (a,b))
20 factor p1 p2 x = (p1 x, p2 x)
```

Lines 1–9 define a function `coequate` that takes two functions $p, q : A \to B$ and lists of elements of A and B as parameters and returns a list of glued points in B.

Let us visualize how `coequate` works with concrete data, namely two functions $f : A_1 \to A_2$ and $h : A_1 \to A_2$ defined in lines 51–56 and 65–70, respectively.

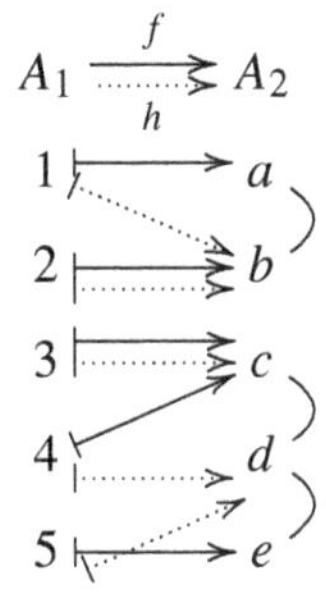

The three small arcs appearing on the right show the relation returned by `coequate`. The minimal information is given as follows.

$$a \sim^0 b, \quad c \sim^0 d, \quad d \sim^0 e$$

Lacking transitivity, we cannot claim $c \sim^0 e$. Try suggested tests.

Definition 3.13 Let J be a discrete category, namely a set. If a J-shaped diagram $F : J \to \mathscr{A}$ has a colimit, it is called a *coproduct* or *direct sum*. Notice that J may be infinite.

Given a cocone from F to X, in other words, a set of morphisms $\{f_j : F(j) \to X\}_{j \in \mathrm{Obj}(J)}$, the diagram

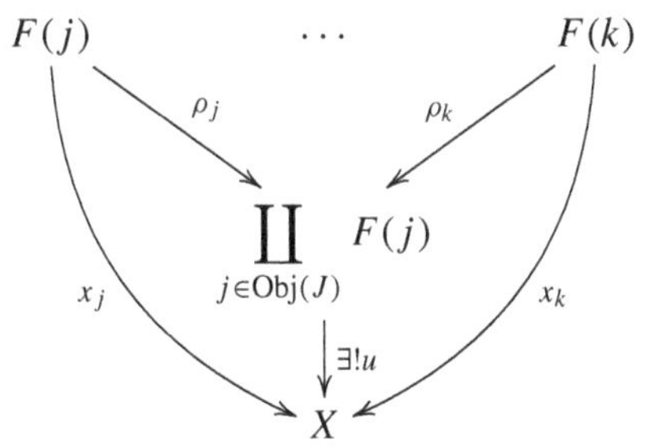

has a mediating morphism u, which we especially denote by

$$u = \coprod_{j \in \mathrm{Obj}(J)} f_j.$$

Proposition 3.6 *We use the notation in Definition 3.13. Given a morphism $g : X \to Y$, we have the following:*

$$g \circ \left(\coprod_{j \in \mathrm{Obj}(J)} x_j \right) = \coprod_{j \in \mathrm{Obj}(J)} (g \circ x_j),$$

which means that operating g from left is commutative with the coproduct operation.

Proof In diagram

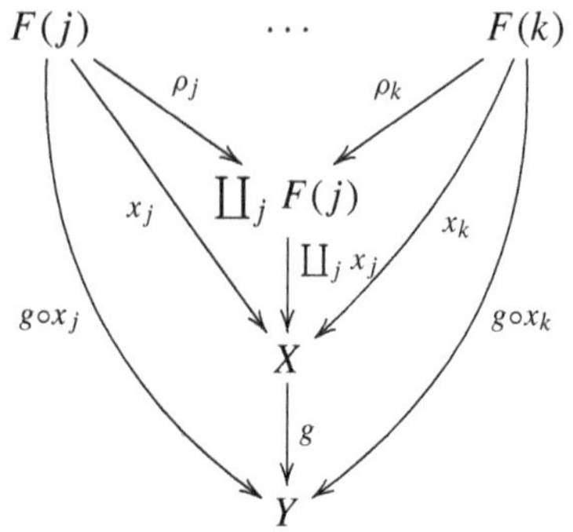

the triangles are all commutative. Larger triangles made by composition are also commutative. Let us consider a cocone $(\{g \circ x_j\}, Y)$. A mediating morphism from a universal cocone to Y is

$$\coprod_{j \in \mathrm{Obj}(J)} (g \circ x_j). \tag{3.6}$$

The morphism

$$g \circ \left(\coprod_{j \in \mathrm{Obj}(J)} x_j \right) \tag{3.7}$$

makes all the triangles

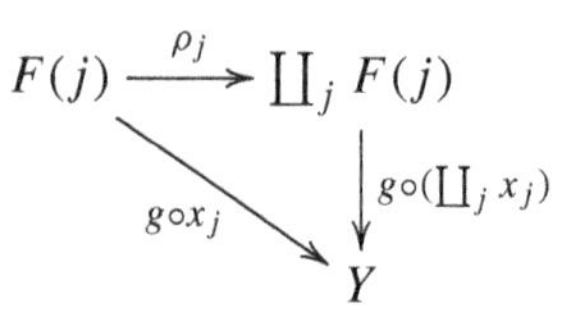

commute. Therefore it is a mediating morphism, which should be unique. Thus, morphisms (3.6) and (3.7) coincide □

3.6 Existence of Limits

Not all J-shaped diagrams $F : J \to \mathscr{A}$ have limits or colimits.

Definition 3.14

- For a fixed index category J, if all J-shaped diagrams in $\mathscr{A}$ have limits[colimits], category $\mathscr{A}$ is said to *have limits[colimits] of shape J*.
- If a category $\mathscr{A}$ have limits[colimits] of shape J for all small and discrete categories J, then $\mathscr{A}$ is said to *have products*[*have coproducts*].

- If all $\bullet \rightrightarrows \bullet$ -shaped diagrams in $\mathscr{A}$ have limits[colimits], category $\mathscr{A}$ is said to *have equalizers*[*have coequalizers*].
- If all $\bullet \leftarrow \bullet \rightarrow \bullet$-shaped diagrams have limits, category $\mathscr{A}$ is said to *have pullbacks*.
- If all $\bullet \rightarrow \bullet \leftarrow \bullet$-shaped diagrams have colimits, category $\mathscr{A}$ is said to *have pushouts*.
- If a category $\mathscr{A}$ has limits[colimits] of shape J for all small categories J, category $\mathscr{A}$ is said to be *complete*[*cocomplete*].

An often used sufficient condition for a category to have limits is presented below. A little preparation is required for this.

Let $\mathscr{A}$ be a category that has equalizers, and J be an index category. We regard $\mathrm{Obj}(J)$ and $\mathrm{Mor}(J)$ are both discrete categories. Suppose category $\mathscr{A}$ has products. There is a universal cone

$$\begin{array}{ccccc} & & \prod\limits_{f \in \mathrm{Mor}(J)} F(\mathrm{cod}(f)) & & \\ & \swarrow^{\pi_f} & & \searrow^{\pi_{f'}} & \\ F(\mathrm{cod}(f)) & & \cdots & & F(\mathrm{cod}(f')) \end{array} \tag{3.8}$$

to a diagram $F \circ \mathrm{cod}$ of shape $\mathrm{Mor}(J)$, namely a product indexed by $f \in \mathrm{Mor}(J)$. The data structure is presented as follows.

$$\left(\prod_{f \in \mathrm{Mor}(J)} F(\mathrm{cod}(f)), \quad \{\pi_f\}_{f \in \mathrm{Mor}(J)}, \quad F \circ \mathrm{cod} \right)$$

Let us consider another cone to $F \circ \mathrm{cod}$

$$\begin{array}{ccccc} & & \prod\limits_{j \in \mathrm{Obj}(J)} F(j) & & \\ & \swarrow^{p_{\mathrm{dom}(f)}} & & \searrow^{p_{\mathrm{dom}(f')}} & \\ F(\mathrm{dom}(f)) & & & & F(\mathrm{dom}(f')) \\ \downarrow{\scriptstyle F(f)} & & & & \downarrow{\scriptstyle F(f')} \\ F(\mathrm{cod}(f)) & & \cdots & & F(\mathrm{cod}(f')) \end{array}$$

with data structure

$$\left(\prod_{j \in \mathrm{Obj}(J)} F(j), \quad \{F(f) \circ p_{\mathrm{dom}(f)}\}_{f \in \mathrm{Mor}(J)}, \quad F \circ \mathrm{cod} \right),$$

where p_j's are projections accompanying the product $\prod_{j \in \mathrm{Obj}(J)} F(j)$. These projections appear in the form of $p_{\mathrm{dom}(f)}$ in the diagram. It often happens that $\mathrm{dom}(f_1) = \mathrm{dom}(f_2)$ for a pair $f_1 \neq f_2$. So, the same $p_{\mathrm{dom}(f)}$ may appear repeatedly in the upper half of the diagram. However, this does not cause any inconvenience since the base of the cone is indexed by f. The universal cone (3.8) gives a unique mediating morphism s that makes all the squares in the diagram commute.

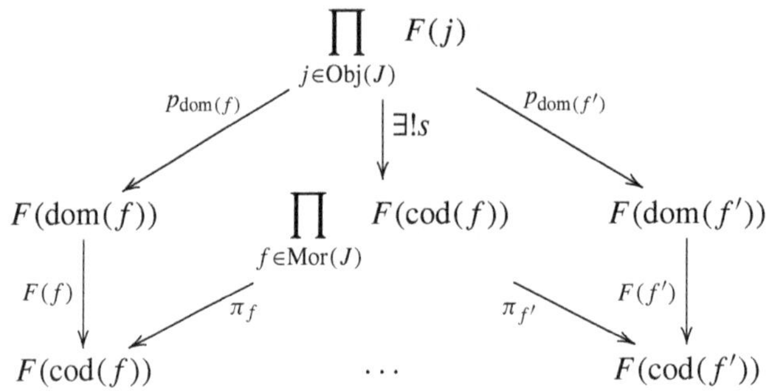

Universality of products allows us to consider the products of morphisms. We often write as

$$s = \prod_{f \in \mathrm{Mor}(J)} F(f) \circ p_{\mathrm{dom}(f)}.$$

We consider yet another cone from $\prod_{j \in \mathrm{Obj}(J)} F(j)$ to $F \circ \mathrm{cod}$, the base of which is indexed by $f \in \mathrm{Mor}(J)$.

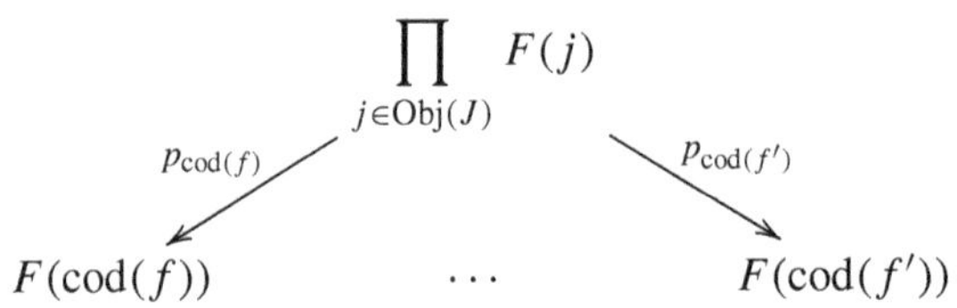

The data structure for this is given as

$$\left(\prod_{j \in \mathrm{Obj}(J)} F(j), \quad \{p_{\mathrm{cod}(f)}\}_{f \in \mathrm{Mor}(J)}, \quad F \circ \mathrm{cod} \right)$$

It is sometimes the case that $\mathrm{cod}(f_1) = \mathrm{cod}(f_2)$ for a pair $f_1 \neq f_2$, which means that some objects repeatedly appear in the base of the cone. Notice that no trouble is caused since the base is indexed by f.

There is a unique morphism t to the universal cone (3.8)

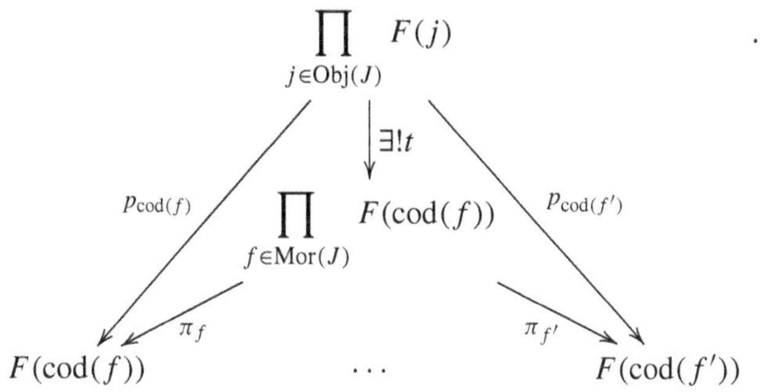

Universality of products allows us to consider the products of morphisms. We often write as

$$t = \prod_{f \in \mathrm{Mor}(J)} p_{\mathrm{cod}(f)}.$$

We got two morphisms s and t

$$\prod_{j \in \mathrm{Obj}(J)} F(j) \overset{s}{\underset{t}{\rightrightarrows}} \prod_{f \in \mathrm{Mor}(J)} F(\mathrm{cod}(f)) \,. \tag{3.9}$$

Now, we can formulate the theorem for the existence of limits.

Theorem 3.1 *Let $\mathscr{A}$ be a category with equalizers and products, and J be an index category. We regard* $\mathrm{Obj}(J)$ *and* $\mathrm{Mor}(J)$ *as discrete categories, over which we may form products. Under these assumptions, we have an equalizer $e : K \to \prod_{j \in \mathrm{Obj}(J)} F(j)$ for two morphisms s and t in* (3.9). *The cone from K to F obtained by composing e with projections p_j's gives a limit for F. We may write as $K = \varprojlim F$. The data structure for the limiting cone is presented as*

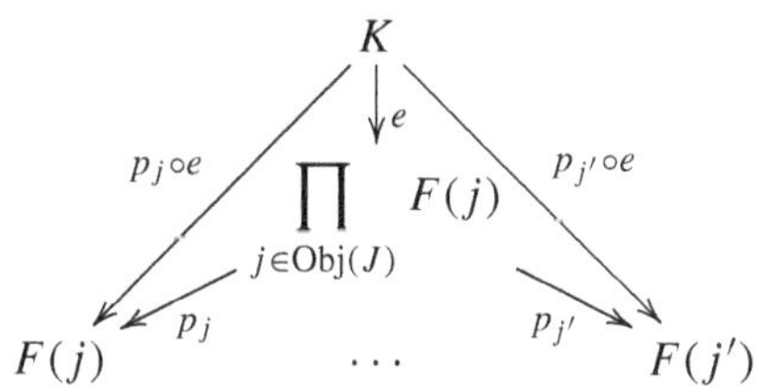

$$(K, \{p_j \circ e\}_{j \in \mathrm{Obj}(J)}, F).$$

Proof Let $(X, x = \{x_j\}_{j \in \mathrm{Obj}(J)}, F)$ be any cone to F. This cone automatically is a cone to $F|_{\mathrm{Obj}(J)}$, where the discrete category $\mathrm{Obj}(J)$ is considered as a subcategory of J. Universality of products gives a unique morphism x in the commutative diagram below.

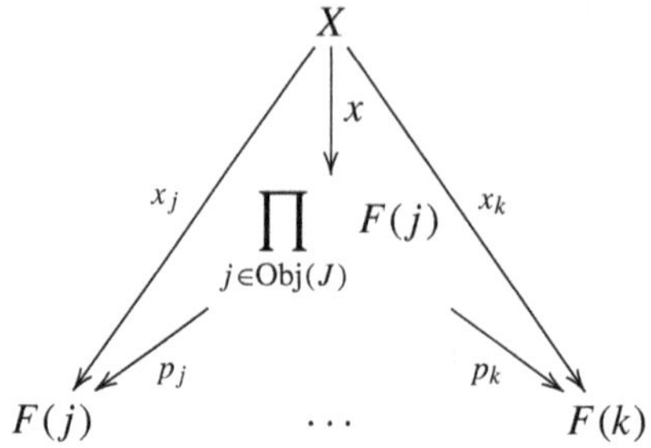

Composing with parallel morphisms s and t, we get

$$X \xrightarrow{x} \prod_{j \in \mathrm{Obj}(J)} F(j) \underset{t}{\overset{s}{\rightrightarrows}} \prod_{f \in \mathrm{Mor}(J)} F(\mathrm{cod}(j)) \, ,$$

where $x = \prod_{j \in \mathrm{Obj}(J)} x_j$.

We now show that $s \circ x = t \circ x$. Rewriting the LHS, we have

$$\begin{aligned} s \circ x &= \left(\prod_{f \in \mathrm{Mor}(J)} F(f) \circ p_{\mathrm{dom}(f)} \right) \circ x \\ &= \prod_{f \in \mathrm{Mor}(J)} \left(F(f) \circ p_{\mathrm{dom}(f)} \circ x \right) \\ &= \prod_{f \in \mathrm{Mor}(J)} \left(F(f) \circ x_{\mathrm{dom}(f)} \right) . \end{aligned}$$

Line 2 is derived by Proposition 3.5 which claims that limits and composition of a morphism commute. Line 3 is obtained directly from the definition of x, namely $p_j \circ x = x_j$ with j replaced by $\mathrm{dom}(f)$.

Rewriting the RHS, we have

$$\begin{aligned} t \circ x &= \left(\prod_{f \in \mathrm{Mor}(J)} p_{\mathrm{cod}(f)} \right) \circ x \\ &= \prod_{f \in \mathrm{Mor}(J)} \left(p_{\mathrm{cod}(f)} \circ x \right) \\ &= \prod_{f \in \mathrm{Mor}(J)} x_{\mathrm{cod}(f)} . \end{aligned}$$

Line 2 is derived from the fact that limits and composition of a morphism commute. Line 3 is obtained directly from the definition of x, namely $p_j \circ x = x_j$ with j replaced by $\mathrm{cod}(f)$.

Note that $\{x_j\}$ is a cone to F. Then, we see that all the triangles

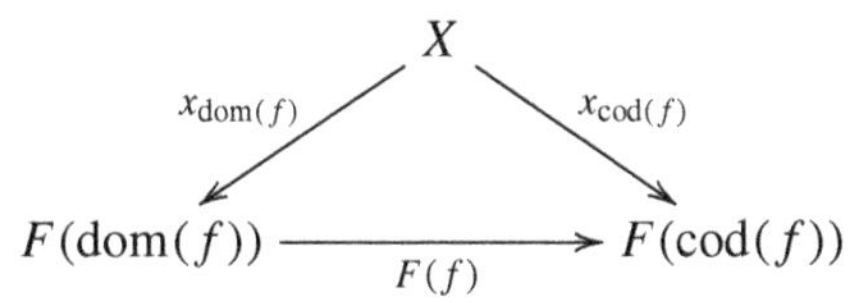

commute. In other words, for all f we have

$$F(f) \circ x_{\mathrm{dom}(f)} = x_{\mathrm{cod}(f)},$$

which means $s \circ x = t \circ x$.

Universality of equalizer e gives a unique mediating morphism $u : X \to K$ that makes the diagram

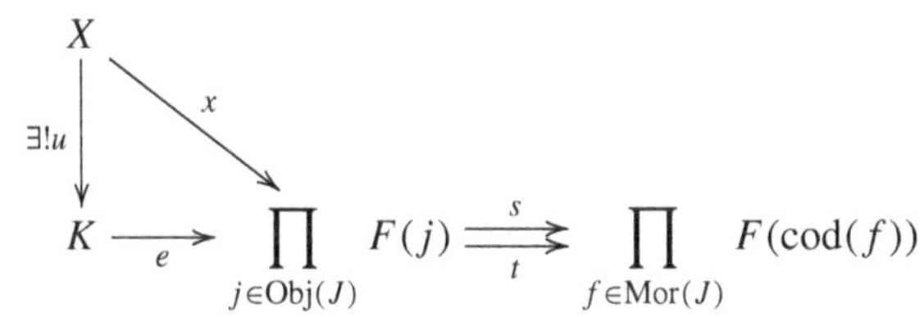

commute. The data $\{p_j \circ e\}$ gives a universal cone to J-shaped diagram F. We have

$$K = \varprojlim F.$$

□

So much for a formal discussion. Let us do it concretely.

Example 3.17 Let J be a category given below.

$$J = (1 \xrightarrow{f} 2).$$

The following two sets

$$\mathrm{Obj}(J) = \{1, 2\}$$
$$\mathrm{Mor}(J) = \{1_1, 1_2, f\}$$

are considered as discrete categories. We assume existence of products. We have the limiting cone below.

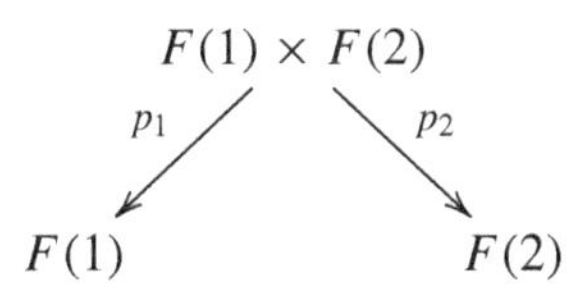

We reindex the base with $\mathrm{Mor}(J)$.

$$\mathrm{cod}(1_1) = 1, \quad \mathrm{cod}(1_2) = \mathrm{cod}(f) = 2$$

Projections indexed by $\mathrm{Mor}(J)$ are μ's below.

$$\mu_{1_1} = p_1, \quad \mu_{1_2} = p_2, \quad \mu_f = p_2$$

We get a cone from $\prod_{j \in \mathrm{Obj}(J)} F(j) = F(1) \times F(2)$ to $\mathrm{Mor}(J)$-shaped diagram $F \circ \mathrm{cod}$

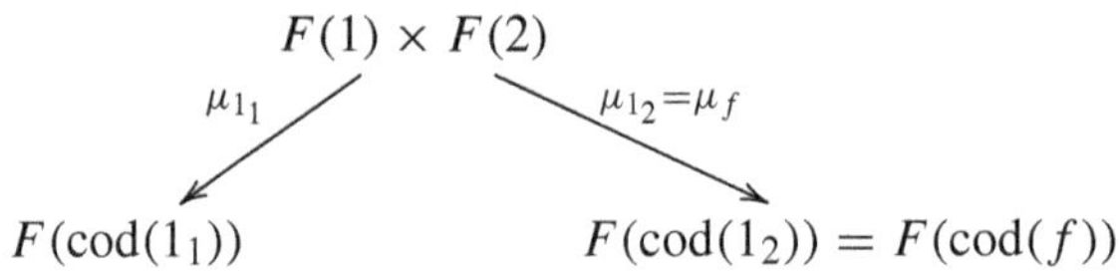

We assume a universal cone to $\mathrm{Mor}(J)$-shaped diagram $F \circ \mathrm{cod}$, which in this case a product:

$$\prod_{f \in \mathrm{Mor}(J)} F(\mathrm{cod}(f)) = F(1) \times F(2) \times F(2).$$

We have the following.

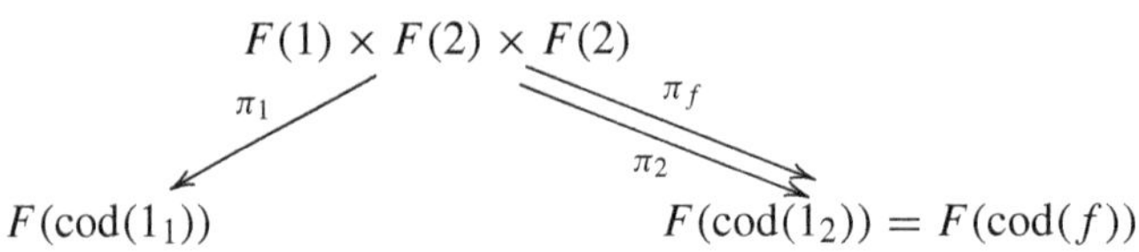

The mediating morphism t makes all the triangles in the following diagram commute.

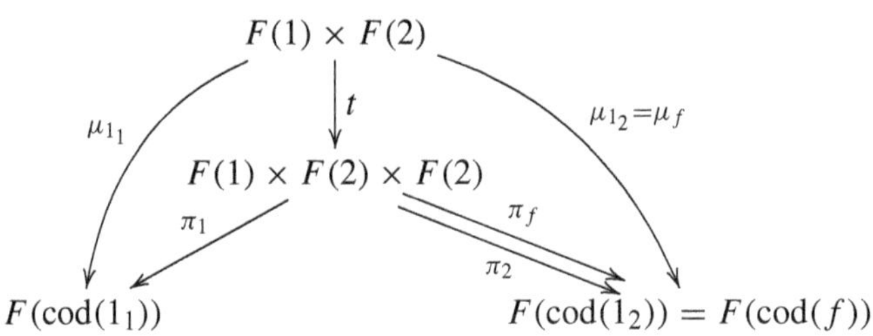

Next, we consider another cone to $\mathrm{Mor}(J)$-shaped diagram $F \circ \mathrm{cod}$

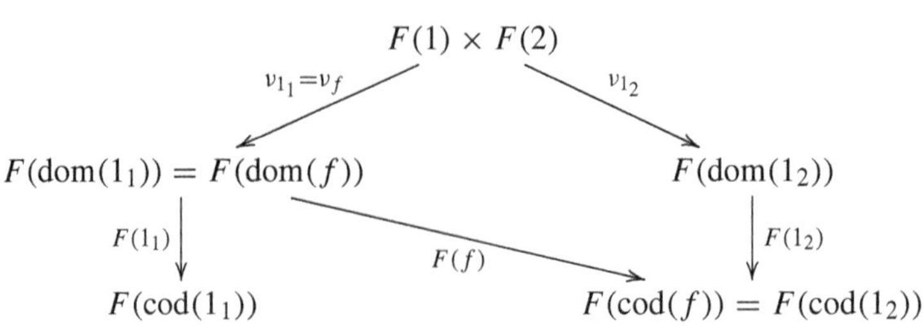

where ν's are projections accompanying the cone from $F(1) \times F(2)$ to $F \circ \text{dom}$ that satisfy

$$\nu_{1_1} = \nu_f = p_1, \quad \nu_{1_2} = p_2.$$

The whole diagram amounts to considering the cone

$$(F(1) \times F(2), \{F(1_1) \circ \nu_{1_1}, F(1_2) \circ \nu_{1_2}, F(f) \circ \nu_f\}, F \circ \text{cod}).$$

The mediating morphism from this cone to the universal cone is obtained as

$$s : F(1) \times F(2) \to F(1) \times F(2) \times F(2).$$

Let us revisit a previous Example 3.17. This time we take $\mathscr{C} = \textbf{Set}$. We will realize that limits in **Set** are subsets of a product.

Example 3.18 Let J be a category

$$J = (1 \xrightarrow{f} 2).$$

We have

$$\begin{aligned} \text{Obj}(J) &= \{1, 2\} \\ \text{Mor}(J) &= \{1_1, 1_2, f\}. \end{aligned}$$

We consider a diagram $F : J \to \textbf{Set}$ as follows. Its function on objects is given by

$$F(1) = \{3, 4, 5\}, \quad F(2) = \{6, 7\}.$$

Since J has only one morphism other than the identities, F's function on morphisms is determined by the image of f. We set

$$\begin{array}{rl} F(f): & F(1) \to F(2) \\ & 3 \mapsto 6 \\ & 4 \mapsto 6 \\ & 5 \mapsto 7 \end{array}$$

We follow the construction in Example 3.17. First, we have

$$\begin{aligned} F(\text{cod}(1_1)) &= F(1) = \{3, 4, 5\} \\ F(\text{cod}(1_2)) &= F(2) = \{6, 7\} \\ F(\text{cod}(f)) &= F(2) = \{6, 7\} \\ \prod_{h \in \text{Mor}(J)} F(\text{cod}(h)) &= \{3, 4, 5\} \times \{6, 7\} \times \{6, 7\} \end{aligned}$$

We construct a mapping t in the diagram

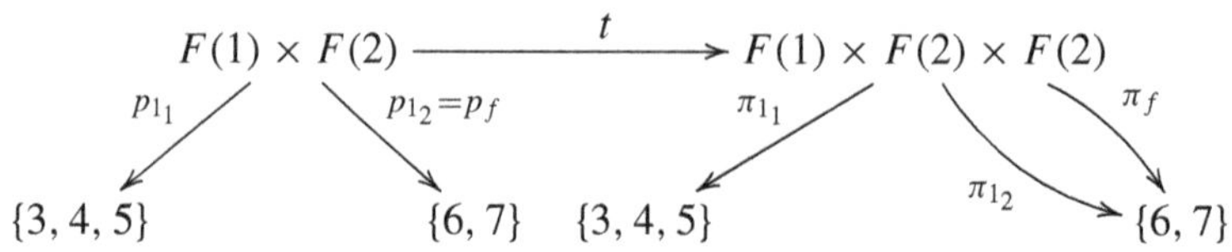

as follows. Notice that the component $F(2)$ is diagonally duplicated in the codomain.

$$
\begin{array}{rcl}
\text{indexed by 1 and 2} & & \text{indexed by } 1_1,\ 1_2,\ \text{and } f \\
t : F(1) \times F(2) & \longrightarrow & F(1) \times F(2) \times F(2) \\
(3,6) & \longmapsto & (3,6,6) \\
(3,7) & \longmapsto & (3,7,7) \\
(4,6) & \longmapsto & (4,6,6) \\
(4,7) & \longmapsto & (4,7,7) \\
(5,6) & \longmapsto & (5,6,6) \\
(5,7) & \longmapsto & (5,7,7)
\end{array}
$$

The correspondence is found in the following way.

- On the left side, list out all the pair from $F(1) \times F(2)$. Note that $F(1)$ and $F(2)$ are indexed by 1 and 2.
- On the right side, prepare three columns that are indexed by 1_1, 1_2, and f. Now, we have totally five columns indexed by 1, 2, 1_1, 1_2, f from left.
- Copy column $1 = \mathrm{cod}(1_1)$ (the first column from left) to column 1_1 (the third column from left).
- Copy column $2 = \mathrm{cod}(1_2)$ (the second column from left) to column 1_2 (the forth column from left).
- Copy column $2 = \mathrm{cod}(f)$ (the second column from left) to column f (the right-most column).

Construction of s needs some detour.

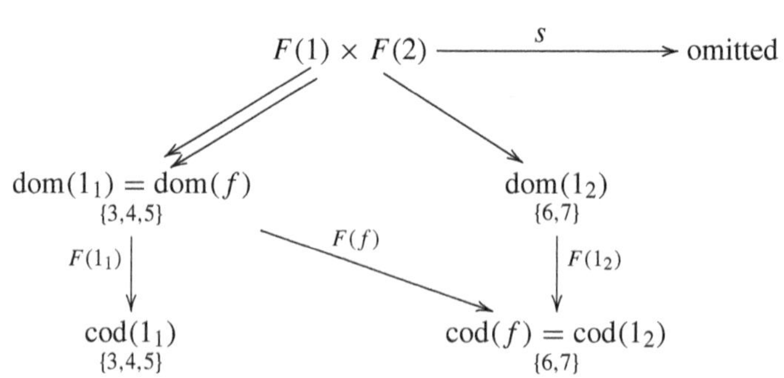

The mapping s is determined as follows.

$$\begin{array}{rcl} \text{indexed by 1 and 2} & & \text{indexed by } 1_1,\ 1_2, \text{ and } f \\ s : F(1) \times F(2) & \longrightarrow & F(1) \times F(2) \times F(2) \\ (3,6) & \longmapsto & (3,6,6) \\ (3,7) & \longmapsto & (3,7,6) \\ (4,6) & \longmapsto & (4,6,6) \\ (4,7) & \longmapsto & (4,7,6) \\ (5,6) & \longmapsto & (5,6,7) \\ (5,7) & \longmapsto & (5,7,7) \end{array}$$

This correspondence is found in the following way. Pairs and triplets are vertically lined up. As a whole they may be viewed as five columns lined up from left to right, being indexed by $1, 2, 1_1, 1_2$, and f.

- Columns 1 and 2 (two columns from the left part) list out all elements of $F(1) \times F(2)$ as pairs.
- Columns 1_1, 1_2, and f (three columns from the right part) are reserved for storing triplets.
- Column 1_1 is filled with images of $F(1_1) = 1_{F(1)}$ of the corresponding element in column $1 = \text{dom}(1_1)$. In this case, obtained by simply copying the whole column 1.
- Column 1_2 is filled with images of $F(1_2) = 1_{F(2)}$ of the corresponding element in column $2 = \text{dom}(1_2)$. In this case, obtained by simply copying the whole column 2.
- Column f is filled with images of $F(f)$ of the corresponding element in column $1 = \text{dom}(f)$. This time, we have to trace exact correspondence by $F(f)$.

This completes the construction of s.

To equalize s and t, we line up the images side by side.

$$\begin{array}{rclc} & & \text{by } t & \text{by } s \\ (3,6) & \longmapsto & (3,6,6) & (3,6,6) \\ (3,7) & \longmapsto & (3,7,7) & (3,7,6) \\ (4,6) & \longmapsto & (4,6,6) & (4,6,6) \\ (4,7) & \longmapsto & (4,7,7) & (4,7,6) \\ (5,6) & \longmapsto & (5,6,6) & (5,6,7) \\ (5,7) & \longmapsto & (5,7,7) & (5,7,7) \end{array}$$

Having observed this data, we find that e is a mapping that embeds

$$\{(3,6), (4,6), (5,7)\}$$

to $\{3, 4, 5\} \times \{6, 7\}$.

Haskell code for Example 3.18 is given below.

Listing 3.7 toylimit.hs

```
-- Category J
---- Obj(J) = {1, 2}
data ObjJ = O1 | O2  deriving (Eq, Enum, Show)

---- Mor(J) = {1_1, 1_2, f}
data MorJ = I1 | I2 | Mf  deriving (Eq, Enum, Show)

---- dom : Mor(J) -> Obj(J)
domJ :: MorJ -> ObjJ
domJ I1 = O1
domJ I2 = O2
domJ Mf = O1

---- cod : Mor(J) -> Obj(J)
codJ :: MorJ -> ObjJ
codJ I1 = O1
codJ I2 = O2
codJ Mf = O2

-- toy sets
---- F(1) = {3,4,5}
data F1 = V3 | V4 | V5  deriving (Eq, Show)

---- F(2) = {6,7}
data F2 = V6 | V7 deriving (Eq, Show)

---- F(1_1)
fmapI1 :: F1 -> F1
fmapI1 x = x

---- F(1_2)
fmapI2 :: F2 -> F2
fmapI2 x = x

---- F(f)
fmapMf :: F1 -> F2
fmapMf V3 = V6
fmapMf V4 = V6
fmapMf V5 = V7

-- t
t :: (F1,F2) -> (F1,F2,F2)
t (x,y) = (x,y,y)

-- s
s :: (F1,F2) -> (F1,F2,F2)
s (x,y) = (fmapI1 x, fmapI2 y, fmapMf x)

-- equalizer
objK = [(x,y) | x <- [V3,V4,V5], y <- [V6,V7],
                s (x,y) == t (x,y) ]
```

Lines 3 and 6 define types for objects and morphisms of J.
Lines 8–18 define the dom and cod functions.
Lines 20–25 define $F(1)$ and $F(2)$ as types.
Lines 27–39 define the function on morphisms of functor F.
Lines 41–47 define morphisms s and t.
Lines 49–51 define an object K for the equalizer $e : K \to F(1) \times F(2)$.
A test can be done as follows.

```
*Main> objK
[(V3,V6),(V4,V6),(V5,V7)]
```

Let us reflect on what cones are in category **Set** before we proceed to the next example. Given a J-shaped diagram $F : J \to$ **Set**, all $F(j)$'s are sets. So, we can talk about their elements. Consider a cone from A to F. We require the following diagram to commute.

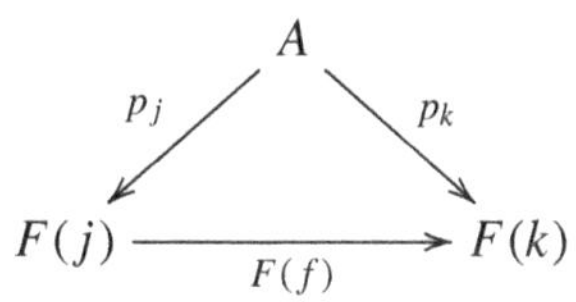

This means that $(F(f))(p_j(a)) = p_k(a)$ for all $a \in A$ and $f : j \to k$.

Example 3.19 Category **Set** has products, which are Cartesian products. Category **Set** also has equalizers. An equalizer of two mappings $f, g : A \to B$ can be defined by a subset $K \subset A$ such that

$$K = \{x \in A \mid f(x) = g(x)\}$$

and its embedding $e : K \to A$.

Since **Set** has products and equalizers, we can claim, by Theorem 3.1, that **Set** has limits.

Limits in **Set** appear in the following manner. Let $F : J \to$ **Set** be a J-shaped diagram. We present an element of the Cartesian product $\prod_{j \in \mathrm{Obj}(J)} F(j)$ as $(a_j)_{j \in \mathrm{Obj}(J)}$. A subset defined as the solution set of the system of equations

$$(F(f))(a_j) = a_k \quad (f \in \mathrm{Mor}(J)),$$

where unknowns are indexed by $\mathrm{Obj}(J)$ and equations are indexed by $\mathrm{Mor}(J)$.

3.7 Existence of Colimits

An often used sufficient condition for a category to have colimits is presented below. The dual of Theorem 3.1 will be given. It is often said that the dual of some theorem can be simply proved by reversing the arrows. When it comes to calculate examples in category **Set**, unfortunately, this is not the case. This is because there is a significant asymmetry in **Set**. Just think about the condition for a correspondence to be a function. There are many things to take care.

We repeat the same preparation as for the limit.

Let $\mathscr{A}$ be a category that has coequalizers, and J be an index category. We regard $\mathrm{Obj}(J)$ and $\mathrm{Mor}(J)$ are both discrete categories. Suppose category $\mathscr{A}$ has coproducts. There is a universal cocone

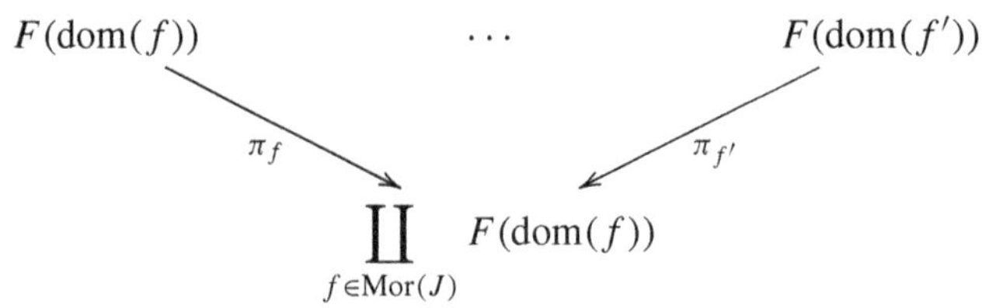

from a diagram $F \circ \mathrm{dom} : \mathrm{Mor}(J) \to \mathscr{A}$ of shape $\mathrm{Mor}(J)$, namely a coproduct indexed by $f \in \mathrm{Mor}(J)$. The data structure is presented as follows.

$$\left(F \circ \mathrm{dom}, \quad \{\pi_f\}_{f \in \mathrm{Mor}(J)}, \coprod_{f \in \mathrm{Mor}(J)} F(\mathrm{dom}(f)) \right),$$

where π_f's are coprojections.

Let us consider another cocone from $F \circ \mathrm{dom}$

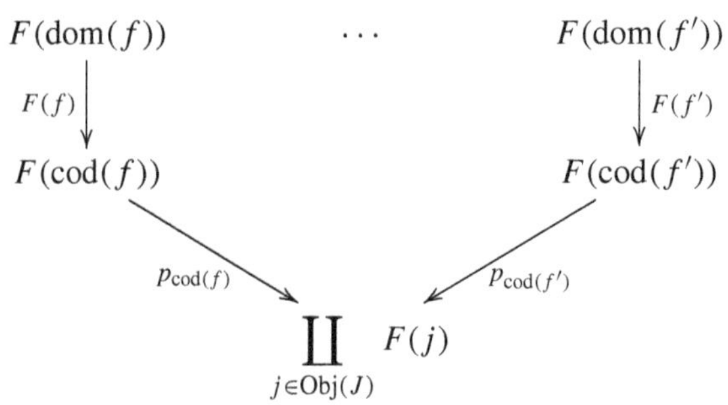

with data structure

$$\left(F \circ \mathrm{dom}, \quad \{p_{\mathrm{cod}(f)} \circ F(f)\}_{f \in \mathrm{Mor}(J)}, \coprod_{j \in \mathrm{Obj}(J)} F(j) \right),$$

where p_j's are coprojections accompanying the coproduct $\coprod_{j \in \mathrm{Obj}(J)} F(j)$. These coprojections appear in the form of $p_{\mathrm{cod}(f)}$ in the diagram. It often happens that $\mathrm{cod}(f_1) = \mathrm{cod}(f_2)$ for a pair $f_1 \neq f_2$. So, the same $p_{\mathrm{cod}(f)}$ may appear repeatedly in the lower half of the diagram. However, this does not cause any inconvenience since the base of the cocone is indexed by f. Therefore, there is a unique mediating morphism s

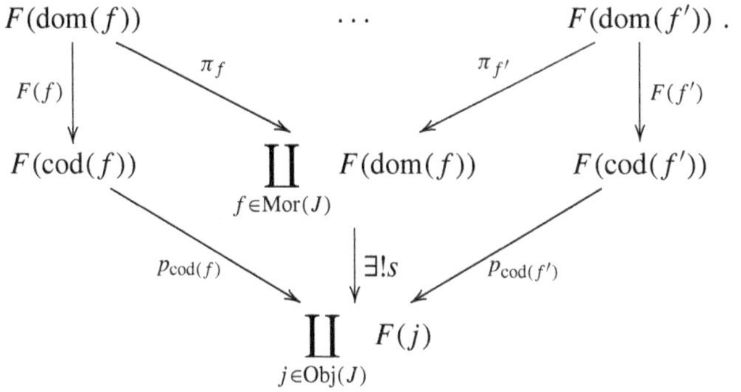

We may write

$$s = \coprod_{f \in \mathrm{Mor}(J)} p_{\mathrm{cod}(f)} \circ F(f).$$

We consider yet another cocone from $\mathrm{Mor}(J)$-shaped diagram $F \circ \mathrm{dom}$ to $\coprod_{j\ \mathrm{Obj}(J)} F(j)$:

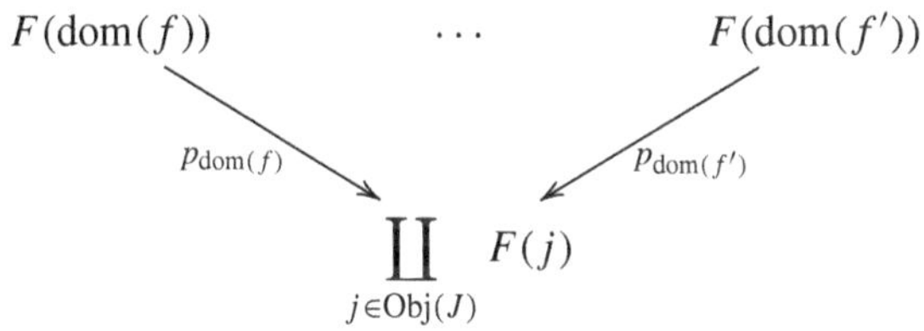

The data structure is given by

$$\left(F \circ \mathrm{dom}, \quad \{p_{\mathrm{dom}(f)}\}_{f \in \mathrm{Mor}(J)}, \coprod_{j \in \mathrm{Obj}(J)} F(j) \right).$$

This cocone has a base indexed by $f \in \mathrm{Mor}(J)$. Therefore, there is a unique mediating morphism t:

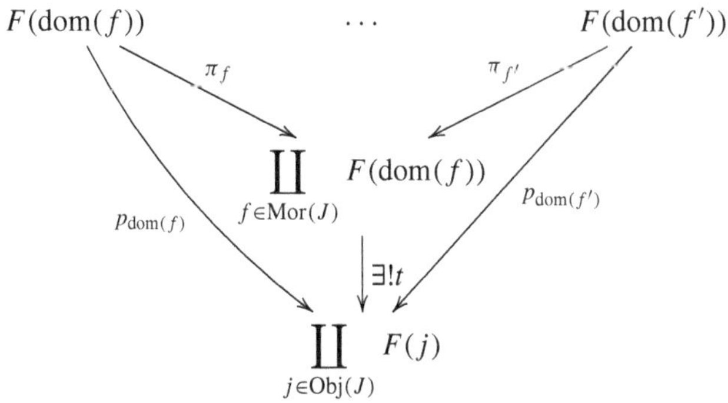

We may write

$$t = \coprod_{f \in \mathrm{Mor}(J)} p_{\mathrm{dom}(f)}.$$

So much for the construction of the parallel morphisms s and t:

$$\coprod_{f \in \mathrm{Mor}(J)} F(\mathrm{dom}(f)) \overset{s}{\underset{t}{\rightrightarrows}} \coprod_{j \in \mathrm{Obj}(J)} F(j) \tag{3.10}$$

Now, it is time to state the theorem.

Theorem 3.2 *Let $\mathscr{A}$ be a category with coequalizers and coproducts, and J be an index category. We regard* $\mathrm{Obj}(J)$ *and* $\mathrm{Mor}(J)$ *as discrete categories, over which we may form coproducts. Under these assumptions, we have a coequalizer $e : \coprod_{j \in \mathrm{Obj}(J)} F(j) \to K$ for two morphisms s and t in* (3.10)*. Then, gives a cocone*

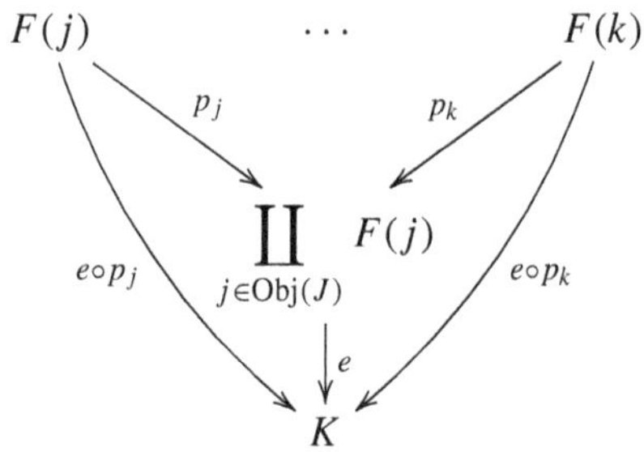

$$(F, \quad \{e \circ p_j\}_{j \in \mathrm{Obj}(J)}, \quad K)$$

that is a limiting cocone of the J-shaped diagram $F : J \to \mathscr{A}$. We may write $K = \varinjlim F$.

Proof Let $(F, \{x_j\}, X)$ be a cocone from F to X. Universality of coproducts gives a unique morphism $x = \coprod_{j \in \mathrm{Obj}(J)} x_j$ that makes the following diagram commute.

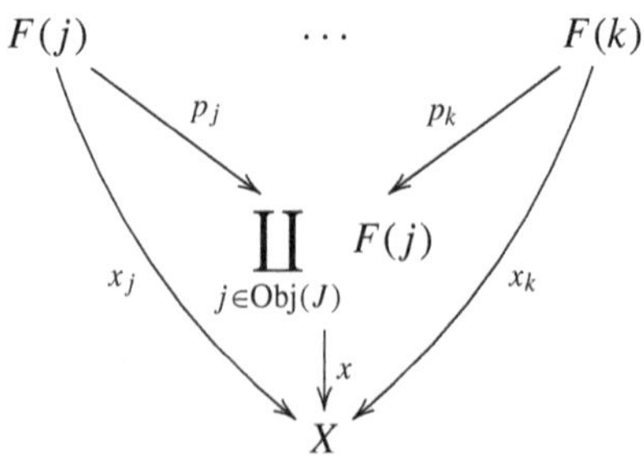

Composing with s and t, we have the following diagram.

$$\coprod_{f \in \mathrm{Mor}(J)} F(\mathrm{dom}(f)) \underset{t}{\overset{s}{\rightrightarrows}} \coprod_{j \in \mathrm{Obj}(J)} F(j) \xrightarrow{x} X$$

We want to show that $x \circ s = x \circ t$. First, we rewrite the LHS.

$$\begin{aligned} x \circ s &= x \circ \left(\coprod_{f \in \mathrm{Mor}(J)} p_{\mathrm{cod}(f)} \circ F(f) \right) \\ &= \coprod_{f \in \mathrm{Mor}(J)} \left(x \circ p_{\mathrm{cod}(f)} \circ F(f) \right) \\ &= \coprod_{f \in \mathrm{Mor}(J)} \left(x_{\mathrm{cod}(f)} \circ F(f) \right) \end{aligned}$$

Line 2 is obtained by the exchange of colimit and composition by a morphism. This exchange is not trivial (Proposition 3.6). Line 3 is obtained by the definition of x, namely $x \circ p_j = x_j$, where j is replaced by $\mathrm{cod}(f)$.

Next, we rewrite the RHS.

$$\begin{aligned} x \circ t &= x \circ \left(\coprod_{f \in \mathrm{Mor}(J)} p_{\mathrm{dom}(f)} \right) \\ &= \coprod_{f \in \mathrm{Mor}(J)} \left(x \circ p_{\mathrm{dom}(f)} \right) \\ &= \coprod_{f \in \mathrm{Mor}(J)} x_{\mathrm{dom}(f)} \end{aligned}$$

Line 2 is obtained by the exchange of colimit and composition by a morphism. Line 3 is obtained by the definition of x. Since x_j's constitute a cocone from F,

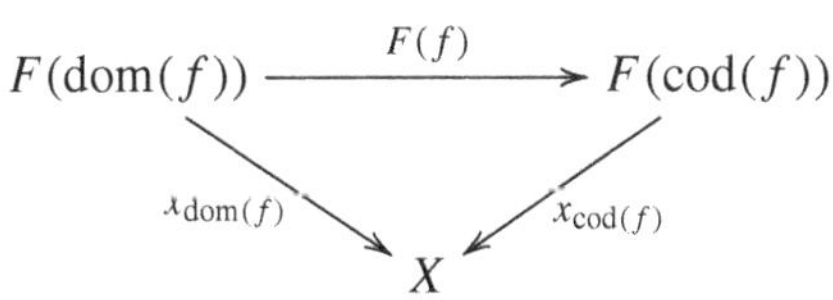

commutes for all f's. Thus, we have

$$x_{\mathrm{cod}(f)} \circ F(f) = x_{\mathrm{dom}(f)}$$

for all $f \in \mathrm{Mor}(J)$. We can conclude that $x \circ s = x \circ t$. Universality of coequalizer e gives us a unique mediating morphism $u : K \to X$ as in the following diagram.

$$\begin{array}{ccccc} \coprod_{f \in \mathrm{Mor}(J)} F(\mathrm{dom}(f)) & \overset{s}{\underset{t}{\rightrightarrows}} & \coprod_{j \in \mathrm{Obj}(J)} F(j) & \xrightarrow{e} & K \\ & & & \searrow^{x} & \downarrow \exists! u \\ & & & & X \end{array}$$

This means that $\{e \circ p_j\}$ is a universal cocone. We may write $K = \varinjlim F$. □

Cocones in category **Set** can be discussed in a similar way to cones. Given a J-shaped diagram $F : J \to$ **Set**, we can talk about elements of $F(j)$. A cocone to A requires the following triangles to commute.

$$\begin{array}{ccc} F(j) & \xrightarrow{F(f)} & F(k) \\ & {}_{p_j}\searrow \quad \swarrow_{p_k} & \\ & A & \end{array}$$

This amounts to taking the union of a family of sets $\{F(j)\}$ pasting $x_j \in F(j)$ and $x_k = F(f)(x_j) \in F(k)$. With this in mind, we proceed to a discussion below.

Example 3.20 Let J be a category

$$J = (1 \xrightarrow{f} 2).$$

We have

$$\begin{aligned} \mathrm{Obj}(J) &= \{1, 2\} \\ \mathrm{Mor}(J) &= \{1_1, 1_2, f\}. \end{aligned}$$

We consider a diagram $F : J \to$ **Set** as follows. Its function on objects is given by

$$F(1) = \{3, 4, 5, 6\}, \quad F(2) = \{7, 8, 9\}.$$

Since J has only one morphism other than the identities, F's function on morphisms is determined by the image of f. We set

$$\begin{aligned} F(f) :\ & F(1) \to F(2) \\ & 3 \mapsto 7 \\ & 4 \mapsto 7 \\ & 8 \\ & 5 \mapsto 9 \\ & 6 \mapsto 9 \end{aligned}$$

We have the following.

$$\begin{aligned}
F(\mathrm{dom}(1_1)) &= F(1) = \{3, 4, 5, 6\} \\
F(\mathrm{dom}(1_2)) &= F(2) = \{7, 8, 9\} \\
F(\mathrm{dom}(f)) &= F(1) = \{3, 4, 5, 6\} \\
\coprod_{f \in \mathrm{Mor}(J)} F(\mathrm{dom}(f)) &= \{3, 4, 5, 6\} \sqcup \{7, 8, 9\} \sqcup \{\mathbf{3}, \mathbf{4}, \mathbf{5}, \mathbf{6}\}
\end{aligned}$$

Operator $\sqcup$ denotes a disjoint union. To avoid overlapping, we use bold face figures in the last operand.

Morphism t is a mapping. We write down the whole correspondence below.

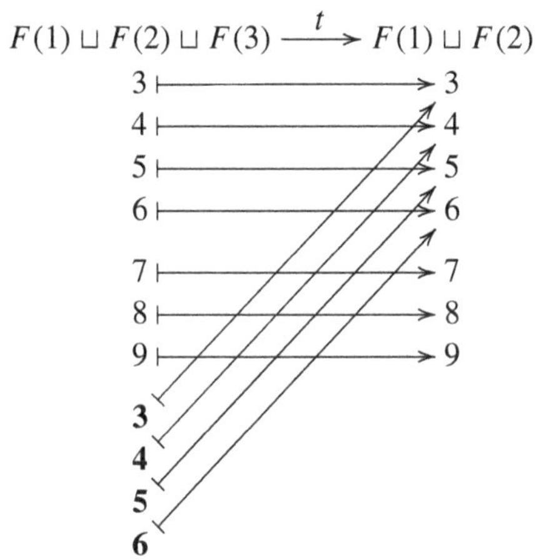

Likewise, morphism s is described as follows.

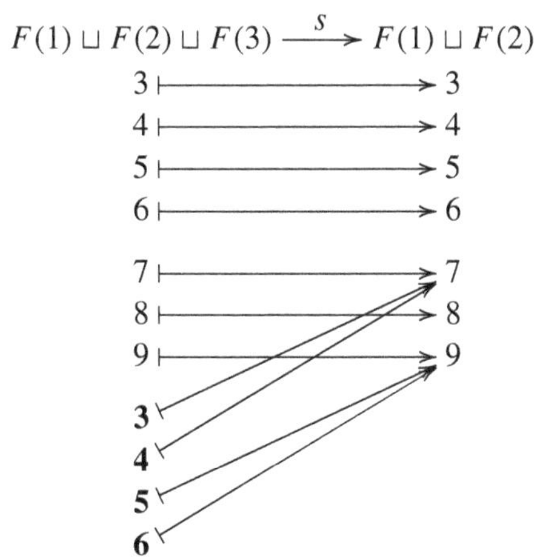

To get a coequalizer for s and t, we have to calculate the quotient $(F(1) \sqcup F(2))/\sim$, where $\sim$ is an equivalence relation in $F(1) \sqcup F(2)$, which is the reflexive, symmetric, transitive closure of a relation $\sim^0$ defined as follows.

Let $b, b' \in F(1) \sqcup F(2)$.

$$b \sim^0 b' \overset{\text{def}}{\iff} \exists a \in F(1) \sqcup F(2) \sqcup F(1); \quad b = s(a) \text{ and } b' = t(a)$$

Refer to the sample code below for an algorithm to collect equivalence classes of $\sim$. The result of calculation is

$$\{\{3, 4, 7\}, \{5, 6, 9\}, \{8\}\}.$$

The coequalizer e is a projection of the quotient.

Listing 3.8 toycolimit.hs

```
import Coequalizer

-- Category J
---- Obj(J) = {1, 2}
data ObjJ = O1 | O2  deriving (Eq, Enum, Show)

---- Mor(J) = {1_1, 1_2, f}
data MorJ = I1 | I2 | Mf  deriving (Eq, Enum, Show)

---- dom : Mor(J) -> Obj(J)
domJ :: MorJ -> ObjJ
domJ I1 = O1
domJ I2 = O2
domJ Mf = O1

---- cod : Mor(J) -> Obj(J)
codJ :: MorJ -> ObjJ
codJ I1 = O1
codJ I2 = O2
codJ Mf = O2

-- toy sets
---- F(1) = {3,4,5,6}
data F1 = V3 | V4 | V5 | V6  deriving (Eq, Show)

---- F(2) = {7,8,9}
data F2 = V7 | V8 | V9  deriving (Eq, Show)

---- F(1_1)
fmapI1 :: F1 -> F1
fmapI1 x = x

---- F(1_2)
fmapI2 :: F2 -> F2
fmapI2 x = x

```

```
---- F(f)
fmapMf :: F1 -> F2
fmapMf V3 = V7
fmapMf V4 = V7
fmapMf V5 = V9
fmapMf V6 = V9

-- F(1) + F(2) + F(1) -- disjoint union
data F121 = P3 | P4 | P5 | P6
            | Q7 | Q8 | Q9
            | R3 | R4 | R5 | R6
  deriving (Eq, Enum, Show)

-- coprojections to F(1)+F(2)+F(1)
sigma1 :: F1 -> F121
sigma1 V3 = P3
sigma1 V4 = P4
sigma1 V5 = P5
sigma1 V6 = P6

sigma2 :: F2 -> F121
sigma2 V7 = Q7
sigma2 V8 = Q8
sigma2 V9 = Q9

sigma'f :: F1 -> F121
sigma'f V3 = R3
sigma'f V4 = R4
sigma'f V5 = R5
sigma'f V6 = R6

-- partial inverses to coprojections
sigma1i :: F121 -> F1
sigma1i P3 = V3
sigma1i P4 = V4
sigma1i P5 = V5
sigma1i P6 = V6

sigma2i :: F121 -> F2
sigma2i Q7 = V7
sigma2i Q8 = V8
sigma2i Q9 = V9

sigma'fi :: F121 -> F1
sigma'fi R3 = V3
sigma'fi R4 = V4
sigma'fi R5 = V5
sigma'fi R6 = V6

-- F(1) + F(2) -- disjoint union
data F12  = S3 | S4 | S5 | S6 | T7 | T8 | T9
  deriving (Eq, Enum, Show)

-- coprojections to F(1)+F(2)
rho1 :: F1 -> F12
```

```
92 rho1 V3 = S3
93 rho1 V4 = S4
94 rho1 V5 = S5
95 rho1 V6 = S6
96
97 rho2 :: F2 -> F12
98 rho2 V7 = T7
99 rho2 V8 = T8
100 rho2 V9 = T9
101
102 -- t: F(1)+F(2)+F(1) -> F(1)+F(2)
103 t :: F121 -> F12
104 t x
105    | x ‘elem‘ [P3 .. P6]  = rho1 (sigma1i x)
106    | x ‘elem‘ [Q7 .. Q9]  = rho2 (sigma2i x)
107    | x ‘elem‘ [R3 .. R6]  = rho1 (sigma’fi x)
108
109 {- suggested test for t
110 map t [P3 .. R6]
111 -}
112
113 -- s: F(1)+F(2)+F(1) -> F(1)+F(2)
114 s :: F121 -> F12
115 s x
116    | x ‘elem‘ [P3 .. P6]  = rho1 (fmapI1 (sigma1i x))
117    | x ‘elem‘ [Q7 .. Q9]  = rho2 (fmapI2 (sigma2i x))
118    | x ‘elem‘ [R3 .. R6]  = rho2 (fmapMf (sigma’fi x))
119
120 {- suggested test for s
121 map s [P3 .. R6]
122 -}
123
124 {- do it
125 rel = coequate s t [P3 .. R6]
126
127 partition rel [S3 .. T9]
128 -}
```

Line 1 imports a helper module Listing 3.6.
All the lines to Line 20 define an index category J.
Lines 22–42 define a diagram $F : J \to \mathbf{Set}$.
Lines 44–84 define

$$\coprod_{f \in \mathrm{Mor}(J)} F(\mathrm{dom}(f)) = F(1_1) \sqcup F(1_2) \sqcup F(f)$$

Especially, lines 50–66 define the coprojections

$$\sigma_{1_1}, \quad \sigma_{1_2}, \quad \sigma_f.$$

Lines 68–84 define inverse partial functions of coprojections:

$$\sigma_{1_1}^{-1}, \quad \sigma_{1_2}^{-1}, \quad \sigma_f^{-1}.$$

Lines 86–100 define

$$\coprod_{j \in \mathrm{Obj}(J)} F(j) = F(1) \sqcup F(1).$$

Especially lines 90–100 define coprojections

$$\rho_1, \quad \rho_2.$$

Lines 102–107 define

$$t : \coprod_{f \in \mathrm{Mor}(J)} F(\mathrm{dom}(f)) \to \coprod_{j \in J} F(j).$$

Lines 113–118 define s in a similar manner.

The tests for t and s can be done as follows.

```
*Main> map t [P3 .. R6]
[S3,S4,S5,S6,T7,T8,T9,S3,S4,S5,S6]
*Main> map s [P3 .. R6]
[S3,S4,S5,S6,T7,T8,T9,T7,T7,T9,T9]
```

Using the imported function `coequate`, we can calculate the quotient to coequate s and t as follows.

```
*Main> rel = coequate s t [P3 .. R6]
*Main> partition rel [S3 .. T9]
[[T8],[S6,T9,S5],[S4,T7,S3]]
```

We sum up the previous examples.

Example 3.21 Category **Set** has coproducts and coequalizers. A coproduct is a *disjoint union*, where elements of those sets that are to be joined are tagged so that tags indicate from which set it comes from. That's why it is called a *tagged union*.

Having coproducts and coequalizers, category **Set** has colimits. See Theorem 3.2.

Colimits can be constructed as follows. Given a J-shaped diagram $F : J \to$ **Set**, we first form a disjoint union $\coprod_{j \in \mathrm{Obj}(J)} F(j)$. Then, take the quotient by an equivalence relation $\sim$ and get

$$L = \coprod_{j \in \mathrm{Obj}(J)} F(j) / \sim$$

that is an apex of a cocone. Let π be the projection to the quotient. The relation $\sim$ is the equivalence closure of $\sim^0$ defined as follows. For $b, b' \in \coprod_{j \in \mathrm{Obj}(J)} F(j)$

$$b \sim^0 b' \overset{\text{def}}{\Longleftrightarrow} \exists (f : j \to k) \in \mathrm{Mor}(J);\ \exists a \in F(j);\quad b = F(f)(\rho_j(a)) \text{ and } b' = \rho_j(a)$$

A universal cocone if given as

$$\left(F,\quad \{\pi \circ \rho_j\}_{j \in \mathrm{Obj}(J)},\ \coprod_{j \in \mathrm{Obj}(J)} F(j)/\sim \right).$$

Functors and Limits

4

Functors map diagrams to diagrams and cones to cones. We are naturally tempted to ask if the image of a limiting cone is also a limiting cone in the target category. We first study some very small examples in which the answers to this question are either yes or no depending on the functor in question. Then, we will study the Hom functors that guarantee the positive answers for all limiting cones.

Suppose a category $\mathscr{A}$ has both products and coproducts. A natural question to pose is whether the following isomorphism exists:

$$\mathscr{A}(A, B \times C) \simeq \mathscr{A}(A, B) \times \mathscr{A}(A, C).$$

If the answer is yes, we may say that the functor $\mathscr{A}(A, -)$ preserves products. More specifically, functor $\mathscr{A}(A, -)$ maps products to products.

We pose another question: if

$$\mathscr{A}(A \sqcup B, C) \simeq \mathscr{A}(A, C) \times \mathscr{A}(B, C)$$

holds. If the answer is positive, we may say that functor $\mathscr{A}(-, C)$ preserves coproducts. But, note that the functor maps coproducts to products due to its variance, namely the functor is contravariant.

We study some sufficient conditions for the preservation of limits in general.

4.1 Functors Map Cones to Cones

We fix an index category J and a J-shaped diagram $D : J \to \mathscr{A}$. Given a functor $F : \mathscr{A} \to \mathscr{B}$ and its composite with D, and get another J-shaped diagram $F \circ D : J \to \mathscr{B}$.

S. Yukita, *Category Theory Using Haskell*, Computer Science Foundations and Applied Logic, https://doi.org/10.1007/978-3-031-68538-5_4

Further, we see that cones from A to D are mapped to cones from $F(A)$ to $F \circ D$.

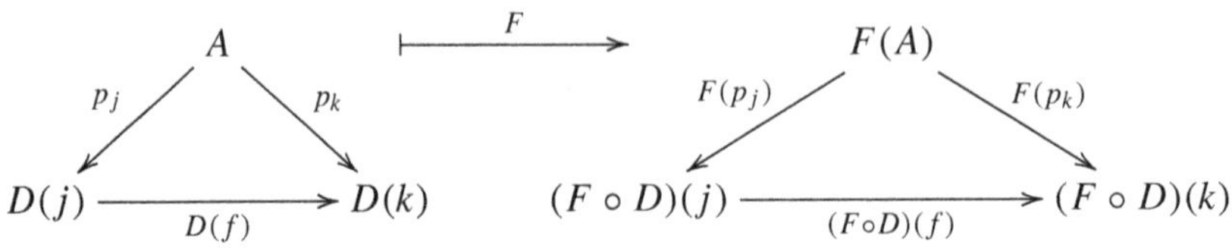

In the same way, we see that cones from D to A are mapped to cones from $F \circ D$ to $F(A)$.

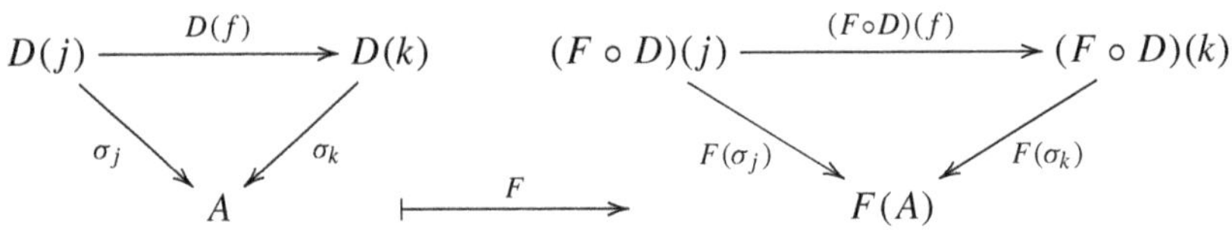

Definition 4.1 A functor $F : \mathscr{A} \to \mathscr{B}$ is said to *preserve limits* of J-shaped diagrams if, for all J-shaped diagrams $D : J \to \mathscr{A}$ that have a limit $\varprojlim D$, there exists $\varprojlim(F \circ D)$ and an isomorphism

$$F(\varprojlim D) \simeq \varprojlim(F \circ D).$$

A functor $F : \mathscr{A} \to \mathscr{B}$ is said to *preserve colimits* of J-shaped diagrams if, for all J-shaped diagrams $D : J \to \mathscr{A}$ that have a colimit $\varinjlim D$, there exists $\varinjlim(F \circ D)$ and an isomorphism

$$F(\varinjlim D) \simeq \varinjlim(F \circ D).$$

Example 4.1 There are cases in which $\varprojlim(F \circ D)$ does not exist even though $\varprojlim D$ exists. Let J be a discrete category with two objects.

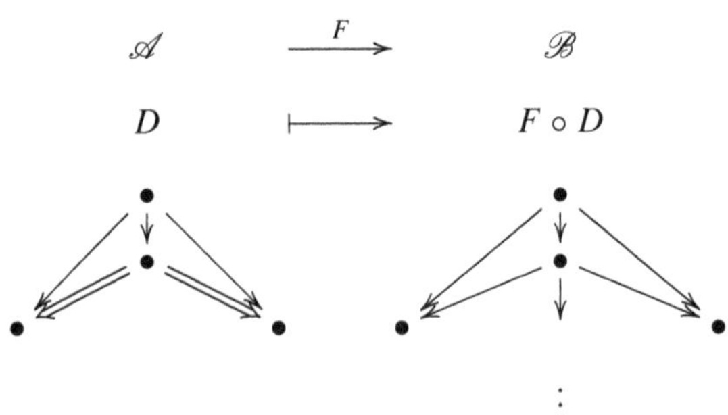

Arrows $\Rightarrow$ constitute a limiting cone in $\mathscr{A}$, but the diagram $F \circ D$ has no limits in $\mathscr{B}$.

Example 4.2 There are cases in which

$$F(\varprojlim D) \simeq \varprojlim(F \circ D)$$

does not hold even though both $\varprojlim D$ and $\varprojlim(F \circ D)$ exist. Let J be a discrete category with two objects.

$$\begin{array}{ccc} \mathscr{A} & \xrightarrow{F} & \mathscr{B} \\ D & \longmapsto & F \circ D \end{array}$$

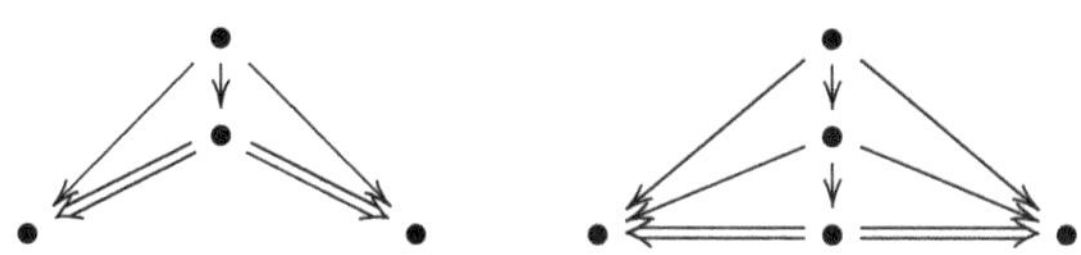

Arrows $\Rightarrow$ in category $\mathscr{A}$ constitute a limiting cone. Arrows $\Rightarrow$ in category $\mathscr{B}$ also constitute a limiting cone. However, the latter is not isomorphic to the functor image of the former. Note that we assumed, in the drawing above, the functor maps objects and morphisms in the left part to those at the same locations on the right.

Example 4.3 If we reverse all the arrows in the previous Examples 4.1 and 4.2, we get examples of colimits that may fail to be preserved.

4.2 Hom Functors and Limits

In this section, we fix a locally small category $\mathscr{A}$. The Hom functor is a functor in two variables.

$$\mathscr{A}(-, -) : \mathscr{A}^{\mathrm{op}} \times \mathscr{A} \to \mathbf{Set}.$$

We fix the first argument and get a covariant functor

$$\mathscr{A}(A, -) : \mathscr{A} \to \mathbf{Set},$$

which turns out to preserve limits.

Theorem 4.1 *Let J be an index category and $\mathscr{A}$ a locally small category. If a J-shaped diagram $D : J \to \mathscr{A}$ has a limit, then J-shaped diagram $\mathscr{A}(A, -) \circ D$ also has a limit. Further, we have a natural isomorphism:*

$$\mathscr{A}(A, \varprojlim D) \simeq \varprojlim(\mathscr{A}(A, -) \circ D). \tag{4.1}$$

Remark 4.1 Equation (4.1) is often written as

$$\mathscr{A}(A, \varprojlim_j D(j)) \simeq \varprojlim_j \mathscr{A}(A, D(j)),$$

making the indexing process explicitly. In the literature, we often encounter presentations such as

$$\mathrm{Hom}_{\mathscr{A}}(A, \varprojlim_j D(j)) \simeq \varprojlim_j \mathrm{Hom}_{\mathscr{A}}(A, D(j)).$$

Before undertaking the proof, we examine the situation in depth. The J-shaped diagram $D : J \to \mathscr{A}$ and its composite with the functor $\mathscr{A}(A, -) : \mathscr{A} \to \mathbf{Set}$ yields another J-shaped diagram:

$$\begin{array}{ccccc}
J & \xrightarrow{D} & \mathscr{A} & \xrightarrow{\mathscr{A}(A,-)} & \mathbf{Set} \\
j & \longmapsto & D(j) & \longmapsto & \mathscr{A}(A, D(j)) \\
{\scriptstyle f}\downarrow & & \downarrow{\scriptstyle D(f)} & & \downarrow{\scriptstyle D(f)\circ-} \\
k & \longmapsto & D(k) & \longmapsto & \mathscr{A}(A, D(k)).
\end{array}$$

This way, we get a functor $\mathscr{A}(A, -) \circ D : J \to \mathbf{Set}$. We consider the following two cones:

$$\begin{array}{ccc}
 & L & \\
{\scriptstyle p_j}\swarrow & & \searrow{\scriptstyle p_k} \\
D(j) & \xrightarrow[D(f)]{} & D(k)
\end{array}
\qquad
\begin{array}{ccc}
 & \mathscr{A}(A, L) & \\
{\scriptstyle \mathscr{A}(A,p_j)}\swarrow & & \searrow{\scriptstyle \mathscr{A}(A,p_k)} \\
\mathscr{A}(A, D(j)) & \xrightarrow[D(f)\circ-]{} & \mathscr{A}(A, D(k)).
\end{array}$$

Let the left diagram be a limiting cone. We will show the right diagram is also a limiting cone.

If we want to claim the naturality of the correspondence, the following formulation is convenient. However, note that the discussion will be slightly restricted to the situation where all the J-shaped diagrams in $\mathscr{A}$ have limits.

- We define limits $\varprojlim$ as a functor. We know that limits are unique up to isomorphisms. This means they are not unique. We choose a representative from each isomorphism class, using the axiom of choice. We can extend the so-obtained function on objects to function on morphisms. See Sect. 4.5 for details.
- We define a functor $\mathscr{A}^{\mathrm{op}} \times [J, \mathscr{A}] \to \mathbf{Set}$ that takes a limit first and then applies Hom as follows:

$$\begin{array}{ccccc}
\mathscr{A}^{\mathrm{op}} \times [J, \mathscr{A}] & \xrightarrow{\mathrm{Id} \times \varprojlim} & \mathscr{A}^{\mathrm{op}} \times \mathscr{A} & \xrightarrow{\mathrm{Hom}} & \mathbf{Set} \\
(A, D) & \longmapsto & (A, \varprojlim D) & \longmapsto & \mathscr{A}(A, \varprojlim D).
\end{array}$$

- We define a functor $\mathscr{A}^{\mathrm{op}} \times [J, \mathscr{A}] \to \mathbf{Set}$ that creates a J-shaped diagram first, and then applies the limit functor as follows:

$$\begin{array}{ccccc} \mathscr{A}^{\mathrm{op}} \times [J, \mathscr{A}] & \longrightarrow & [J, \mathbf{Set}] & \xrightarrow{\varprojlim} & \mathbf{Set} \\ (A, D) & \longmapsto & \mathscr{A}(A, -) \circ D & \longmapsto & \varprojlim(\mathscr{A}(A, -) \circ D). \end{array}$$

- We define a natural isomorphism between the two functors above. To be more specific, we examine how the outputs change under the influence of morphisms $f \in \mathscr{A}(A, A')$ and $\alpha \in [J, \mathscr{A}](D, D')$. For example, the latter functor needs calculation with the following morphisms at its first step.

$$\begin{array}{cccc} A & \mathscr{A}(A, -) \circ D & D & \mathscr{A}(A, -) \circ D \\ {\scriptstyle f}\downarrow & \uparrow{\scriptstyle \mathscr{A}(f, -) \circ D} & {\scriptstyle \alpha}\downarrow & \downarrow{\scriptstyle \mathscr{A}(A, \alpha)} \\ A' & \mathscr{A}(A', -) \circ D & D' & \mathscr{A}(A, -) \circ D \end{array}$$

  In the above diagrams, we adopt the convention of drawing for contravariant functors. See Remark 4.1.

Proof (Theorem 4.1)
We leave the naturality proof to the reader. Arguments similar to those in Chap. 7 will do.

Universality of limits gives

$$\mathscr{A}(A, \varprojlim D) \simeq \mathrm{Cones}(A, D). \tag{4.2}$$

This is the definition of limits itself. See Remark 3.1.

Since $\varprojlim(\mathscr{A}(A, -) \circ D)$ is a limit in **Set**, its existence is guaranteed by Theorem 3.1 and an argument in Example 3.19. All the objects involved can be studied using their elements. Let x be an element of $\varprojlim(\mathscr{A}(A, -) \circ D)$. A limiting cone in **Set** can be seen as a set of cones from a point to the base:

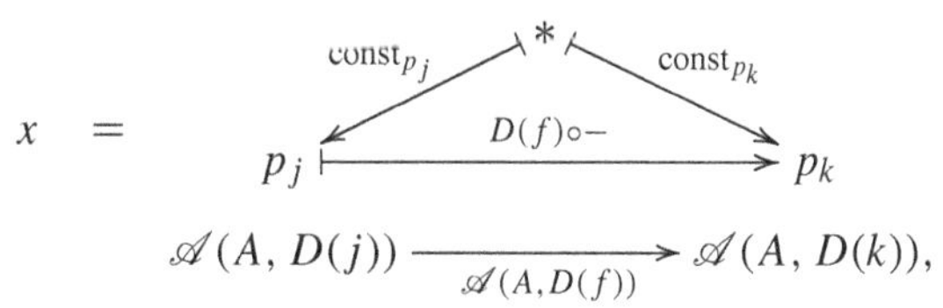

where const_{p_j} is a function from a singleton set $\{*\}$ to $\mathscr{A}(A, D(j))$ that has a single value p_j. The data representation of this cone is

$$(\{*\}, \{\mathrm{const}_{p_j}\}_{j \in \mathrm{Obj}(J)}, \mathscr{A}(A, -) \circ D).$$

Such a cone is in one-to-one correspondence with $(\{p_j\}, D)$, an element of Cones (A, D). Thus, we have

$$\mathrm{Cones}(A, D) = \varprojlim(\mathscr{A}(A, -) \circ D).$$

Combining this equation and (4.2), we get the conclusion. □

4.3 Hom Functors and Colimits

We fix a locally small category $\mathscr{A}$. The Hom functor is a functor in two variables.

$$\mathscr{A}(-, -) : \mathscr{A}^{\mathrm{op}} \times \mathscr{A} \to \mathbf{Set}$$

We fix the second argument to get a contravariant functor

$$\mathscr{A}(-, A) : \mathscr{A}^{\mathrm{op}} \to \mathbf{Set},$$

which turns out to preserve limits, but due to contravariance both limits and colimits play the roles.

Theorem 4.2 *Let J be an index category and $\mathscr{A}$ a locally small category. If a J-shaped diagram $D : J \to \mathscr{A}$ has a* colimit, *then J-shaped diagram $\mathscr{A}(-, A) \circ D$ also has a* limit. *Further, we have a natural isomorphism:*

$$\mathscr{A}(\varinjlim D, A) \simeq \varprojlim(\mathscr{A}(-, A) \circ D). \tag{4.3}$$

Remark 4.2 We cannot compose $\mathscr{A}(-, A) : \mathscr{A}^{\mathrm{op}} \to \mathbf{Set}$ and $D : J \to \mathscr{A}$ as they are. Taking the opposites $D : J^{\mathrm{op}} \to \mathscr{A}^{\mathrm{op}}$ solves the problem. We can compose two functors as follows:

$$J^{\mathrm{op}} \xrightarrow{D} \mathscr{A}^{\mathrm{op}} \xrightarrow{\mathscr{A}(-,A)} \mathbf{Set}.$$

We get a diagram that appears in the RHS of Eq. (4.3).

Before undertaking the proof, we examine the situation in depth. The J-shaped diagram $D : J^{\mathrm{op}} \to \mathscr{A}^{\mathrm{op}}$ and its composite with the functor $\mathscr{A}(-, A) : \mathscr{A}^{\mathrm{op}} \to \mathbf{Set}$ yields another J-shaped diagram:

$$\begin{array}{ccccc}
J^{\mathrm{op}} & \xrightarrow{D} & \mathscr{A}^{\mathrm{op}} & \xrightarrow{\mathscr{A}(-,A)} & \mathbf{Set} \\
\\
j & \longmapsto & D(j) & \longmapsto & \mathscr{A}(D(j), A) \\
{\scriptstyle f}\big\downarrow & & \big\downarrow{\scriptstyle D(f)} & & \big\uparrow{\scriptstyle -\circ D(f)} \\
k & \longmapsto & D(k) & \longmapsto & \mathscr{A}(D(k), A).
\end{array}$$

This way, we get a functor $\mathscr{A}(-, A) \circ D : J^{\mathrm{op}} \to \mathbf{Set}$, or We consider the following cocone and cone:

$$\begin{array}{ccccc} D(j) & \xrightarrow{D(f)} & D(k) \\ & {\scriptstyle p_j}\searrow \quad \swarrow{\scriptstyle p_k} & \\ & L & \end{array} \qquad \begin{array}{ccc} \mathscr{A}(D(j), A) & \xleftarrow{-\circ D(f)} & \mathscr{A}(D(k), A) \\ {\scriptstyle \mathscr{A}(p_j, A)}\nwarrow & & \nearrow{\scriptstyle \mathscr{A}(p_k, A)} \\ & \mathscr{A}(L, A) & \end{array}$$

Let the left diagram be a limiting *cocone*. We will show the right diagram is a limiting *cone*. Notice that arrows are reversed in the right diagram due to contravariance.

If we want to claim the naturality of the correspondence, the following formulation is convenient. However, note that the discussion will be slightly restricted to the situation where all the J-shaped diagrams in $\mathscr{A}$ have colimits.

- We define colimits $\varinjlim$ as a functor. We know that colimits are unique up to isomorphisms. This means they are not unique. We choose a representative from each isomorphism class, using the axiom of choice. We can extend the so-obtained function on objects to function on morphisms. See Sect. 4.5 for details.
- We define a functor $[J, \mathscr{A}]^{\mathrm{op}} \times \mathscr{A} \to \mathbf{Set}$ that takes a colimit first and then applies Hom as follows:

$$\begin{array}{ccccc} [J, \mathscr{A}]^{\mathrm{op}} \times \mathscr{A} & \xrightarrow{\varinjlim^{\mathrm{op}} \times \mathrm{Id}} & \mathscr{A}^{\mathrm{op}} \times \mathscr{A} & \xrightarrow{\mathrm{Hom}} & \mathbf{Set} \\ (D, A) & \longmapsto & (\varinjlim D, A) & \longmapsto & \mathscr{A}(\varinjlim D, A) \end{array}$$

- We define a functor $[J, \mathscr{A}]^{\mathrm{op}} \times \mathscr{A} \to \mathbf{Set}$ that creates a J-shaped diagram first, and then applies the limit functor as follows:

$$\begin{array}{ccccc} [J, \mathscr{A}]^{\mathrm{op}} \times \mathscr{A} & \longrightarrow & [J, \mathbf{Set}] & \xrightarrow{\varprojlim} & \mathbf{Set} \\ (D, A) & \longmapsto & \mathscr{A}(-, A) \circ D & \longmapsto & \varprojlim(\mathscr{A}(-, A) \circ D) \end{array}$$

- We define a natural isomorphism between the two functors above.

Colimits and limits play their roles in the same theater.

Proof (Theorem 4.2)
We leave the naturality proof to the reader. Arguments similar to those in Chap. 7 will do.

Universality of colimits gives

$$\mathscr{A}(\varinjlim D, A) \simeq \mathrm{Cones}(D, A).$$

This is the definition of colimits itself. See Remark 3.1.

Given a cocone $(D, \{x_j\}, X)$ we have a mediating morphism $t : \varinjlim D \to X$ as in the following diagram:

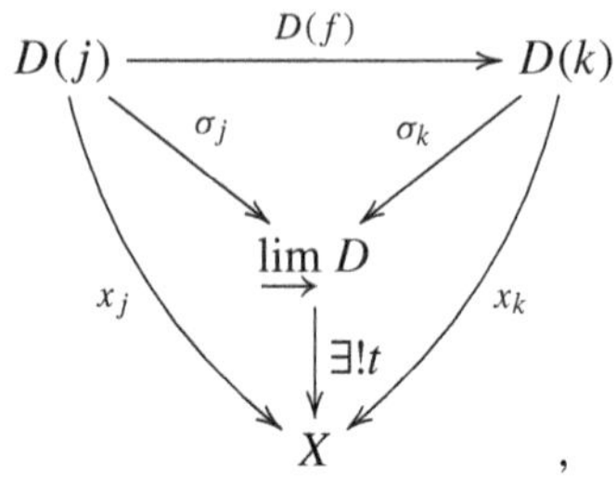

where all the triangles commute and $f : j \to k$ is an arbitrary morphism in J. We hand it to the contravariant functor $\mathscr{A}(-, A)$ and get the following.

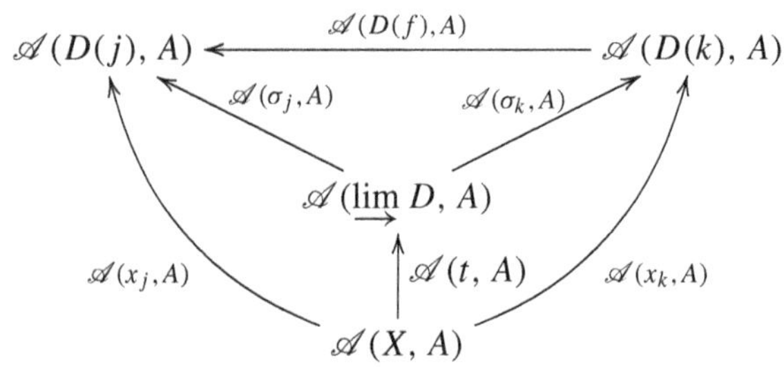

Due to functoriality, all the triangles commute. We want to know whether $\mathscr{A}(\varinjlim D, A)$ is a limit or not. In any case **Set** guarantees the existence of limits. Let L be a limit and u a mediating morphism as follows.

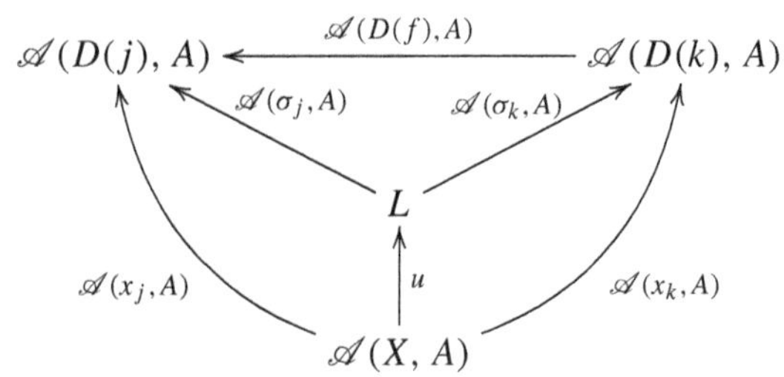

We use the same notation for projections. All the objects involved can be studied using their elements. Let x be an element of L. A limiting cone in **Set** can be seen as a set of cones from a point $*$ to the base:

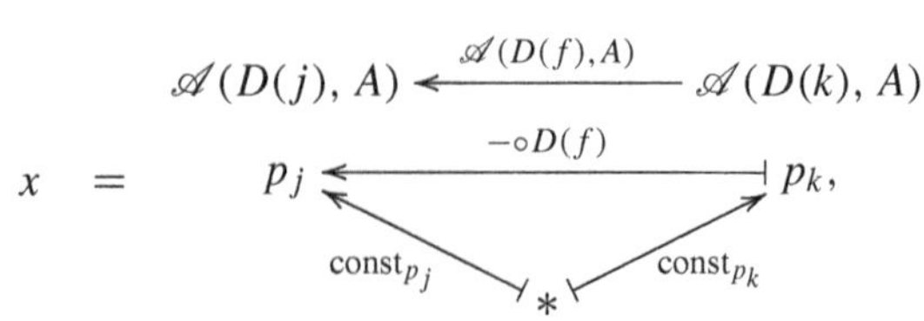

where const_{p_j} is a function from a singleton set $\{*\}$ to $\mathscr{A}(D(j), A)$ that has a single value p_j. The data representation of this J^{op}-shaped cone is

$$(\{*\}, \{\mathrm{const}_{p_j}\}_{j \in \mathrm{Obj}(J)}, \mathscr{A}(-, A) \circ D).$$

Such a cone is in one-to-one correspondence with $(D, \{p_j\})$, an element of Cones (D, A). Thus, we have

$$\varprojlim(\mathscr{A}(-, A) \circ D) \simeq \mathrm{Cones}(D, A)$$

and get to the conclusion. □

4.4 Executable Examples

We will see interactions between Hask, Hom functors, `( -> )`, `( , )`, and `Either`.

Listing 4.1 homlimit.hs

```
data XA = A1 | A2 | A3
  deriving (Show, Eq, Enum)

data XB = B1 | B2
  deriving (Show, Eq, Enum)

data XC = C1 | C2 | C3 | C4 | C5 | C6
  deriving (Show, Eq, Enum)

-- Hom(XC, (XA times XB)) ~= Hom(XC, XA) times Hom(XC, XB)
alpha1 :: (c -> (a,b)) -> (c -> a, c -> b)
alpha1 f
  = (fst . f, snd . f)

alpha2 :: (c -> a, c -> b) -> (c -> (a,b))
alpha2 (f,g)
  = \c -> (f c, g c)

-- Hom(XA + XB, XC) ~= Hom(XA, XC) times Hom(XB, XC)
beta1 :: (Either a b -> c) -> (a -> c, b -> c)
beta1 f
  = (f . Left, f . Right)

beta2 :: (a -> c, b -> c) -> (Either a b -> c)
beta2 (f,g)
  = h
    where
      h (Left a)  = f a
      h (Right b) = g b

-- test data for
-- alpha1 . alpha2 = id    and
-- alpha2 . alpha1 = id

f1 :: XC -> (XA,XB)
f1 C1 = (A1,B1)
f1 C2 = (A1,B2)
f1 C3 = (A2,B1)
f1 C4 = (A2,B2)
f1 C5 = (A3,B1)
f1 C6 = (A3,B2)

```

```
f2 :: XC -> XA
f2 C1 = A1
f2 C2 = A1
f2 C3 = A2
f2 C4 = A2
f2 C5 = A3
f2 C6 = A3

g2 :: XC -> XB
g2 C1 = B1
g2 C2 = B1
g2 C3 = B1
g2 C4 = B2
g2 C5 = B2
g2 C6 = B2

{- executable tests
map ((alpha2 . alpha1) f1) [C1 .. C6]
map f1 [C1 .. C6]

map (fst ((alpha1 . alpha2) (f2,g2)) ) [C1 .. C6]
map f2 [C1 .. C6]
map (snd ((alpha1 . alpha2) (f2,g2)) ) [C1 .. C6]
map g2 [C1 .. C6]
-}

-- test data for
-- beta1 . beta2 = id    and
-- beta2 . beta1 = id

f3 :: Either XA XB -> XC
f3 (Left A1)  = C1
f3 (Left A2)  = C2
f3 (Left A3)  = C3
f3 (Right B1) = C4
f3 (Right B2) = C5

d1 :: [Either XA XB]
d1 = (map Left [A1 ..]) ++ (map Right [B1 ..])

f4 :: XA -> XC
f4 A1 = C1
f4 A2 = C2
f4 A3 = C3

g4 :: XB -> XC
g4 B1 = C4
g4 B2 = C5

{- executable tests
map ((beta2 . beta1) f3) d1
map f3 d1

map (fst ((beta1 . beta2) (f4,g4))) [A1 ..]
map f4 [A1 ..]

map (snd ((beta1 . beta2) (f4,g4))) [B1 ..]
map g4 [B1 ..]
-}
```

Lines 1–8 define three finite sets. All the type names begin with "X." We will drop this letter when we talk about objects in the context of category theory.

Lines 11–17 define a pair of functions that establish the isomorphism:

$$\mathrm{Hom}(C, A \times B) \underset{\alpha_2}{\overset{\alpha_1}{\rightleftarrows}} \mathrm{Hom}(C, A) \times \mathrm{Hom}(C, B).$$

Lines 20–29 define a pair of functions that establish the isomorphism:

$$\mathrm{Hom}(A \sqcup B, C) \underset{\beta_2}{\overset{\beta_1}{\rightleftarrows}} \mathrm{Hom}(A, C) \times \mathrm{Hom}(B, C).$$

Notice that the contravariant functor exchanges the sorts of limits.

Lines 35–57 provide test data for checking if α_1 and α_2 are mutual inverses. We run the test code described in the commented-out area lines 60–66. For example, to test

$$\alpha_2 \circ \alpha_1 = 1_{\mathrm{Hom}(C, A \times B)}$$

run the test code as follows.

```
*Main> :t alpha2 . alpha1
alpha2 . alpha1 :: (c -> (a, b)) -> c -> (a, b)
*Main> map ((alpha2 . alpha1) f1) [C1 .. C6]
[(A1,B1),(A1,B2),(A2,B1),(A2,B2),(A3,B1),(A3,B2)]
*Main> map f1 [C1 .. C6]
[(A1,B1),(A1,B2),(A2,B1),(A2,B2),(A3,B1),(A3,B2)]
```

To test

$$\alpha_1 \circ \alpha_2 = 1_{\mathrm{Hom}(C, A) \times \mathrm{Hom}(C, B)}$$

run the test code as follows.

```
*Main> :t alpha1 . alpha2
alpha1 . alpha2 :: (c -> a, c -> b) -> (c -> a, c -> b)
*Main> map (fst ((alpha1 . alpha2) (f2,g2)) ) [C1 .. C6]
[A1,A1,A2,A2,A3,A3]
*Main> map f2 [C1 .. C6]
[A1,A1,A2,A2,A3,A3]
*Main> map (snd ((alpha1 . alpha2) (f2,g2)) ) [C1 .. C6]
[B1,B1,B1,B2,B2,B2]
*Main> map g2 [C1 .. C6]
[B1,B1,B1,B2,B2,B2]
```

Lines 73–90 provide test data for checking if β_1 and β_2 are mutual inverses. We run the test code described in the commented-out area lines 93–100. For example, to test

$$\beta_2 \circ \beta_1 = 1_{\mathrm{Hom}(A \sqcup B, C)}$$

run the test code as follows.

```
*Main> :t beta2 . beta1
beta2 . beta1 :: (Either a b -> c) -> Either a b -> c
*Main> map ((beta2 . beta1) f3) d1
[C1,C2,C3,C4,C5]
*Main> map f3 d1
[C1,C2,C3,C4,C5]
```

To test

$$\beta_1 \circ \beta_2 = 1_{\mathrm{Hom}(A,C) \times \mathrm{Hom}(B,C)}$$

run the test code as follows.

```
*Main> :t beta1 . beta2
beta1 . beta2 :: (a -> c, b -> c) -> (a -> c, b -> c)
*Main> map (fst ((beta1 . beta2) (f4,g4))) [A1 ..]
[C1,C2,C3]
*Main> map f4 [A1 ..]
[C1,C2,C3]
*Main> map (snd ((beta1 . beta2) (f4,g4))) [B1 ..]
[C4,C5]
*Main> map g4 [B1 ..]
```

4.5 Limit Functors

In this section, we define limits and colimits as functors from category $[J, \mathscr{A}]$ to $\mathscr{A}$. We assume that J is as small as before and $\mathscr{A}$ is locally small. We also assume that all J-shaped diagrams have limits.

Remark 4.3 This is a continued comment of Remark 4.3. We always assume that J, an index category, is small and $\mathscr{A}$ locally small. This makes the functor category $[J, \mathscr{A}]$ locally small. We briefly see why.

Let D and D' be any pair of objects in $[J, \mathscr{A}]$. We have to show that $[J, \mathscr{A}]$ is small. Since $\mathscr{A}$ is locally small, $\mathscr{A}(D(j), D'(j))$ is small for all $j \in \mathrm{Obj}(J)$. Therefore, their product $\prod_{j \in \mathrm{Obj}(J)} \mathscr{A}(D(j), D'(j))$ is also small. A natural transformation from D to D' can be seen as an element of this set. Thus, we conclude that $[J, \mathscr{A}](D, D')$ is small.

4.5.1 Limits as a Functor

We assume that all J-shaped diagrams in $\mathscr{A}$ have a limit. For each diagram $F : J \to \mathscr{A}$, we choose a representative from the isomorphism class of limits and assign the representative to $\varprojlim F$. We consider this assignment as a *function on objects* of the functor we are going to define.

Next, we construct the function on morphisms. Given a morphism $\alpha : F \to G$ in $[J, \mathscr{A}]$, consider the following diagram.

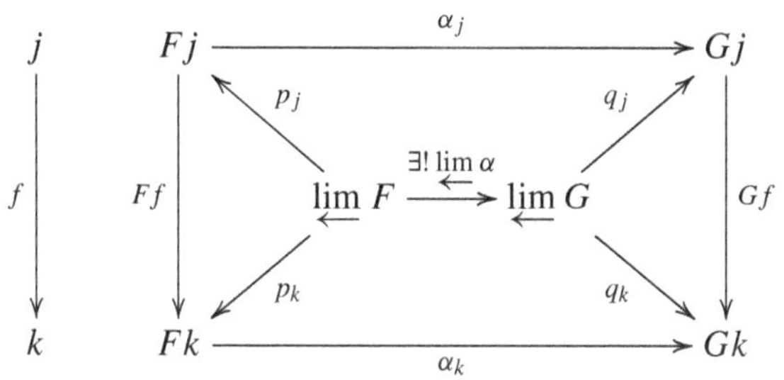

Universality of $\varprojlim G$ gives a unique mediating morphism $\varprojlim \alpha$ from the cone $(\varprojlim F, \{\alpha_j \circ p_j\}, G)$.

Given another morphism $\beta : G \to H$ in $[J, \mathscr{A}]$, we can draw a diagram:

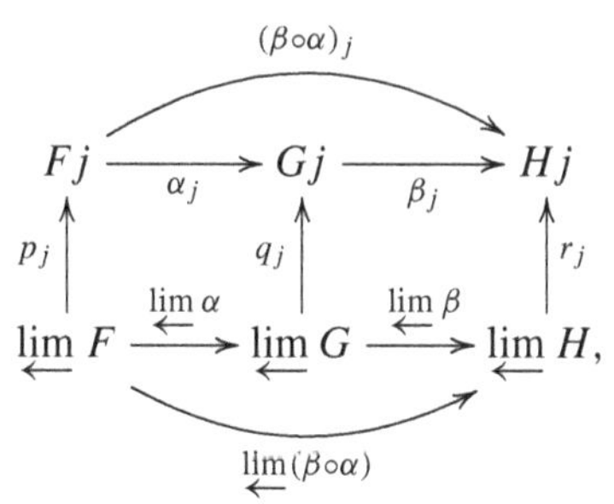

(4.4)

in which we want to have

$$\varprojlim \beta \circ \varprojlim \alpha = \varprojlim(\beta \circ \alpha).$$

In diagram (4.4), we find the followings:

(1) The top triangle commutes simply because it is the composition of natural transformations.

(2) The left square commutes since universality of $\varprojlim G$ gives a unique mediating morphism $\varprojlim \alpha$ for the cone $(\varprojlim F, \{\alpha_j \circ p_j\}, G)$.
(3) The right square commutes since universality of $\varprojlim H$ gives a unique mediating morphism $\varprojlim \beta$ for the cone $(\varprojlim G, \{\beta_j \circ q_j\}, H)$.
(4) (2) and (3) tell us that $\varprojlim \beta \circ \varprojlim \alpha$ makes the large square commute. Cone $(\varprojlim F, \{\beta_j \circ \alpha_j \circ p_j\}, H)$ and $\varprojlim H$ should give a unique mediating morphism, which should be $\varprojlim \beta \circ \varprojlim \alpha$ with no other choice.
(5) Cone $(\varprojlim F, \{(\beta \circ \alpha)_j \circ p_j\}, H)$ and $\varprojlim H$ give a unique mediating morphism $\varprojlim(\beta \circ \alpha)$.

From (1) we see that cones $(\varprojlim F, \{\beta_j \circ \alpha_j \circ p_j\}, H)$ and $(\varprojlim F, \{(\beta \circ \alpha)_j \circ p_j\}, H)$ are identical. Therefore, mediating morphisms given by (4) and (5) must coincide.

This concludes the construction of the functor $\varprojlim : [J, \mathscr{A}] \to \mathscr{A}$.

4.5.2 Colimits as a Functor

Let us continue a parallel discussion for colimits. We assume that all J-shaped diagrams in $\mathscr{A}$ have a colimit. For each diagram $F : J \to \mathscr{A}$, we choose a representative from the isomorphism class of colimits and assign the representative to $\varinjlim F$. We consider this assignment as a *function on objects* of the functor we are going to define.

Next, we construct the function on morphisms. Given a morphism $\alpha : F \to G$ in $[J, \mathscr{A}]$, consider the following diagram.

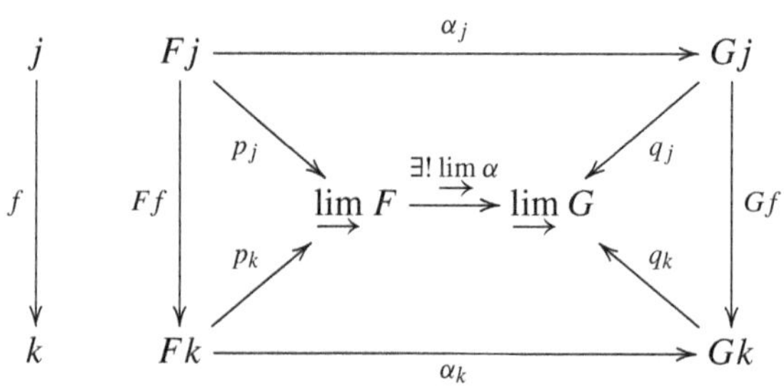

Universality of $\varinjlim F$ gives a unique mediating morphism $\varinjlim \alpha$ to the cocone $(F, \{q_j \circ \alpha_j\}, \varinjlim G)$.

Given another morphism $\beta : G \to H$ in $[J, \mathscr{A}]$, we can draw a diagram:

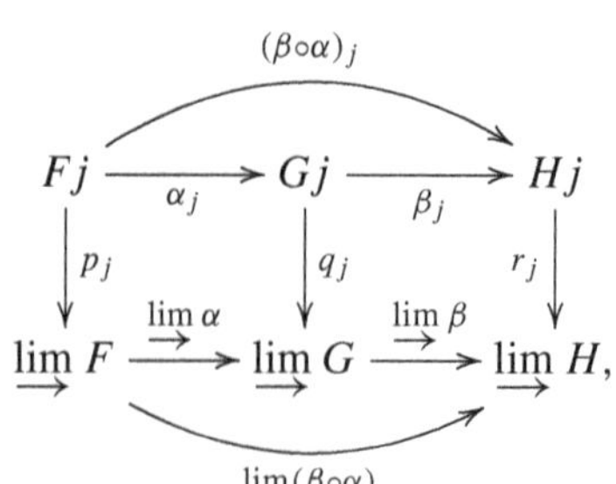

(4.5)

in which we want to have

$$\varinjlim \beta \circ \varinjlim \alpha = \varinjlim (\beta \circ \alpha).$$

In diagram (4.5), we find the followings:

(1) The top triangle commutes simply because it is the composition of natural transformations.
(2) The left square commutes since universality of $\varinjlim F$ gives a unique mediating morphism $\varinjlim \alpha$ for the cocone $(F, \{q_j \circ \alpha_j\}, \varinjlim G)$.
(3) The right square commutes since universality of $\varprojlim G$ gives a unique mediating morphism $\varinjlim \beta$ for the cocone $(G, \{r_j \circ \beta_j\}, \varinjlim H)$.
(4) (2) and (3) tell us that $\varinjlim \beta \circ \varinjlim \alpha$ makes the large square commute. Cocone $(F, \{r_j \circ \beta_j \circ \alpha_j\}, \varinjlim H)$ and $\varinjlim F$ should give a unique mediating morphism, which should be $\varinjlim \beta \circ \varinjlim \alpha$ with no other choice.
(5) Cocone $(F, \{r_j \circ (\beta \circ \alpha)_j\}, \varinjlim H)$ and $\varinjlim F$ give a unique mediating morphism $\varinjlim (\beta \circ \alpha)$.

From (1) we see that cocones $(F, \{r_j \circ \beta_j \circ \alpha_j\}, \varinjlim H)$ and $(F, \{r_j \circ (\beta \circ \alpha)_j\}, \varinjlim H)$ are identical. Therefore, mediating morphisms given by (4) and (5) must coincide.

This concludes the construction of the functor $\varinjlim : [J, \mathscr{A}] \to \mathscr{A}$.

Adjoints 5

We will see various phenomena involving adjunction, universal arrows, equivalence of categories, adjoint equivalence, and their interrelationship. Adjunction appears in everyday life in mathematics, computer science, and theoretical physics.

Let V be a finite-dimensional real vector space with inner product $\langle -, - \rangle$. Linear transformation A and its transpose ${}^tA$ are, by definition, related as follows: for any pair $x, y \in V$, equation

$$\langle Ax, y \rangle = \langle x, {}^tAy \rangle$$

holds. We often represent linear transformations as matrices, where transposition is really matrix transposition. If we work with the complex numbers, we usually introduce Hermitian inner products and Hermitian conjugates. If you are familiar with these examples, you are sure to become familiar with the notion of adjunction.

We will study the adjunction of functors

$$\mathscr{A}(FX, A) \cong \mathscr{X}(X, UA).$$

5.1 Adjunctions

Let $\mathscr{X}$ and $\mathscr{A}$ be locally small categories, F and U be functors. We will study various phenomena in which arrows $h : FX \to A$ and $f : X \to UA$ are in one-to-one correspondence as in the following diagram:

S. Yukita, *Category Theory Using Haskell*, Computer Science Foundations and Applied Logic, https://doi.org/10.1007/978-3-031-68538-5_5

$$\mathscr{A} \underset{F}{\overset{U}{\rightleftarrows}} \mathscr{X}$$

$$\begin{array}{ccc} FX & \overset{F}{\longleftarrow\!\shortmid} & X \\ {\scriptstyle h}\downarrow & & \downarrow{\scriptstyle f} \\ A & \underset{U}{\longmapsto} & UA, \end{array}$$

for all pairs $X \in \mathrm{Obj}(\mathscr{X})$ and $A \in \mathrm{Obj}(\mathscr{A})$.

Definition 5.1 (*Adjunctions*) If there are bijections

$$\phi_{X,A} : \mathscr{A}(FX, A) \cong \mathscr{X}(X, UA)$$

for all pairs (X, A) depending naturally on X and A, then we say that F is a *left adjoint* of G and that G is a *right adjoint* of F. We denote this situation by $F \dashv U$. Naturality is explained later.

When $h : FX \to A$ and $f : X \to UA$ are in the above correspondence, namely, $\phi_{X,A}(h) = f$, we write a diagram like

$$\frac{FX \overset{h}{\to} A}{X \underset{f}{\to} UA}.$$

These two morphisms are called mutual *transposition* or *adjunct*. We call the part over the horizontal bar the *numerator* and the part under the bar the *denominator* in analogy with fractions.

We say that the family of bijections $\phi_{X,A}$ is natural in X if

$$\frac{FX' \overset{Fx}{\longrightarrow} FX \overset{h}{\to} A}{X' \underset{x}{\to} X \underset{f}{\to} UA}$$

for any morphism $x : X' \to X$ in $\mathscr{X}$. We may write as $\phi_{X',A}(h \circ Fx) = f \circ x$.

We say that the family of bijections $\phi_{X,A}$ is natural in A if

$$\frac{FX \overset{h}{\to} A \overset{a}{\to} A'}{X \underset{f}{\to} UA \underset{Ua}{\longrightarrow} UA'}$$

for any morphism $a : A \to A'$ in $\mathscr{A}$. We may write as $\phi_{X,A'}(a \circ h) = Ua \circ f$.

Naturality in X and A can be presented in a single diagram:

$$\frac{FX' \overset{Fx}{\longrightarrow} FX \overset{h}{\to} A \overset{a}{\to} A'}{X' \underset{x}{\to} X \underset{f}{\to} UA \underset{Ua}{\longrightarrow} UA'}. \tag{5.1}$$

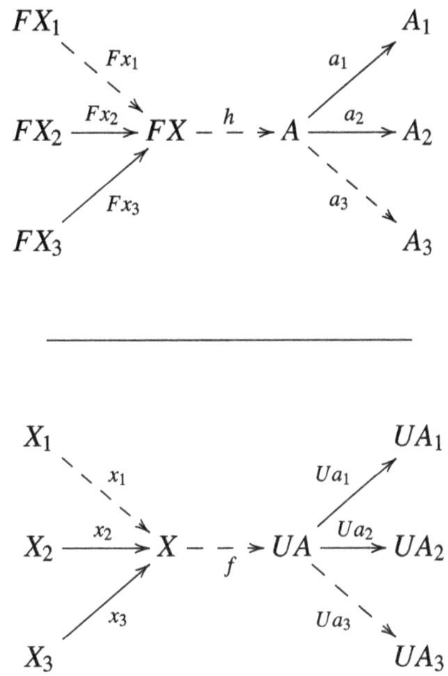

Fig. 5.1 Adjunct pair

Pivoting the adjunct pair h and f, the extension to the left is supported by functor F and to the right by U as in the following diagram:

$$\frac{\text{by functor } F}{\text{extended to left}} \frac{\to FX \xrightarrow{h} A \to}{\to X \underset{f}{\to} UA \to} \frac{\text{extended to right}}{\text{by functor } U}.$$

We will introduce the concept of the universal arrow later. As a preparation, we give a bit crowded drawing Fig. 5.1. Corresponding pairs of paths by the vertical translation, for example, two dashed paths, are adjuncts.

5.2 Unit and Counit

Let us specialize in an adjunct pair

$$\frac{FX \xrightarrow{h} A}{X \underset{f}{\to} UA}$$

by replacing A with FX and h with $1_{FX} : FX \to FX$. We denote its adjunct by η_X. We have

$$\frac{FX \xrightarrow{1_{FX}} FX}{X \underset{\eta_X}{\longrightarrow} UFX}.$$

Another specialization is to replace X with UA and f with $1_{UA} : UA \to UA$. We denote its adjunct by ε_A. We have

$$\frac{FUA \xrightarrow{\varepsilon_A} A}{UA \underset{1_{UA}}{\longrightarrow} UA}.$$

The family of morphisms η_X is said to be the *unit* of the adjunction. The family of morphisms ε_A is said to be the *counit* of the adjunction.

Proposition 5.1 (Natural transformations) *The assignments $X \mapsto \eta_X$ and $A \mapsto \varepsilon_A$ define natural transformations $\eta : \mathrm{Id}_{\mathscr{X}} \to UF$ and $\varepsilon : FU \to \mathrm{Id}_{\mathscr{A}}$.*

Proof First, we make clear the goal. An assignment $\eta : X \mapsto \eta_X$ is said to be a natural transformation if the following diagram commutes

$$\begin{array}{ccc} X_1 & \xrightarrow{\eta_{X_1}} & UFX_1 \\ {\scriptstyle f}\downarrow & \circlearrowleft & \downarrow{\scriptstyle UFf} \\ X_2 & \underset{\eta_{X_2}}{\longrightarrow} & UFX_2 \end{array}$$

for any pair of objects $X_1, X_2 \in \mathrm{Obj}(\mathscr{X})$ and a morphism $f : X_1 \to X_2$.

Recall that the above diagram is a part of the following diagram:

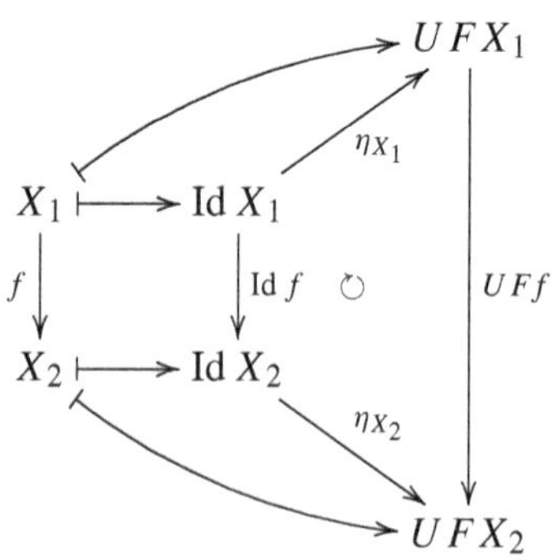

which might convey more intuition to the reader. However, to save space we will not stick to this luxurious drawing style.

We can put the goal in other words: The two paths

$$X_1 \xrightarrow[\eta_{X_1}]{} UFX_1 \xrightarrow[UFf]{} UFX_2 \tag{5.2}$$

and

$$X_1 \xrightarrow[f]{} X_2 \xrightarrow[\eta_{X_2}]{} UFX_2 \tag{5.3}$$

must coincide.

Both are paths in $\mathscr{X}$. We put them in the denominator positions of the fractional diagrams.

Let us begin with path (5.2). We place its adjunct in the numerator:

$$\frac{FX_1 \xrightarrow{1_{FX_1}} FX_1 \xrightarrow{Ff} FX_2}{X_1 \xrightarrow[\eta_{X_1}]{} UFX_1 \xrightarrow[UFf]{} UFX_2}. \tag{5.4}$$

Next, we place path (5.3) in the denominator and its adjunct in the numerator:

$$\frac{FX_1 \xrightarrow{Ff} FX_2 \xrightarrow{1_{FX_2}} FX_2}{X_1 \xrightarrow[f]{} X_2 \xrightarrow[\eta_{X_2}]{} UFX_2}. \tag{5.5}$$

We see that numerators in (5.4) and (5.5) are both

$$FX_1 \xrightarrow{Ff} FX_2.$$

Therefore, by adjunction, the denominators must coincide.

Next we show that the assignment $\varepsilon : A \mapsto \varepsilon_A$ is a natural transformation. We have to show that the following diagram commutes:

$$\begin{array}{ccc} FUA_1 & \xrightarrow{\varepsilon_{A_1}} & A_1 \\ {\scriptstyle FUf}\downarrow & \circlearrowright & \downarrow{\scriptstyle f} \\ FUA_2 & \xrightarrow[\varepsilon_{A_2}]{} & A_2 \end{array}$$

for any pair of objects $A_1, A_2 \in \mathrm{Obj}(\mathscr{A})$ and a morphism $f : A_1 \to A_2$.

We can put the goal in other words: The two paths

$$FUA_1 \xrightarrow{\varepsilon_{A_1}} A_1 \xrightarrow{f} A_2 \tag{5.6}$$

and

$$FUA_1 \xrightarrow{FUf} FUA_2 \xrightarrow{\varepsilon_{A_2}} A_2 \tag{5.7}$$

must coincide.

Both are paths in $\mathscr{A}$. We put them in the numerator positions of the fractional diagrams.

Let us begin with path (5.6). We place its adjunct in the denominator:

$$\frac{FUA_1 \xrightarrow{\varepsilon_{A_1}} A_1 \xrightarrow{f} A_2}{UA_1 \xrightarrow[1_{UA}]{} UA_1 \xrightarrow[Uf]{} UA_2}. \tag{5.8}$$

Next, we place the path (5.7) in the numerator and place its adjunct in the denominator:

$$\frac{FUA_1 \xrightarrow{FUf} FUA_2 \xrightarrow{\varepsilon_{A_2}} A_2}{UA_1 \xrightarrow[Uf]{} UA_2 \xrightarrow[1_{UA_2}]{} UA_2}. \tag{5.9}$$

We see that denominators in (5.8) and (5.9) are both

$$UA_1 \xrightarrow{Uf} UA_2.$$

Therefore, by adjunction, the numerators must coincide. □

Definition 5.2 Natural transformations $\eta : \mathrm{Id}_{\mathscr{X}} \to UF$ and $\varepsilon : FU \to \mathrm{Id}_{\mathscr{A}}$ derived from an adjunction $F \dashv U$ are called the *unit* and *counit* of the adjunction, respectively.

5.3 Triangle Identities

Proposition 5.2 (Triangle identities) *Given an adjunction $F \dashv U$ with unit $\eta : \mathrm{Id}_{\mathscr{X}} \to UF$ and counit $\varepsilon : FU \to \mathrm{Id}_{\mathscr{A}}$, we have two commutative diagrams called the* triangle identities.

$$\begin{array}{ccccc} U & \xrightarrow{\eta U} & UFU & \qquad FUF & \xleftarrow{F\eta} & F \\ & \searrow^{=} & \downarrow{\scriptstyle U\varepsilon} & \qquad {\scriptstyle \varepsilon F}\downarrow & \swarrow_{=} & \\ & & U & \qquad F & & \end{array} \tag{5.10}$$

Let us first check the types of objects, morphisms, functors, and natural transformations and see if they are properly fit in the diagrams (5.10).

ηU, $U\varepsilon$, $F\eta$, and εF are combinations of functors and natural transformations. We have to make clear in what way they are concatenated.

Let us put (5.10) into the component-based style as follows.

$$\begin{array}{ccccc} UA & \xrightarrow{\eta_{UA}} & UFUA & \qquad FUFX & \xleftarrow{F\eta_X} & FX \\ & \searrow^{=} & \downarrow{\scriptstyle U\varepsilon_A} & \qquad {\scriptstyle \varepsilon_{FX}}\downarrow & \swarrow_{=} & \\ & & UA & \qquad FX & & \end{array}$$

In this diagram, the players are only objects and morphisms. So, the correctness of wiring depends simply on domain-codomain matching. We may proceed in this style in the rest of our discussion. However, component-free presentation is seen everywhere in various articles, papers, and books. Let us immerse ourselves in this style.

We begin with ηU. We can read it as η is concatenated with U as in

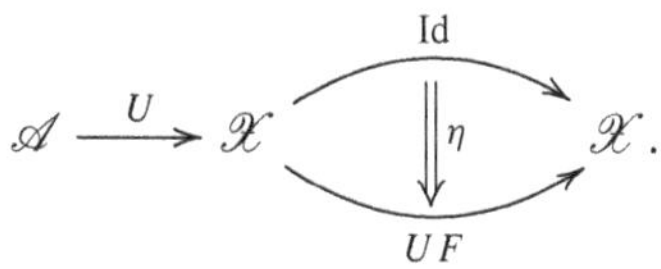

Horizontal composition gives us another natural transformation

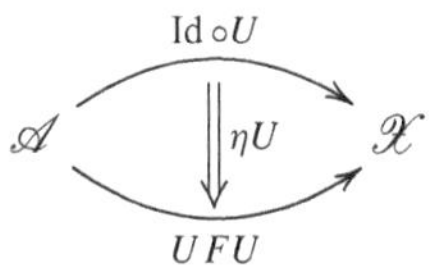

namely

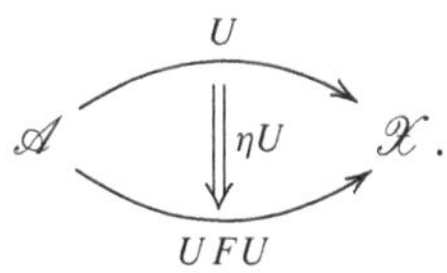

The component of ηU at $A \in \mathrm{Obj}(\mathscr{A})$ is given by

$$(\eta U)_A \stackrel{\text{def}}{=} \eta_{UA} : UA \to UFUA.$$

Next, we analyze $U\varepsilon$. We can read it as ε is prepended by U as in

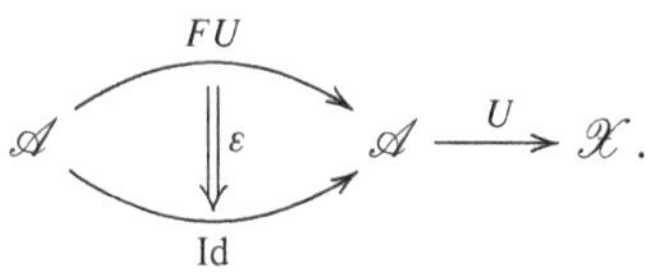

Horizontal composition gives us another natural transformation

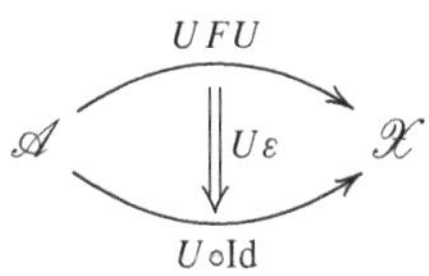

namely

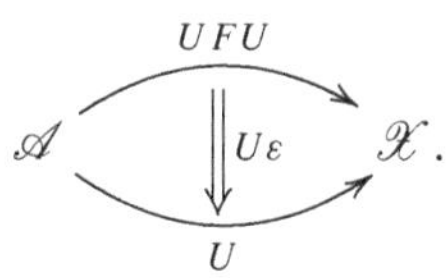

The component of $U\varepsilon$ at $A \in \mathrm{Obj}(\mathscr{A})$ is given by

$$(U\varepsilon)_A \stackrel{\text{def}}{=} U(\varepsilon_A) : UFUA \to UA.$$

We omit the discussion on horizontal compositions of $F\eta$ and εF. We just give their components. The component of $F\eta : F \to FUF$ at X is given by

$$(F\eta)_X \stackrel{\text{def}}{=} F(\eta_X) : FX \to FUFX.$$

The component of $\eta F : FUF \to F$ at X is given by

$$(\varepsilon F)_X \overset{\text{def}}{=} \varepsilon_{FX} : FUFX \to FX.$$

Proof (Proposition 5.2)

We first prove

$$UA \underset{\eta_{UA}}{\longrightarrow} UFUA \underset{U\varepsilon_A}{\longrightarrow} UA \quad = \quad UA \underset{1_{UA}}{\longrightarrow} UA.$$

The both sides of the equation are paths in $\mathscr{X}$. We put them in the denominators with their adjuncts in the numerators as follows.

$$\frac{FUA \xrightarrow{1_{FUA}} FUA \xrightarrow{\varepsilon_A} A}{UA \underset{\eta_{UA}}{\longrightarrow} UFUA \underset{U\varepsilon_A}{\longrightarrow} UA} \tag{5.11}$$

$$\frac{FUA \xrightarrow{\varepsilon_A} A}{UA \underset{1_{UA}}{\longrightarrow} UA} \tag{5.12}$$

The numerators in (5.11) and (5.12) are equal. Therefore, the denominators must coincide.

Next, we prove

$$FX \xrightarrow{F\eta_X} FUFX \xrightarrow{\varepsilon_{FX}} FX \quad = \quad FX \xrightarrow{1_{FX}} FX.$$

The both sides of the equation are paths in $\mathscr{A}$. We put them in the numerators with their adjuncts in the denominators as follows.

$$\frac{FX \xrightarrow{F\eta_X} FUFX \xrightarrow{\varepsilon_{FX}} FX}{X \underset{\eta_X}{\longrightarrow} UFX \underset{1_{FX}}{\longrightarrow} UFX} \tag{5.13}$$

$$\frac{FX \xrightarrow{1_{FX}} FX}{X \underset{\eta_X}{\longrightarrow} UFX} \tag{5.14}$$

The denominators in (5.13) and (5.14) are equal. Therefore, the numerators must coincide. □

5.4 Universal Arrows and Adjunctions

We will explain adjunctions from the view point of universal arrows. In Sect. 5.1, the naturality of adjunction is described by the correspondence

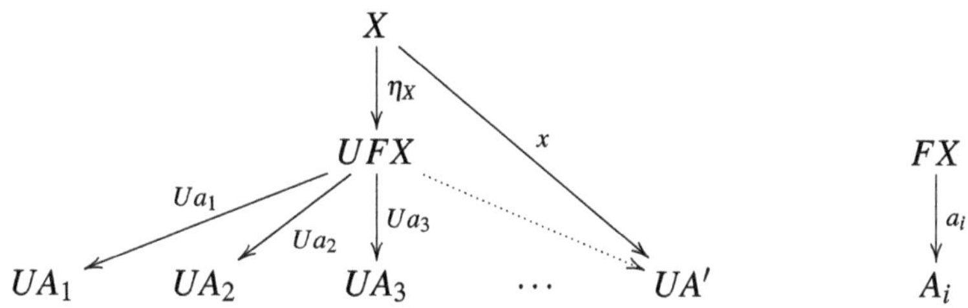

Fig. 5.2 Factorization of $x : FX \to UA'$

$$\frac{FX' \xrightarrow{Fx} FX \xrightarrow{h} A \xrightarrow{a} A'}{X' \underset{x}{\to} X \underset{f}{\to} UA \underset{Ua}{\longrightarrow} UA'}. \tag{5.15}$$

We replace A with FX, h with 1_{FX}, and one with another among various morphisms $a_i : FX \to A_i$ in the adjunction above as follows:

$$\frac{\cdots \to FX \xrightarrow{1_{FX}} FX \xrightarrow{a_i} A_i}{\cdots \to X \underset{\eta_X}{\longrightarrow} UFX \underset{Ua_i}{\longrightarrow} UA_i},$$

where a_i with index i ranges over all the morphisms that have FX as their domain. There may exist an A' such that $\mathscr{A}(FX, A')$ is empty, which we exclude from our discussion. We extract the denominator and rotate 90°. Then, we get Fig. 5.2. Given an arbitrary morphism $x : X \to UA'$, we pose the question of whether or not x factors through η_X with a dashed arrow Ua' (with $a' : FX \to A'$) in Fig. 5.2. If this is the case, η_X is called a *universal arrow* from X to functor U. Further, if this is the case for all $X \in \mathrm{Obj}(\mathscr{X})$, then natural transformation $\eta : \mathrm{Id}_{\mathscr{X}} \to UF$ is said to be *universal to* functor U. We will return to this problem shortly.

Let us replace X with UA and f with 1_{UA}, and one with another among various morphisms $x_i : X_i \to UA$ in adjunction (5.15) to get the following:

$$\frac{FX_i \xrightarrow{Fx_i} FUA \xrightarrow{\varepsilon_A} A \to \cdots}{X_i \xrightarrow{x_i} UA \underset{1_{UA}}{\longrightarrow} UA \to \cdots},$$

where x_i with index i ranges over all the morphisms that have UA as their codomain. There may exist an X' such that $\mathscr{A}(X', UA)$ is empty, which we exclude from our discussion. We extract the numerator and rotate 90°. Then, we get Fig. 5.3. Given an arbitrary morphism $a : FX' \to A$, we pose the question of whether or not a factors through ε_A with a dashed arrow Fx' (with $x' : X' \to UA$) in Fig. 5.3. If this is the case, ε_A is called a *universal arrow* from functor F to A. Further, if this is the case for all $A \in \mathrm{Obj}(\mathscr{A})$, then natural transformation $\varepsilon : FU \to \mathrm{Id}_{\mathscr{A}}$ is said to be *universal from* functor F. We will return to this problem shortly.

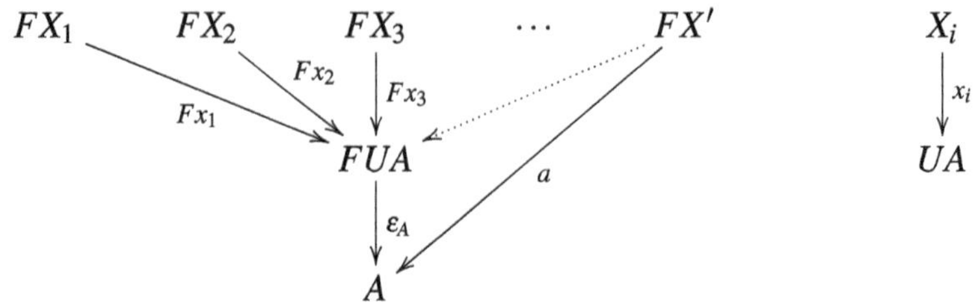

Fig. 5.3 Factorization of $a : FX' \to A$

5.5 Equivalent Formulations of Adjunction

We will show that adjunctions, universal arrows, and triangle identities are all equivalent, which allows us to choose the most appropriate definition depending on the situations.

Theorem 5.1 *Given a pair of functors $U : \mathscr{A} \to \mathscr{X}$ and $F : \mathscr{X} \to \mathscr{A}$, the following four conditions are equivalent.*

1. *$F \dashv U$.*
2. *A pair of natural transformations $\eta : \mathrm{Id}_{\mathscr{X}} \to UF$ and $\varepsilon : FU \to \mathrm{Id}_{\mathscr{A}}$ makes the following two triangles commute*

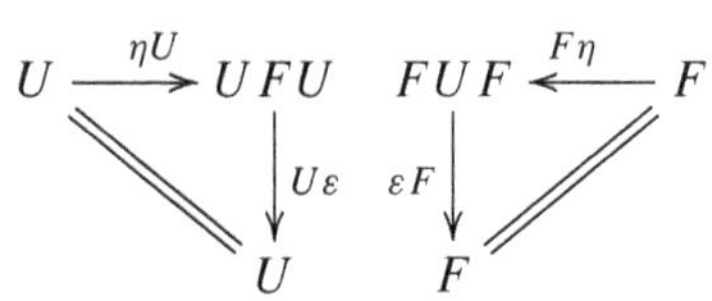

3. *There exists a universal arrow $\eta : \mathrm{Id}_{\mathscr{X}} \to UF$ to functor U, where universal means that for all morphism $f : X \to UA$ there exists a unique morphism $h : FX \to A$ that makes the following triangle commute.*

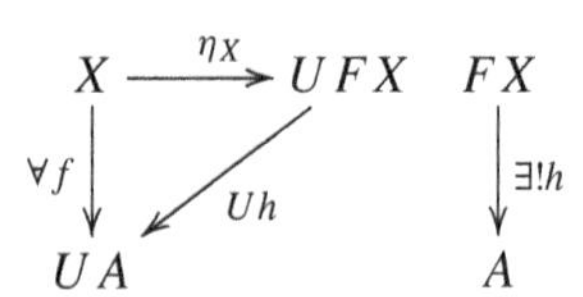

4. *There exists a universal arrow $\varepsilon : FU \to \mathrm{Id}_{\mathscr{A}}$ from functor F, where universal means that for all morphism $h : FX \to A$ there exists a unique morphism $f : X \to UA$ that makes the following triangle commute.*

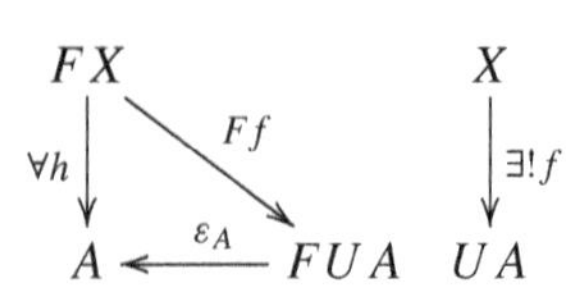

We adopt a strategy as follows.

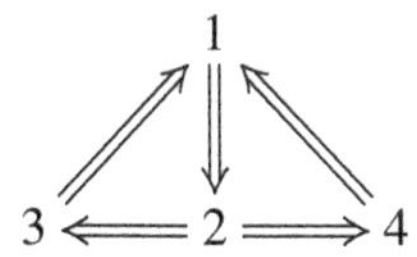

Note that we have already shown $1 \Rightarrow 2$ in Proposition 5.2.

Proof ($2 \Rightarrow 3$ in Theorem 5.1)

We have to find a morphism $UFX \overset{?}{\to} UA$ in the following diagram.

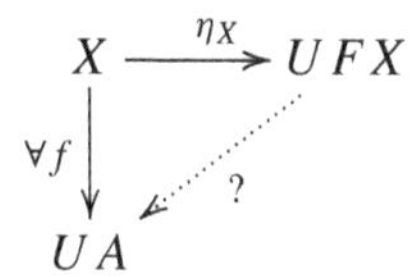

We can think of only one morphism $UFf : UFX \to UFUA$ that originates from UFX. Thanks to the naturality of η we obtain a commutative square

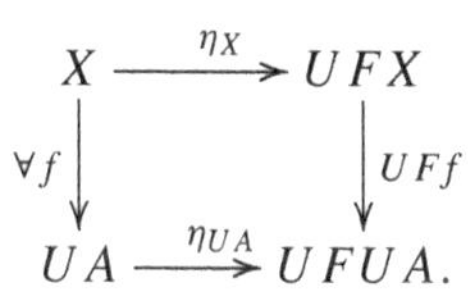

If we search for other materials that can be fit to this square, we find only one thing in the triangle identities. Let them combine.

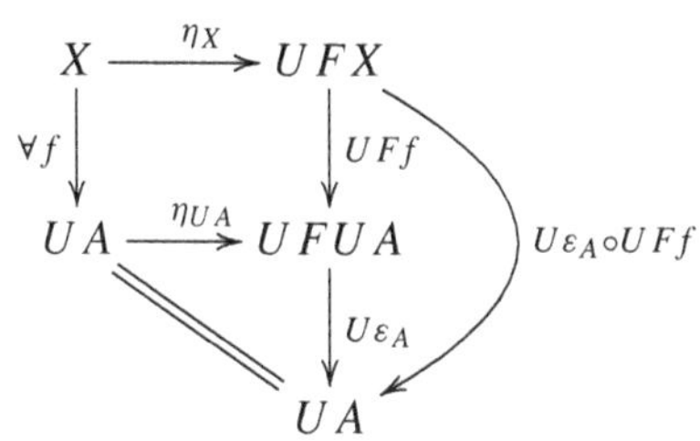

In this diagram, the triangle, square, and consequently trapezoid standing on its toe UA are all commutative. If we put $h = \varepsilon_A \circ Ff$, we have the following.

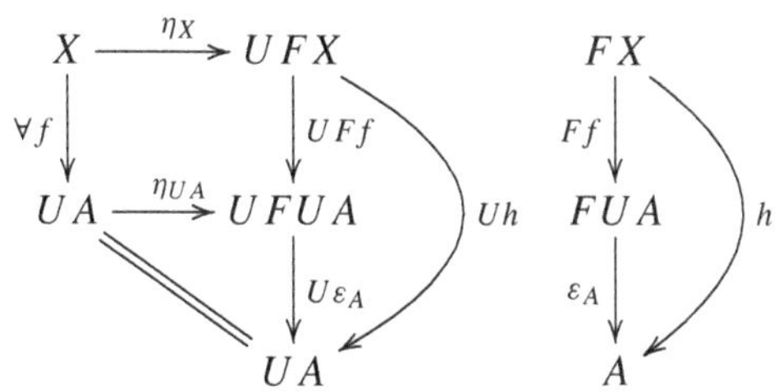

We want to show that such a morphism h is unique. Note that when we apply functor U to morphism h, we can distribute U over the composition $\circ$, namely

$$Uh = U(\varepsilon_A \circ Ff) = U(\varepsilon_A) \circ U(Ff) = U\varepsilon_A \circ UFf.$$

Suppose we have another $h' : FX \to A$ such that $Uh \circ \eta_X = Uh' \circ \eta_X$. Let us draw a square of natural transformation ε with a triangle that fit to the square.

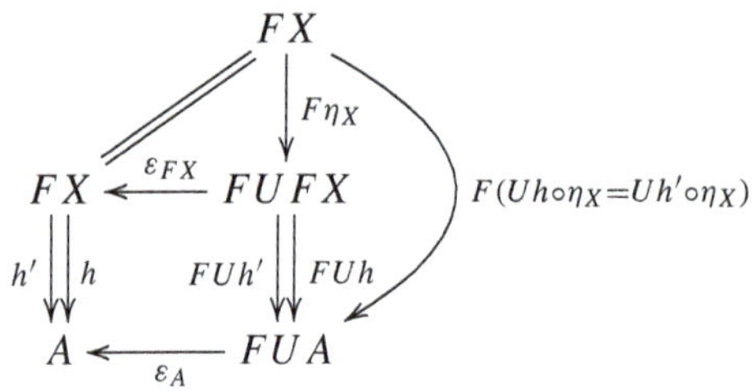

This gives

$$h' = h = \varepsilon_A \circ F(Uh \circ \eta_X = Uh' \circ \eta_X).$$

□

Proof (2 ⇒ 4 in Theorem 5.1)

We have to find a morphism $FX \overset{?}{\to} FUA$ in the following diagram.

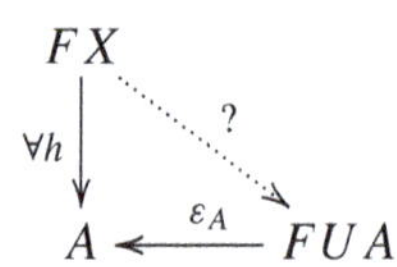

We can think of only one morphism $FUh : FUFX \to FUA$ that terminates at FUA. Thanks to the naturality of ε we obtain a commutative square

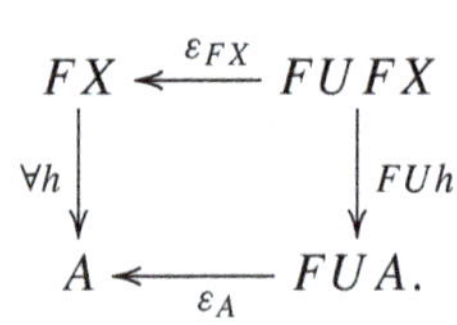

If we search for other materials that can be fit to this square, we find only one thing in the triangle identities. Let them combine.

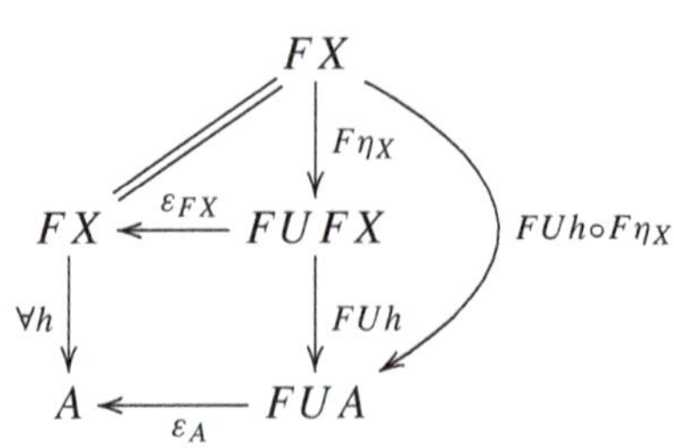

In this diagram, the triangle, square, and consequently trapezoid with a slant roof are all commutative. If we put $f = Uh \circ \eta_X$, we have the following.

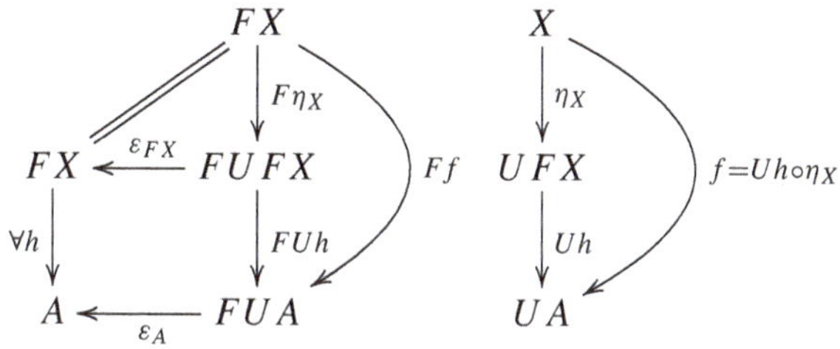

We want to show that such a morphism f is unique.

Suppose we have another $f' : X \to UA$ such that $\varepsilon_X \circ Ff = \varepsilon_X \circ Ff'$. Let us draw a square of natural transformation η with a triangle that fit into the square.

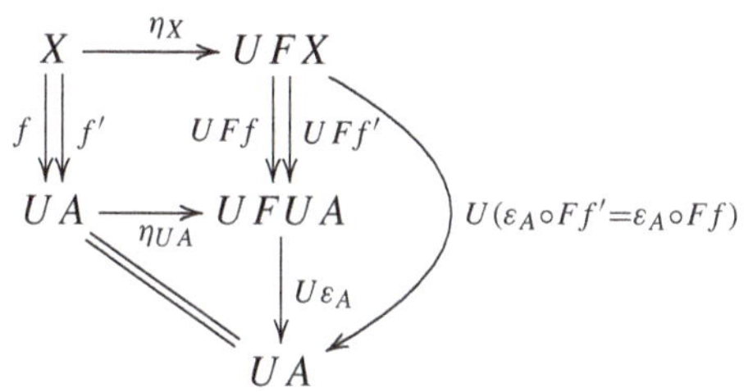

This gives

$$f' = f = U(\varepsilon_A \circ Ff' = \varepsilon_A \circ Ff).$$

□

Proof (3 ⇒ 1 in Theorem 5.1)

By assumption, for any $f : X \to UA$ there exists a unique morphism $h : FX \to A$ that makes the following triangle commute.

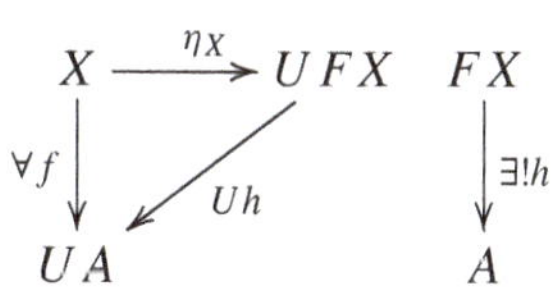

Conversely, given $h : FX \to A$, we can put $f = Uh \circ \eta_X$ to make the diagram above commute. Thus, we obtained a bijection

$$\mathscr{X}(X, UA) \to \mathscr{A}(FX, A).$$

Let us first show naturality in X. Take any morphism $a : X' \to X$.

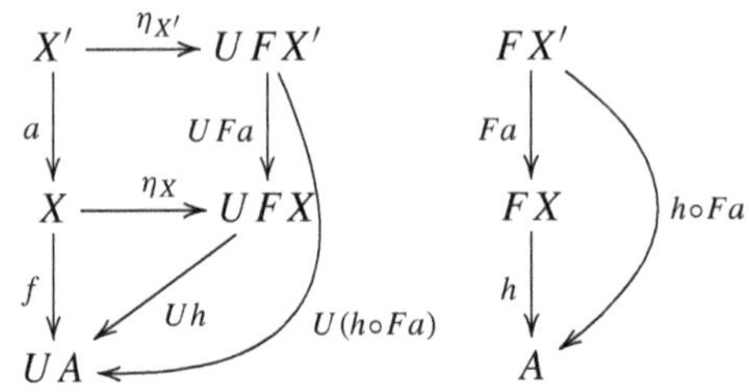

Commutativity in this diagram is as follows.

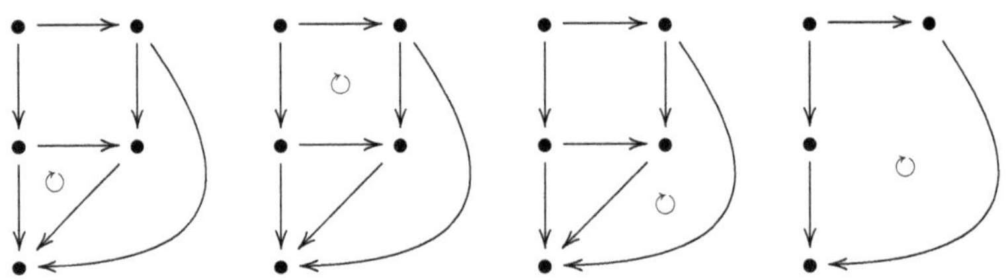

(1) (2) (3) (4)

(1) is given by universal arrow η to functor U.
(2) by natural transformation $\eta : \mathrm{Id}_{\mathscr{X}} \to UF$.
(3) by functor U.
(4) by the combination of (1), (2), and (3).

Universal arrow η to U and (4) establishes

$$\frac{FX' \xrightarrow{Fa} FX \xrightarrow{h} A}{X' \underset{a}{\to} X \underset{f}{\to} UA}.$$

This shows the bijection is natural in X.

Next, we show naturality in A. Take any morphism $b : A \to A'$.

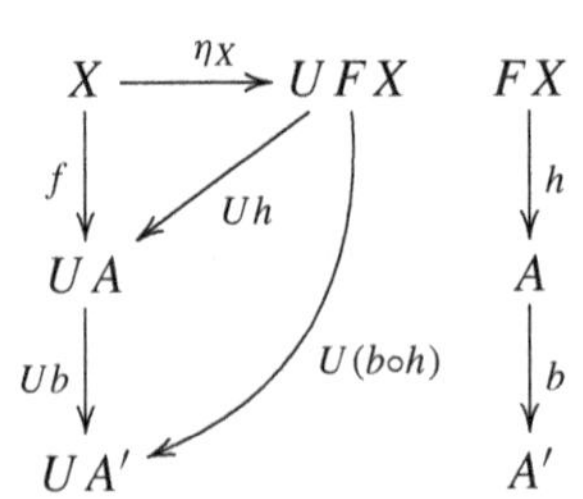

Commutativity in this diagram is as follows.

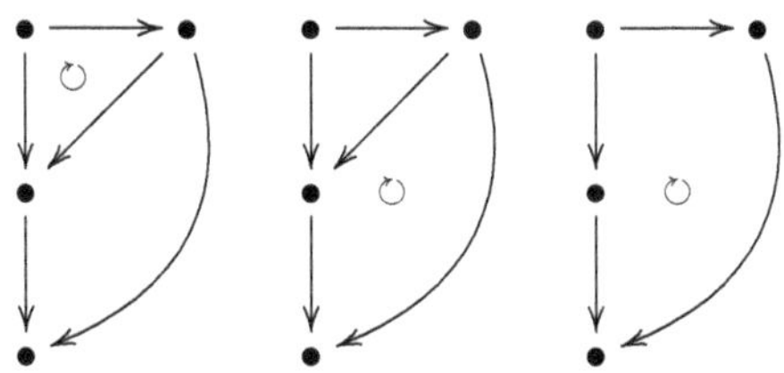

(1) (2) (3)

(1) is given by universal arrow η to functor U.
(2) by functor U.
(3) by the combination of (1) and (2).

Universal arrow η to U and (3) establishes

$$\frac{FX \xrightarrow{h} A \xrightarrow{b} A'}{X \xrightarrow[f]{} UA \xrightarrow[Ub]{} UA'}.$$

This shows the bijection is natural in A. □

Proof (4 ⇒ 1 in Theorem 5.1)

By assumption, for any $h : FX \to A$ there exists a unique morphism $f : X \to UA$ that makes the following triangle commute.

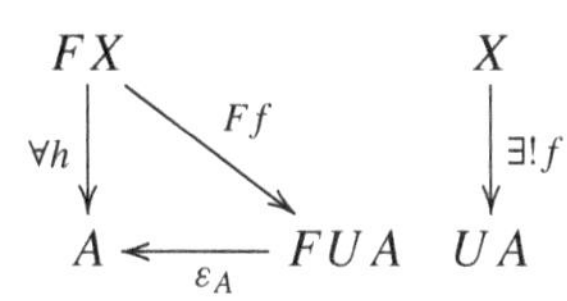

Conversely, given $f : X \to UA$, we can put $h = \varepsilon_A \circ Ff$ to make the diagram above commute. Thus, we obtained a bijection

$$\mathscr{A}(FX, A) \to \mathscr{X}(X, UA).$$

What remains to show is the naturality in X and A.

Let us first show naturality in X. Take any morphism $a : X' \to X$.

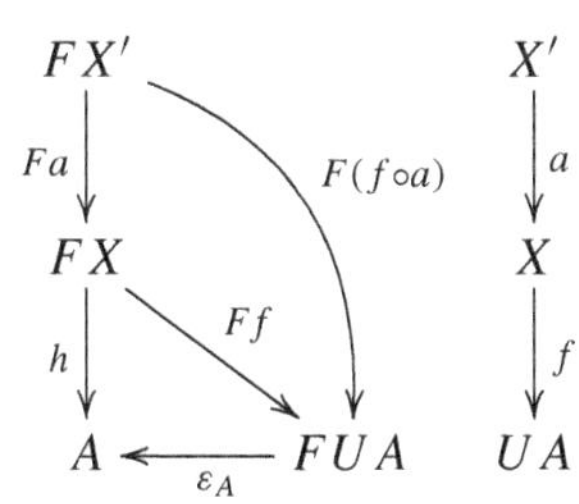

Commutativity in this diagram is as follows.

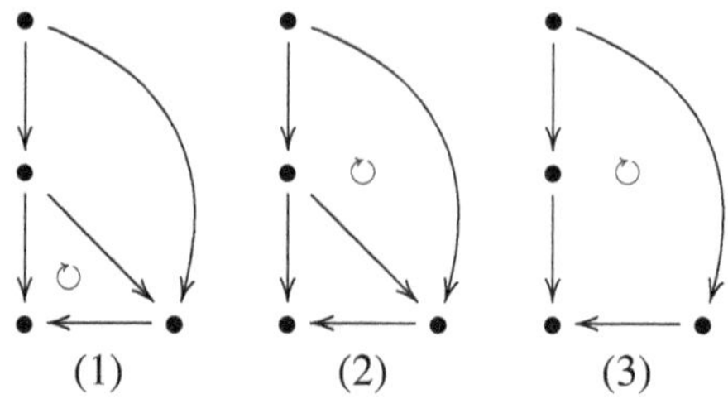

(1) is given by universal arrow ε from functor F.
(2) by functor F.
(3) by the combination of (1) and (2).

Universal arrow ε from F and (3) establishes

$$\frac{FX' \xrightarrow{Fa} FX \xrightarrow{h} A}{X' \underset{a}{\rightarrow} X \underset{f}{\rightarrow} UA}.$$

This shows the bijection is natural in X.

Next, we show naturality in A. Take any morphism $b : A \rightarrow A'$.

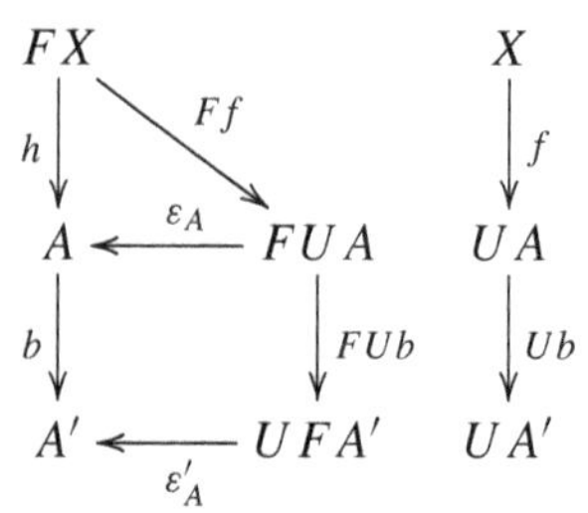

Commutativity in this diagram is as follows.

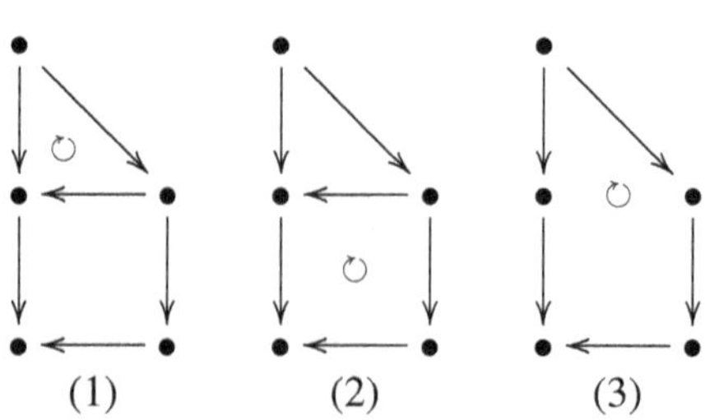

(1) is given by universal arrow ε from functor F.
(2) by natural transformation ε.
(3) by the combination of (1) and (2).

Universal arrow ε from F and (3) establishes

$$\frac{FX \xrightarrow{h} A \xrightarrow{b} A'}{X \xrightarrow[f]{} UA \xrightarrow[Ub]{} UA'}.$$

This shows the bijection is natural in A. □

The concept of *free object* is often used as in the following theorem.

Theorem 5.2 *Given a functor $U : \mathscr{A} \to \mathscr{X}$, the following two conditions are equivalent.*

1. *There exists another functor $F : \mathscr{X} \to \mathscr{A}$ that yields $F \dashv G$.*
2. *For any $X \in \mathrm{Obj}(\mathscr{X})$ there exists a free object $F_0X \in \mathrm{Obj}(\mathscr{A})$. We say that $F_0X \in \mathrm{Obj}(\mathscr{A})$ is a free object if there exists a morphism $\eta_X : X \to UF_0X$ such that for all $f : X \to UA$ there exists a unique $f^\sharp : F_0X \to A$ that makes the following diagram commute.*

$$\begin{array}{ccccc} X & \xrightarrow{\eta_X} & UF_0X & & F_0X \\ & \searrow^{\forall f} & \downarrow{\scriptstyle Uf^\sharp} & & \downarrow{\scriptstyle \exists! f^\sharp} \\ & & UA & & A \end{array} \tag{5.16}$$

In condition 2, F_0 is just a function on objects. We will show that F_0 is eventually extended to a functor F.

Example 5.1 Let $\mathscr{X} = \mathbf{Set}$ and $\mathscr{A} = \mathrm{Vect}_{\mathbb{C}}$. Let $U : \mathrm{Vect}_{\mathbb{C}} \to \mathbf{Set}$ be a forgetful functor. $F : \mathbf{Set} \to \mathrm{Vect}_{\mathbb{C}}$ takes a set X and returns a complex vector space FX that has X as its basis. We give functions on objects in a rough informal style.

$$FX = \bigoplus_{x \in X} \mathbb{C}\vec{b}_x$$

$$UFX = \{\sum_{x \in X} c_x \vec{b}_x \;\Big|\; c_x \in \mathbb{C}\} \simeq \{\{c_x\}_{x \in X}\}$$

$$\begin{aligned} \eta_X : & X \to UFX \\ & x \mapsto \vec{b}_x \end{aligned}$$

Proof (Theorem 5.2)

We prove 1 ⇒ 2. Given $f : X \to UA$, we have a unique adjunct $f^\sharp : FX \to A$:

$$\frac{FX \xrightarrow{\exists! f^\sharp} A}{X \underset{\forall f}{\longrightarrow} UA}.$$

By Theorem 5.1, $\eta_X : X \to UFX$ is a universal arrow from X to functor U. If we replace F_0 with F, we see that diagram (5.16) commutes.

Next, we prove 2 ⇒ 1. The assumption gives a function on objects F_0. We extend it to a functor $F : \mathscr{X} \to \mathscr{A}$.

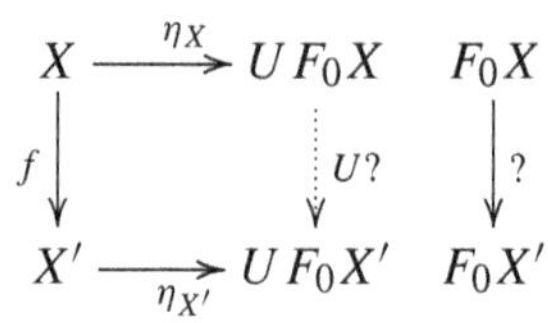

We seek a morphism $? : F_0X \to F_0X'$ for $f : X \to X'$. We apply $\sharp$ operation on

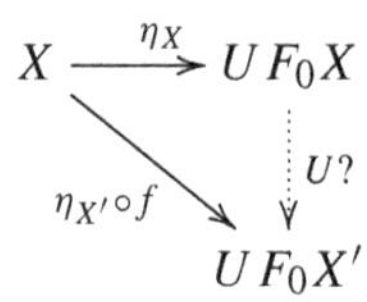

to obtain

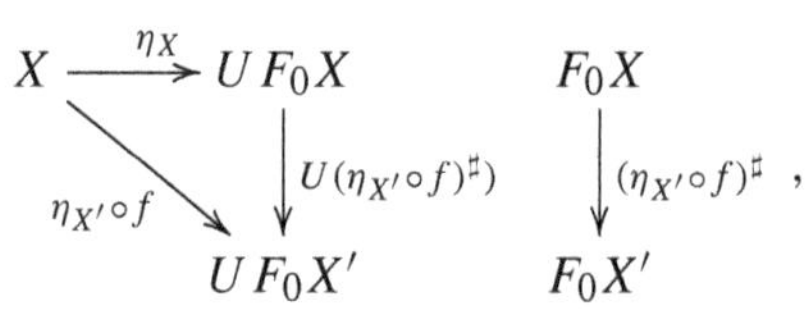

where the dashed morphism $U(\eta_{X'} \circ f)^\sharp$ is uniquely determined.

We are tempted to define a function on morphisms as follows. From now on we write F instead of F_0.

$$Ff = (\eta_{X'} \circ f)^\sharp.$$

We must ask ourselves if this assignment complies with the functor axioms. Two tests must be passed.

(1) $F1_X = 1_{FX}$
(2) For any pair of composable morphisms, $f : X \to X'$ and $g : X' \to X''$ the equality $Fg \circ Ff = F(g \circ f)$ must hold.

First, we examine (1). We have a diagram

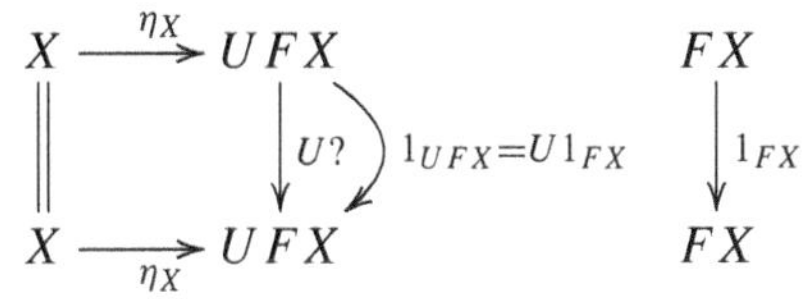

Compare the two squares on the left.

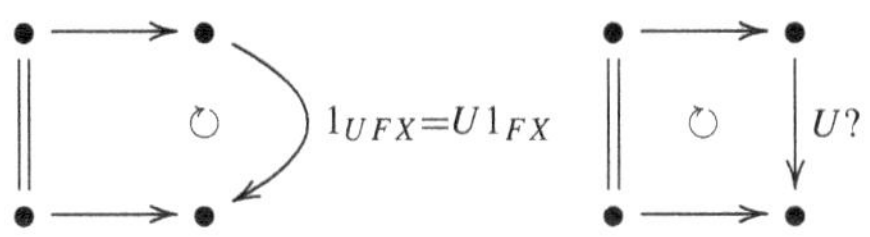

(a) (b)

The square with a round edge (a) is commutative because of a chain of mappings of 1_X by functors F and U. The $U_?$ in rectangle (b) must be unique by the assumption. Thus, we have $F1_X = 1_{FX}$.

Next, we examine axiom (2).

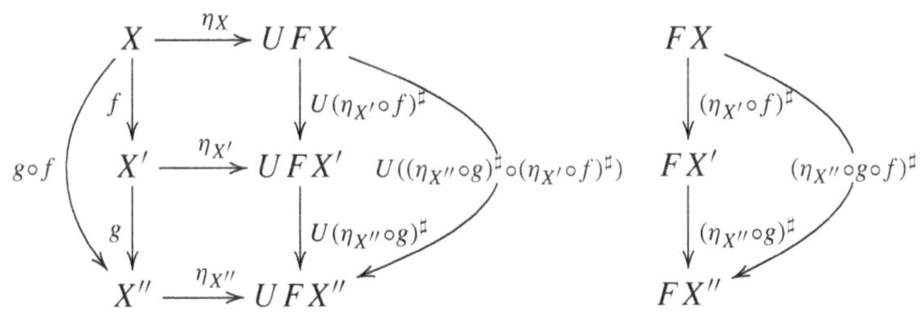

Our definition of a function on morphisms gives

$$F(g \circ f) = (\eta_{X''} \circ g \circ f)^\sharp$$
$$Fg \circ Ff = (\eta_{X''} \circ g)^\sharp \circ (\eta_{X'} \circ f)^\sharp$$

By uniqueness in the assumption guarantees commutativity for all squares and triangles in the diagram. Therefore, we have

$$(\eta_{X''} \circ g \circ f)^\sharp = (\eta_{X''} \circ g)^\sharp \circ (\eta_{X''} \circ f)^\sharp.$$

This shows $F(g \circ f) = Fg \circ Ff$, completing the proof that F is really a functor.

We have shown that the collection of η_X comprises a natural transformation $\eta : \mathrm{Id}_{\mathscr{X}} \to UF$, and that η is a universal arrow to U. By Theorem 5.1, we conclude that $F \dashv U$. □

5.6 Adjunction and Category Equivalence

We will see relations between adjunction and category equivalence.

Definition 5.3 A morphism $f : A \to B$ is called a *monomorphism* if for any pair of morphisms $C \overset{g}{\underset{g'}{\rightrightarrows}} A$ the following implication holds:

$$f \circ g = f \circ g' \implies g = g'.$$

We also say "f is *monic*."

Remark 5.1 In **Set**, monomorphism=injection. We use the same notation in Definition 5.3.

A function $f : A \to B$ is called an injection if the following condition holds:

$$\forall a, a' \in A \ . \ \text{if } f(a) = f(a') \text{ then } a = a'.$$

Suppose f is an injection. If $f \circ g = f \circ g'$, then $f(g(c)) = f(g'(c))$ for all $c \in C$. Since f is injective, we have $g(c) = g'(c)$ for all $c \in C$. We can conclude that $g = g'$ and thus f is monic.

Conversely, suppose f is monic. Since C is arbitrary, we can put $C = \{*\}$, a singleton set. Let $a, a' \in A$ be arbitrary pair of elements, and consider two functions $g : * \mapsto a$ and $g' : * \mapsto a'$. If $f(a) = f(a')$, then $f \circ g = f \circ g'$. Since f is monic, we have $g = g'$, therefore $a = a'$. Thus, f is injective.

Definition 5.4 A morphism $f : A \to B$ is called a *epimorphism* if for any pair of morphisms $B \overset{g}{\underset{g'}{\rightrightarrows}} C$ the following implication holds:

$$g \circ f = g' \circ f \implies g = g'.$$

We also say "f is *epic*."

Remark 5.2 In **Set**, epimorphism=surjection. We use the same notation in Definition 5.4.

A function $f : A \to B$ is called a surjection if the following condition holds:

$$\forall b \in B \ . \ \exists a \in A \ . \ b = f(a).$$

Suppose f is a surjection. For all $b \in B$ there exist $a \in A$ such that $b = f(a)$. If $g \circ f = g' \circ f$, then

$$g(b) = g(f(a)) = g'(f(a)) = g'(b)$$

for all $b \in B$. We have $g = g'$, and thus f is epic.

Conversely suppose f is an epimorphism. We suppose also f is not surjective. Then, we have

$$\exists b \in B \ . \ \forall a \in A \ . \ b \neq f(a).$$

Let $C' = \{1, 2\} \sqcup C$. We define two functions $\bar{g}, \bar{g'} : B \to C'$ as follows.

$$\bar{g}(x) = \begin{cases} 1 & (x = b) \\ g(x) & (x \neq b) \end{cases}$$

$$\bar{g'}(x) = \begin{cases} 2 & (x = b) \\ g'(x) & (x \neq b) \end{cases}$$

We have $\bar{g} \circ f = \bar{g'} \circ f$ but $\bar{g} \neq \bar{g'}$. This contradicts the assumption that f is an epimorphism. Therefore, f must be surjective.

Example 5.2 In **Top** there is a non-surjective epimorphism. Let $i :: \mathbb{Q} \to \mathbb{R}$ an embedding. It is clearly non-surjective, but an epimorphism.

Lemma 5.1 *Given two morphisms $e : A \to B$ and $s : B \to A$ with $e \circ s = 1_B$, then we know that e is an epimorphism and s is a monomorphism.*

Remark 5.3 If we require $f \circ g = 1_B$ at the same time, f and g are isomorphisms.

Proof Let us first prove f is epic. Assume $a \circ e = b \circ e$. If we compose s from the right, we have

$$a \circ e \circ s = b \circ e \circ s,$$

and since $e \circ s = 1_B$ we further have $a = b$, which means that e is epic.

Next, we prove s is monic. Assume $s \circ a = s \circ b$. If we compose e from the left, we have

$$e \circ s \circ a = e \circ s \circ b,$$

and since $e \circ s = 1_B$ we further have $a = b$, which means that s is monic. □

Definition 5.5 Given two morphisms $e : A \to B$ and $s : B \to A$ with $e \circ s = 1_B$, morphism e is called a *split epimorphism*, and s is called a *split monomorphism*. Relations between e and s and relations between B and A are often phrased:

- $s : B \to A$ is a *section* of $e : A \to B$.
- $e : A \to B$ is a *retraction* of $s : B \to A$.
- B is a *retract* of A.

Proposition 5.3 *Given two functors $U : \mathscr{A} \to \mathscr{X}$ and $F : \mathscr{X} \to \mathscr{A}$ with $F \dashv U$, we have the following:*

(a) *U is faithful* $\Leftrightarrow \varepsilon_A : FUA \to A$ *is an epimorphism for any* $A \in \mathrm{Obj}(\mathscr{A})$.
(b) *U is full* $\Leftrightarrow \varepsilon_A : FUA \to A$ *is a split monomorphism for all* $A \in \mathrm{Obj}(\mathscr{A})$.
(c) *U is full and faithful* $\Leftrightarrow \varepsilon_A : FUA \to A$ *is an isomorphism for all* $A \in \mathrm{Obj}(\mathscr{A})$.

Proof The adjunction is depicted as follows.

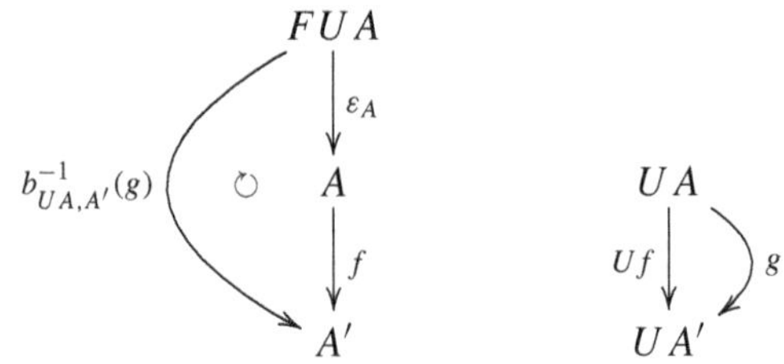

We consider the composition $U_{A,A'}$ with $b^{-1}_{UA,A'}$.

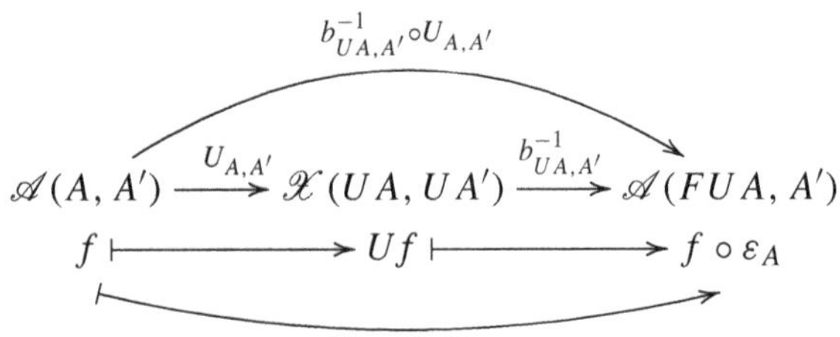

Since $b^{-1}_{UA,A'}$ is bijective, we have the following equivalences.

$$\begin{aligned} U_{A,A'} \text{ is injective } &\Leftrightarrow f \mapsto f \circ \varepsilon_A \text{ is injective} \\ U_{A,A'} \text{ is surjective } &\Leftrightarrow f \mapsto f \circ \varepsilon_A \text{ is surjective} \end{aligned}$$

We prove (a).

$$\begin{aligned} &U \text{ is faithful} \\ \Leftrightarrow\ &\text{For any pair } (A, A')\,,\ U_{A,A'} \text{ is injective} \\ \Leftrightarrow\ &\text{For any pair } (A, A')\,,\ f \mapsto f \circ \varepsilon_A \text{ is injective}(\mathscr{A}(A, A') \to \mathscr{A}(FUA, A')) \end{aligned}$$

That $f \mapsto f \circ \varepsilon_A$ is injective means for any pair f_1, f_2 if $f_1 \circ \varepsilon_A = f_2 \circ \varepsilon_A$ then $f_1 = f_2$, therefore means ε_A is an epimorphism.

We prove (b).

$$\begin{aligned} &U \text{ is full} \\ \Leftrightarrow\ &\text{For any pair } (A, A')\,,\ U_{A,A'} \text{ is surjective} \\ \Leftrightarrow\ &\text{For any pair } (A, A')\,,\ f \mapsto f \circ \varepsilon_A \text{ is surjective}(\mathscr{A}(A, A') \to \mathscr{A}(FUA, A')) \end{aligned}$$

Suppose $f \circ \varepsilon_A = 1_{FUA}$ is surjective for any pair (A, A'). We specialize the case as $A' = FUA$ and $f \circ \varepsilon_A = 1_{FUA}$.

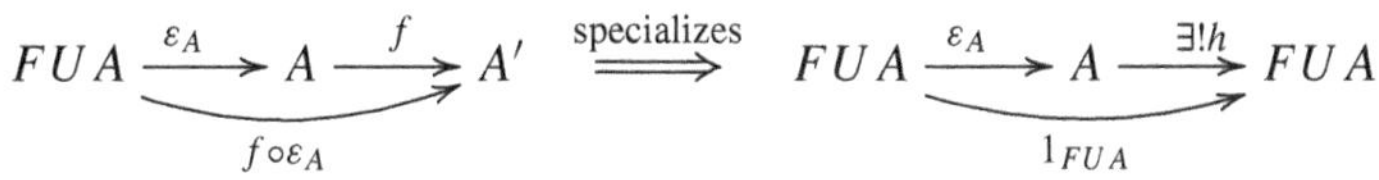

By surjectivity, there exists $h : A \to FUA$ such that $h \circ \varepsilon_A = 1_{FUA}$, which shows that ε_A is a split monomorphism.

Conversely, we assume that ε_A is a split monomorphism. There exists $h : A \to FUA$ such that $h \circ \varepsilon_A = 1_{FUA}$. Given $g : FUA \to A'$, we put $f = g \circ h$. Then, we have

$$f \circ \varepsilon_A = g \circ h \circ \varepsilon_A = g \circ 1_{FUA},$$

which shows that $f \mapsto f \circ \varepsilon_A$ is surjective. The diagram

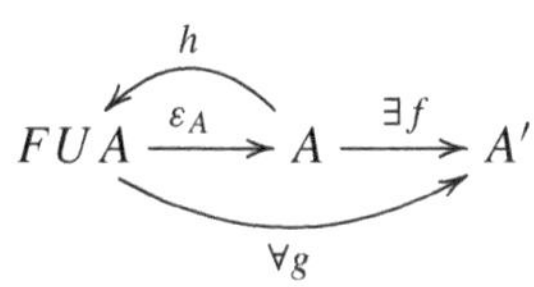

may be a help to get the idea.

We prove (c).

Having proved (a) and (b), we need only apply the following general fact:

$$\text{isomorphism} \Leftrightarrow \text{epimorphism and split monomorphism.}$$

We briefly reproduce the proof for ε_A.
($\Rightarrow$) is trivial.
($\Leftarrow$) Suppose ε_A is a split monomorphism. There exists $u : A \to FUA$ such that $u \circ \varepsilon_A = 1_{FUA}$. Applying $\varepsilon_A \circ$ from the left to the both sides, we get

$$\varepsilon_A \circ u \circ \varepsilon_A = \varepsilon_A = 1_A \circ \varepsilon_A.$$

Since ε_A is an epimorphism, we conclude that $\varepsilon_A \circ u = 1_A$. Thus, ε_A is an isomorphism. □

Proposition 5.4 *Given two functors $U : \mathscr{A} \to \mathscr{X}$ and $F : \mathscr{X} \to \mathscr{A}$, the following two conditions are equivalent:*

1. *Functors U and F give an equivalence of $\mathscr{A}$ and $\mathscr{X}$. To be more specific, there exist natural isomorphisms $\alpha : \mathrm{Id}_{\mathscr{X}} \to UF$ and $\beta : FU \to Id_{\mathscr{A}}$.*
2. *$F \dashv G$, and unit $\eta : \mathrm{Id}_{\mathscr{X}} \to UF$ and counit $\varepsilon : FU \to \mathrm{Id}_{\mathscr{A}}$ are natural isomorphisms.*

Remark 5.4 When condition 2 is satisfied, functors U and F are said to give an *adjoint equivalence* of categories $\mathscr{A}$ and $\mathscr{X}$. The proposition claims that the concepts of category equivalence and adjoint equivalence are equivalent. Note that α and β in condition 1 are not necessarily a unit and counit of the adjunction.

Proof $2 \Rightarrow 1$ is easy. Set $\alpha = \eta$ and $\beta = \varepsilon$, and forget about the fact that they came from an adjunction.

We prove $1 \Rightarrow 2$. We construct an adjunction with the following strategy:

- construct a correspondence between $\mathscr{A}(FX, A)$ and $\mathscr{X}(X, UA)$ natural in X and A,
- construct η and ε out of α and β,
- prove that η and ε are both natural isomorphisms.

Recall that F and G are full and faithful functors, and that they both are essentially surjective on objects. See Proposition 2.1 and Remark 2.3.

Take an arbitrary pair of $A \in \mathrm{Obj}(\mathscr{A})$ and $X \in \mathrm{Obj}(\mathscr{X})$. Given a morphism $f : X \to UA$, we construct a morphism $f^\sharp : FX \to A$. Let us consider the following diagram.

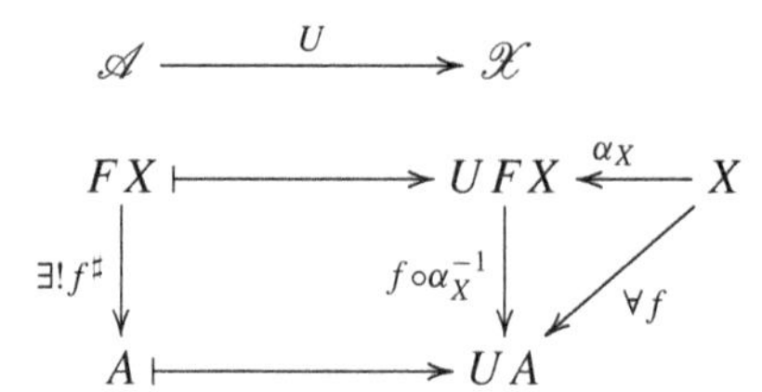

Since U is full and faithful,

$$f^\sharp = U^{-1}_{FX,A}(f \circ \alpha_X^{-1})$$

exists. We have $Uf^\sharp = f \circ \alpha_X^{-1}$. This is the only candidate available for an adjunct.

$$\frac{FX \xrightarrow{f^\sharp} A}{X \underset{f}{\to} UA} = \frac{FX \xrightarrow{1_{FX}} FX \xrightarrow{f^\sharp} A}{X \underset{\alpha_X}{\longrightarrow} UFX \underset{Uf^\sharp}{\longrightarrow} UA}.$$

We still have to show naturality in X and A. From now on we write simply U^{-1} for the inverse function on morphisms.

We prove naturality in X. We have to show for any morphism $X' \to X$

$$\frac{FX' \xrightarrow{Fx} FX \xrightarrow{f^\sharp} A}{X' \underset{x}{\to} X \overset{U}{\underset{f}{\to}} A}$$

holds. In other words,

$$(f \circ x)^\sharp = f^\sharp \circ Fx.$$

The situation is concisely described by the following diagram.

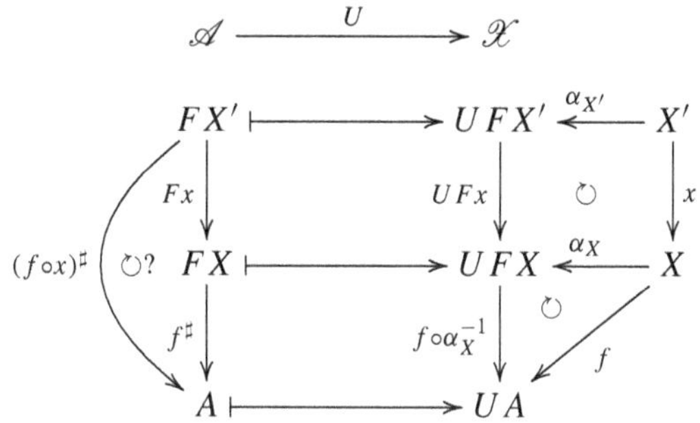

The square on the right is commutative due to the natural transformation α. The triangle below commutes since it is just a composition of morphisms. We show the triangle on the left commutes as follows.

$$\begin{aligned}(f \circ x)^\sharp &= U^{-1}(f \circ x \circ \alpha_{X'}^{-1})\\ &= U^{-1}(f \circ \alpha_X^{-1} \circ UFx)\\ &= U^{-1}(f \circ \alpha_X^{-1}) \circ U^{-1}(UFx)\\ &= f^\sharp \circ Fx\end{aligned}$$

Thus, we established naturality in X.

Next, we prove naturality in A. We have to show for any morphism $a : A \to A'$

$$\frac{FX \xrightarrow{f^\sharp} A \xrightarrow{a} A'}{X \underset{f}{\rightarrow} UA \underset{Ua}{\longrightarrow} UA'}$$

holds. In other words,

$$(Ua \circ f)^\sharp = a \circ f^\sharp.$$

The situation is concisely described by the following diagram.

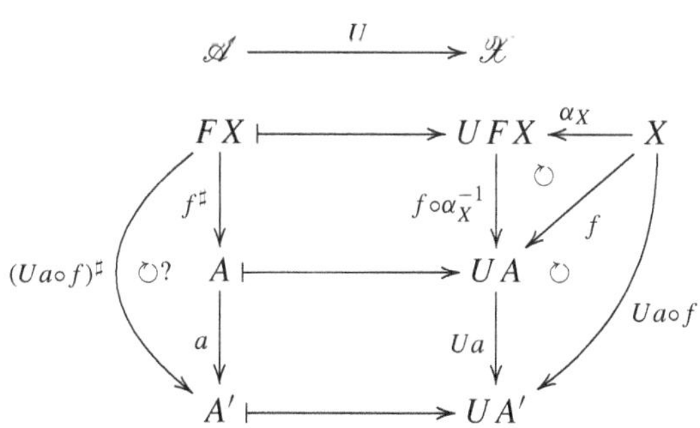

The two triangles on the right are commutative since both are just a composition of morphisms. We show the triangle on the left commute. Let us trace the rewriting:

$$\begin{aligned}(Ua \circ f)^\sharp &= U^{-1}(Ua \circ f \circ \alpha_X^{-1}) \\ &= a \circ U^{-1}(f \circ \alpha_X^{-1}) \\ &= a \circ f^\sharp\end{aligned}$$

Thus, we get naturality in A.

We want to make clear the relation between the pair of natural transformations $\alpha : \mathrm{Id}_{\mathscr{X}} \to UF$ and $\beta : FU \to \mathrm{Id}_{\mathscr{A}}$ and the pair of η and ε.

Let us consider the diagram:

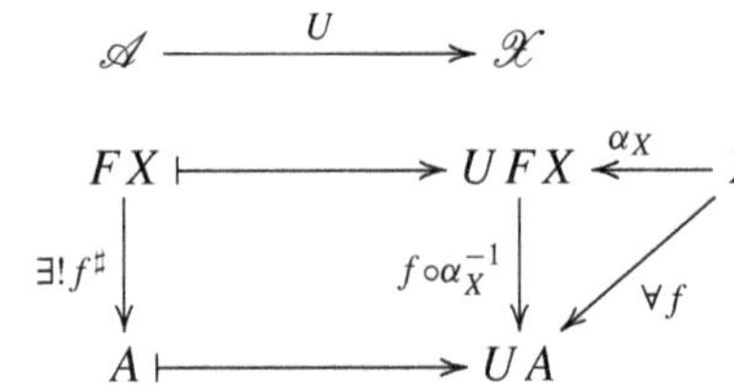

We assign $A = FX$ and $f^\sharp = 1_{FX}$ and rewrite this diagram.

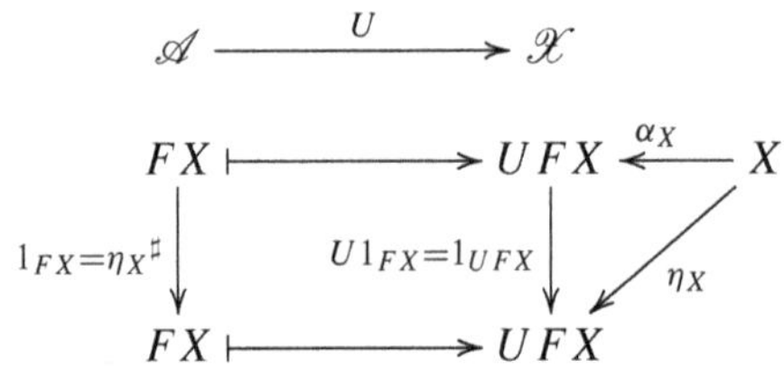

This shows $\eta = \alpha$.

Next, we study ε.

Let us consider the diagram:

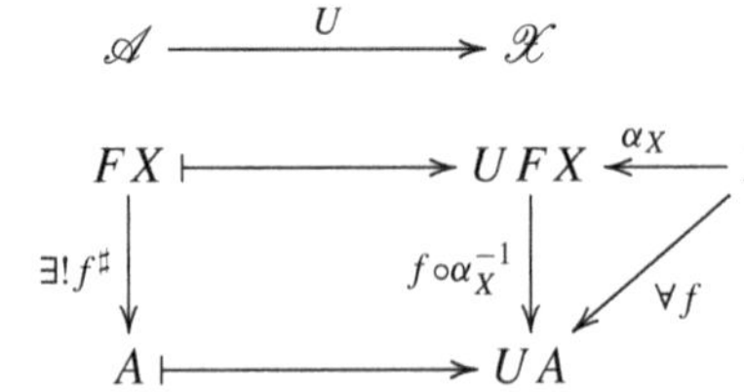

We assign $X = UA$ and $f = 1_{UA}$ and rewrite this diagram.

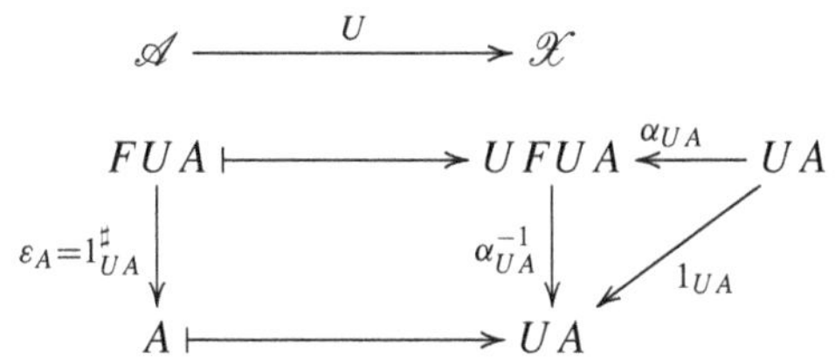

This shows

$$\varepsilon_A = U^{-1}(\alpha_{UA}^{-1}).$$

□

5.7 Global Naturality of Adjunction

Some people feel that the naturality explained so far is a little bit unnatural. Let $\mathscr{A}$ and $\mathscr{X}$ be locally small categories. Consider the adjunction

$$\mathscr{A}(FX, A) \cong \mathscr{X}(X, UA).$$

LHS can be seen as a functor that sends (X, A) to a set:

$$\mathscr{X}^{\mathrm{op}} \times \mathscr{A} \to \mathbf{Set}$$

RHS can also be seen as a functor that sends (X, A) to a set:

$$\mathscr{X}^{\mathrm{op}} \times \mathscr{A} \to \mathbf{Set}$$

We show that both are really functors that are naturally isomorphic. We see this natural isomorphism establishes an adjunction.

(1) We prove $\mathscr{A}(F-, -) : \mathscr{X}^{\mathrm{op}} \times \mathscr{A} \to \mathbf{Set}$ is a functor.
The function on objects is given by

$$\mathscr{A}(F-, -)(X, A) = \mathscr{A}(FX, A)$$

for all $(X, A) \in \mathrm{Obj}(\mathscr{X}^{\mathrm{op}} \times \mathscr{A})$.
Let (X, A) and (X', A') be any pair of objects in $\mathscr{X}^{\mathrm{op}} \times \mathscr{A}$. A morphism from (X, A) to (X', A') is a pair (x, a) where $x : X' \to X$ is a morphism in $\mathscr{X}$ and $a : A \to A'$ is a morphism in $\mathscr{A}$. We write it as

$$(x, a) : (X, A) \to (X', A').$$

We want to send this morphism to a function in a way

$$(x,a) \overset{\mathscr{A}(F-,-)}{\longmapsto} \left(\begin{array}{ccc} \mathscr{A}(FX,A) & \longrightarrow & \mathscr{A}(FX',A') \\ f & \longmapsto & a \circ f \circ Fx \end{array} \right)$$

We have to check if this definition conforms to the axioms of functors.
First we check the image of $(1_X, 1_A)$, the identity morphism in $\mathscr{X}^{\text{op}} \times \mathscr{A}$. We have

$$f = 1_A \circ f \circ F(1_X).$$

since $F(1_X) = 1_{FX}$, which shows

$$\mathscr{A}(F-,-)(1_X, 1_A) = 1_{\mathscr{A}(FX,A)}.$$

Next, we check if the following triangle commutes:

$$(X,A) \xrightarrow{(x,a)} (X',A') \xrightarrow{(x',a')} (X'',A''), \qquad (5.17)$$

with the lower arrow $(X,A) \to (X'',A'')$ labelled $(x \circ x', a' \circ a)$,

where $x : X' \to X$:, $x' : X'' \to X'$, $a : A \to A'$, $a' : A' \to A''$. We apply $\mathscr{A}(F-,-)$ to this triangle to get the following:

$$\mathscr{A}(FX,A) \xrightarrow{f \mapsto a \circ f \circ Fx} \mathscr{A}(FX',A') \xrightarrow{f \mapsto a' \circ f \circ Fx'} \mathscr{A}(FX'',A'').$$

with the lower arrow $\mathscr{A}(FX,A) \to \mathscr{A}(FX'',A'')$ labelled $f \mapsto (a' \circ a) \circ f \circ F(x \circ x')$

We have to see if this triangle commutes. The composition of the upper two edges is as follows.

$$a' \circ a \circ f \circ Fx \circ Fx'$$

Since $Fx \circ Fx' = F(x \circ x')$, we see that the triangle commutes. This completes the proof that $\mathscr{A}(F-,-)$ is a functor.

(2) We prove $\mathscr{X}(-,U-) : \mathscr{X}^{\text{op}} \times \mathscr{A} \to \textbf{Set}$ is a functor.
The function on objects is given by

$$\mathscr{X}(-,U-)(X,A) = \mathscr{X}(X,UA)$$

for all $(X,A) \in \text{Obj}(\mathscr{X}^{\text{op}} \times \mathscr{A})$.
Let (X,A) and (X',A') be any pair of objects in $\mathscr{X}^{\text{op}} \times \mathscr{A}$. A morphism from (X,A) to (X',A') is a pair (x,a) where $x : X' \to X$ is a morphism in $\mathscr{X}$ and $a : A \to A'$ is a morphism in $\mathscr{A}$. We write it as

$$(x,a) : (X,A) \to (X',A').$$

We want to send this morphism to a function in a way

$$(x,a) \overset{\mathscr{X}(-,U-)}{\longmapsto} \left(\begin{array}{ccc} \mathscr{X}(X,UA) & \longrightarrow & \mathscr{X}(X',UA') \\ f & \longmapsto & Ua \circ f \circ x \end{array} \right)$$

Identity morphisms are mapped to identity morphisms. We apply $\mathscr{X}(-,U-)$ to the triangle 5.17 to get the following:

$$\mathscr{X}(X,UA) \xrightarrow{f \mapsto Ua \circ f \circ x} \mathscr{X}(X',UA') \xrightarrow{f \mapsto Ua' \circ f \circ x'} \mathscr{X}(X'',UA''), \qquad \mathscr{X}(X,UA) \xrightarrow{f \mapsto U(a' \circ a) \circ f \circ (x \circ x')} \mathscr{X}(X'',UA'').$$

Since $Ua' \circ Ua = U(a' \circ a)$, we see that this triangle commutes. This completes the proof that $\mathscr{X}(-,U-) : \mathscr{X}^{\mathrm{op}} \times \mathscr{A} \to \mathbf{Set}$ is a functor.

(3) $\alpha : \mathscr{A}(F-,-) \to \mathscr{X}(-,U)$ is a natural isomorphism if the square on the right commutes and all $\alpha_{(X,A)}$ are isomorphisms in the following diagram.

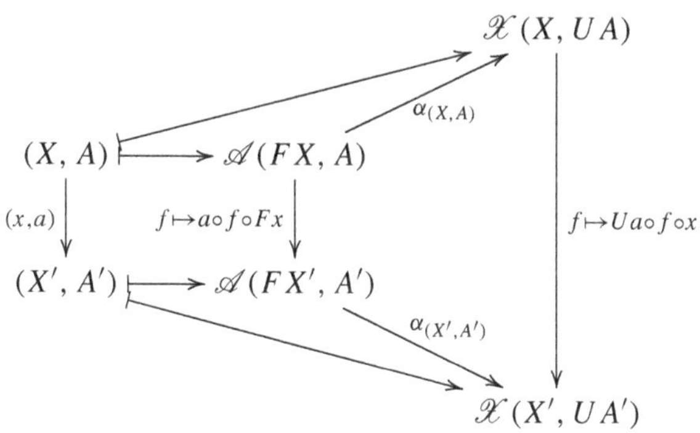

The above properties are guaranteed by Eq. (5.1).

Thus, $F \dashv U$ is equivalent to the existence of a natural isomorphism

$$\alpha \in [\mathscr{X}^{\mathrm{op}} \times \mathscr{A}, \mathbf{Set}](\mathscr{A}(F-,-), \mathscr{X}(-,U-)).$$

We consumed many words and sentences to define the functors involved. Such efforts let us formulate the naturality with a few sentences.

5.8 Adjunctions and Limits

We study the relationships between adjunctions and limits. We showed the operation of taking limits $\varprojlim$ can be seen as a functor $[J,\mathscr{A}] \to \mathscr{A}$ in Sect. 4.5.

Definition 5.6 Let J be a small category, and $\mathscr{A}$ a category. Given $A \in \mathrm{Obj}(\mathscr{A})$, we define a functor ($J$-shaped diagram) $\Delta_A : J \to \mathscr{A}$ as follows.

- the function on objects maps all the objects of J to A.
- the function on morphisms maps all the morphisms of J to 1_A.

Such a functor Δ_A is called a *diagonal functor.*

From now on, we assume $\mathscr{A}$ is locally small to deal with functor category $[J, \mathscr{A}]$. We extend the function

$$\begin{aligned} \Delta : \mathrm{Obj}(\mathscr{A}) &\to \mathrm{Obj}([J, \mathscr{A}]) \\ A &\mapsto \Delta_A \end{aligned}$$

to a functor $\mathscr{A} \to [J, \mathscr{A}]$ by defining functions on morphisms. Given a morphism $a : A \to A'$ in $\mathscr{A}$, we assign a morphism $\Delta_A \to \Delta_{A'}$ in the functor category $[J, \mathscr{A}]$, namely a natural transformation $\alpha(a) : \Delta_A \to \Delta_{A'}$ defined by

$$\alpha(a)_j : \Delta_A(j) \to \Delta_{A'}(j),$$

where $\alpha(a)_j = a$ regardless of j. Such an assignment $\alpha(a)$ is a natural transformation since the square on the right commutes in the following diagram.

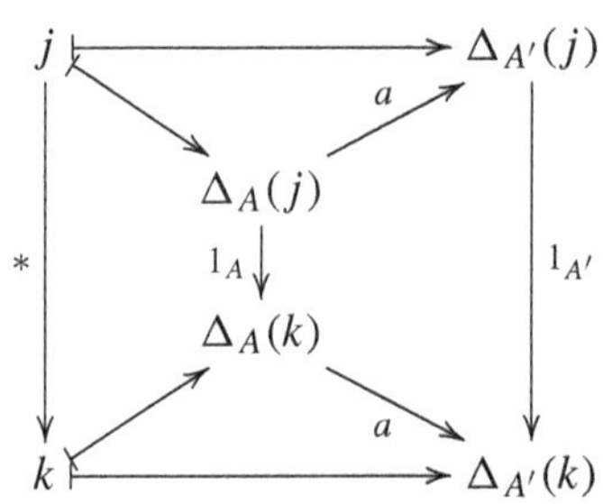

Thus, $A \mapsto \Delta_A$ is extended to a functor from $\mathscr{A}$ to $[J, \mathscr{A}]$.

Proposition 5.5 *We have the following two natural isomorphisms.*

$$\mathscr{A}(A, \varprojlim F) \simeq [J, \mathscr{A}](\Delta_A, F)$$
$$\mathscr{A}(\varinjlim F, A) \simeq [J, \mathscr{A}](F, \Delta_A)$$

Remark 5.5 The limit functor is a right adjoint of Δ_{-}. The colimit functor is a left adjoint of Δ_{-}. Some authors use these facts as definitions of limits and colimits.

Remark 5.6 Let us sum up all the foregoing discussions. A natural transformation from Δ_A to F is equivalent to a cone from A to F.

$$\mathscr{A}(A, \varprojlim F) \simeq [J, \mathscr{A}](\Delta_A, F) \simeq \text{Cones}(A, F)$$

A natural transformation from F to Δ_A is equivalent to a cocone from F to A.

$$\mathscr{A}(\varinjlim F, A) \simeq [J, \mathscr{A}](F, \Delta_A) \simeq \text{Cones}(F, A)$$

We omit the proof since it is just the rephrasing of the definitions.

Monads

6

Monads appear in various disciplines, especially in the science of computing. Understanding phenomena in computing requires more than a simple scheme of input and output correspondence. Non-deterministic computing, stateful computing, and computing that may fail are all everyday life in computer science. Moggi's ideas are lucid in formulating such uncertain mathematical objects in category theory. The Kleisli triple is the main target of this chapter.

6.1 Kleisli Triples

Moggi [1] regards a computer program as a process TB of computing a value of type B on input of type A, where, in our case, A and B are types such as Int, Float, or String, T is a functor such as the List or Maybe functors. The List functor realizes non-deterministic computation. The Maybe functor realizes computation that may fail. Monads or Kleisli triples come into play when we want to concatenate pieces of computation.

Definition 6.1 Given a category $\mathscr{C}$, we consider a triple $(T, \eta, (-)^\sharp)$, where

- T is a function that sends each $A \in \mathrm{Obj}(\mathscr{C})$ to $TA \in \mathrm{Obj}(\mathscr{C})$.
- η is a family of morphisms $\{\eta_A : A \to TA \mid A \in \mathrm{Obj}(\mathscr{C})\}$.
- $(-)^\sharp_{A,B}$ is a function

$$(-)^\sharp_{A,B} : \mathscr{C}(A, TB) \to \mathscr{C}(TA, TB)$$

  for all pairs of $A, B \in \mathrm{Obj}(\mathscr{C})$. Since there is no possibility of confusion, we omit the subscripts and write as $(-)^\sharp$.

S. Yukita, *Category Theory Using Haskell*, Computer Science Foundations and Applied Logic, https://doi.org/10.1007/978-3-031-68538-5_6

Such a triple is called a *Kleisli triple on* $\mathscr{C}$ if it satisfies the following three conditions.

$$(1)\ \ f^\sharp \circ \eta_A = f \quad (2)\ \ \eta_A^\sharp = 1_{TA} \qquad (3)\ \ g^\sharp \circ f^\sharp = (g^\sharp \circ f)^\sharp$$

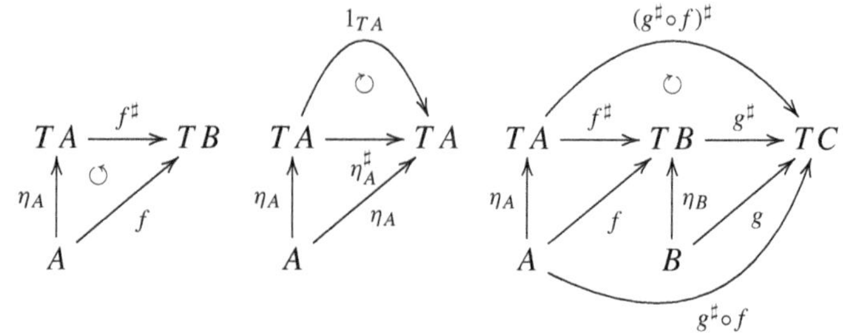

Remark 6.1 Condition (2) can be obtained from condition (1) by setting $f = \eta_A$ in the accompanying diagram. We call a functor from a category $\mathscr{C}$ to itself an *endofunctor*. We will show:

- A function on objects T can be extended to an endofunctor by augmenting a function on morphisms.
- Having established T as a functor, the family of morphisms η turns out to be a natural transformation.

We will further show that the concept of Kleisli triples is equivalent to that of Monads.

In the following sections, we discuss models for

- non-deterministic computation,
- computation with output,
- computation by continuation passing, and
- computation that may fail

in order.

Proposition 6.1 *The function* T *appearing in a Kleisli triple* $(T, \eta, (-)^\sharp)$ *can be extended to an endofunctor* $T : \mathscr{C} \to \mathscr{C}$ *by augmenting a function on morphisms.*

Proof We first seek a function on morphisms compatible with T. For any $f \in \mathscr{C}(A, B)$ we define $Tf \in \mathscr{C}(TA, TB)$ by $Tf = (\eta_B \circ f)^\sharp$ that makes the two triangles commute and thereby does the square in the following diagram.

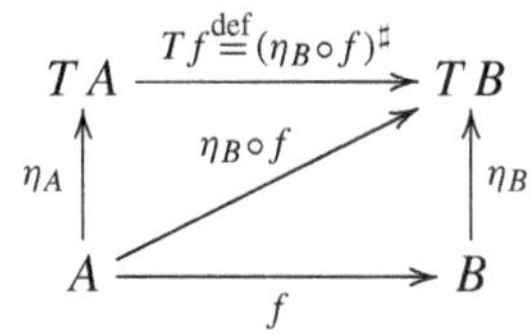

We show that $T1_A = 1_{TA}$, which is in the first place required for T to be a functor. Setting $B = A$ and $f = 1_A$ in the diagram above, we have the following.

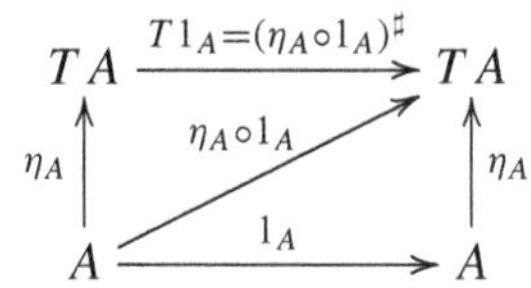

Using condition (2) of Definition 6.1, we have

$$T1_A = \eta_A^\sharp = 1_{TA}.$$

Next, we show that $T(g \circ f) = Tg \circ Tf$. We add two morphisms f and g to the diagram accompanying the condition (3) in Definition 6.1 and obtain the following.

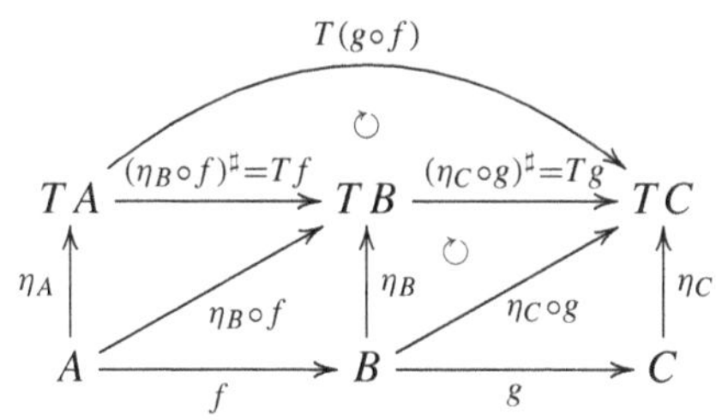

The goal is to show the commutativity of the triangle above that looks like a crescent moon. We can do it as follows.

$$\begin{aligned} Tg \circ Tf &= (\eta_C \circ g)^\sharp \circ (\eta_B \circ f)^\sharp && \text{definition of} T \\ &= ((\eta_C \circ g)^\sharp \circ \eta_B \circ f)^\sharp && \text{condition (3) for } \sharp \\ &= (\eta_C \circ g \circ f)^\sharp && \text{commutativity of the 2nd triangle from right} \\ &= T(g \circ f) && \text{definition of } T \end{aligned}$$

We conclude that the so-extended T is a functor. □

From now on, we regard T appearing in a Kleisli triple as an endofunctor.

Proposition 6.2 *Given a Kleisli triple $(T, \eta, (-)^\sharp)$, the family of morphisms η is a natural transformation from the identity functor* Id $: \mathscr{C} \to \mathscr{C}$ *to $T : \mathscr{C} \to \mathscr{C}$. Namely, we have a natural transformation* Id $\to T$.

Proof The promotion of T to a functor tells us almost completely the naturality of η. But, let us try to put it in a commutative diagram below.

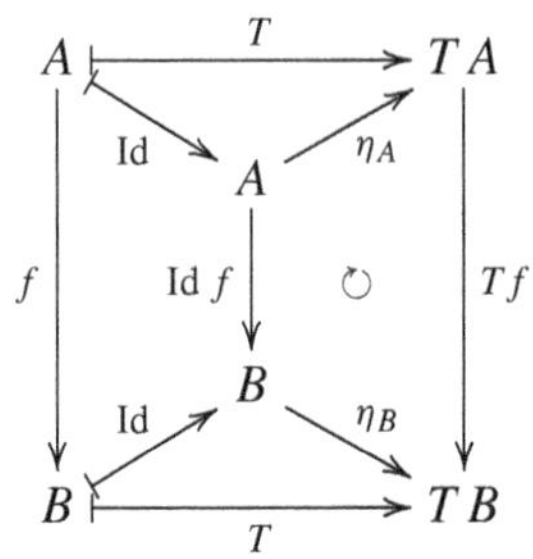

The commutativity of the trapezoid at right comes from the definition of T, which is exactly the naturality of $\eta : \mathrm{Id} \to T$. □

6.2 Moggi' Theory

We will see Moggi's ideas of computation in the following order.

- non-deterministic computation,
- computation with output,
- computation by continuation passing, and
- computation that may fail

The $\sharp$ operator in a Kleisli triple corresponds to

```
(=<<): Monad m => (a -> m b) -> m a -> m b
```

in Haskell. It is a flipped version of the more familiar `bind` operator

```
(>>=): Monad m => m a -> (a -> m b) ->  m b
```

We will discuss Monads later.

6.2.1 Non-determinism and the List Functor

We will embed the list functor into a Kleisli triple.

Listing 6.1 List as a Kleisli triple

```
module NonDeterministic where
-- The List functor

```

```
type T = ([])

-- Definition of
---- eta :: A -> TA
-- plays the role of
---- return :: (Monad m) => a -> m a
---- where m = T.

eta :: a -> T a
eta a = [a]

-- Definition of
---- (-)^# :: Hom(A,TB) -> Hom(TA,TB)
-- plays the role of
---- (=<<) :: (Monad m) => (a -> m a) -> m a -> m b
---- where m = T.

sharp :: (a -> T b) -> T a -> T b
sharp f []     = []
sharp f (x:xs) = f x ++ sharp f xs

-------------
-- Test data
-------------

testFuncs = [(\x -> [0,x,0]),
             (\x -> [0,x,x,0]),
             (\x -> [0,x,x,x,0])]

-- Test1 for
---- f^# o eta_A = f
-- the theoretical goal in code
---- sharp f . eta == f
-- written extensionally
---- (sharp f . eta) a == f a

test1 = map (\f ->(sharp f . eta) 10 == f 10) testFuncs
test1detail = map (\f -> (sharp f. eta) 10) testFuncs


-- Test2 for
---- eta_A^# = id_(TA)
-- the theoretical goal in code
---- sharp eta == id
-- written extensionally
---- sharp eta (ta::T a) == (ta::T a)

testLists = [[],[1],[2,3],[4,5,6]]

test2 = map (\xs -> sharp eta xs == xs) testLists
test2detail = map (\xs -> sharp eta xs) testLists


-- Test3 for
---- g^# o f^# = (g^# o f)^#
```

```
59 -- the theoretical goal in code
60 ---- sharp g . sharp f == sharp (sharp g . f)
61 -- written extensionally
62 ---- (sharp g . sharp f) (ta::T a) ==
63 ---- sharp (sharp g . f) (ta::T a)
64
65 test3cases = [(g,f,d) | g <- testFuncs,
66                         f <- testFuncs,
67                         d <- testLists]
68
69 test3 = map (\(g,f,d) ->
70                 (sharp g . sharp f) d ==
71                 (sharp (sharp g . f)) d)
72             test3cases
```

Line 4 gives an alias to the list functor.

Lines 6-23 implement η and $\sharp$.

Test data and code follow. Successful construction must satisfy the following conditions:

(1) $f^\sharp \circ \eta_A = f$
(2) $\eta_A^\sharp = 1_{TA}$
(3) $g^\sharp \circ f^\sharp = (g^\sharp \circ f)^\sharp$

Lines 29-31 define test data. `test1` in lines 40-41, `test2` in lines 51-54, and `test3` in lines 69-72 are tests for conditions (1), (2), and (3), respectively.

We do not provide proof, instead perform tests enough to give us some confidence. After loading the program, we carry out tests as follows.

```
*NonDeterministic> test1
[True,True,True]
*NonDeterministic> test2
[True,True,True,True]
*NonDeterministic> test3
[True,True,True,True, ... ,True,True]
```

6.2.2 Computation with Output

In the following sample code, the results of consecutive computation are appended to an accumulator string in order.

Listing 6.2 CalcWithOutput.hs

```
module CalcWithOutput where
-- Building a Kleisli triple
-- for a program with simple output
```

```
newtype T a = T {unT :: String -> (a,String)}

-- Definition of
---- (-)^# :: Hom(A,TB) -> Hom(TA,TB)
-- plays the role of
---- return :: (Monad m) => a -> m a
---- where m = T.

eta :: a -> T a
eta x = T (\s -> (x, s))

-- Definition of
----    (-)^# :: Hom(A,TB) -> Hom(TA,TB)
-- plays the role of
---- (=<<) :: (Monad m) => (a -> m a) -> m a -> m b
---- where m = T.

sharp :: (a -> T b) -> (T a -> T b)
sharp f h =
    T (\st -> let
                   (y,st2) = unT h st
                   (z,st3) = unT (f y) st2
              in
                   (z,st3))

-------------
-- Test data
-------------

f1 :: Integer -> T Integer
f1 x = T (\s -> (x^2,
                  s ++ show x ++ "^2=" ++ show (x^2) ++
                  ".\n"))

f2 :: Integer -> T Float
f2 x = T (\s -> (sqrt(fromInteger x),
                  s ++
                  "sqrt of " ++ show x ++
                  " is " ++ show (sqrt(fromInteger x)) ++
                  ".\n"))

testArgs = [(v, s) | v <- [10,100],
                      s <- ["", "---", "hello"]]

test2f1 = [ unT (sharp eta (f1 v)) s == unT (f1 v) s |
              (v,s) <- testArgs ]

test2f2 = [ unT (sharp eta (f2 v)) s == unT (f2 v) s |
              (v,s) <- testArgs ]


-- Test1 for
---- f^# o eta_A = f
-- the theoretical goal in code
```

```
59 ---- sharp f . eta = id
60 -- written extensionally
61 ---- (sharp f . eta) a == f a
62 -- fully extensionally
63 ---- unT ((sharp f . eta) (x::Int) ) (s::String) ==
64 ---- unT (f (x::Int)) (s::String)
65
66 test11 = unT ((sharp f1 . eta) 10) "" == unT (f1 10) ""
67 test11detail = unT ((sharp f1 . eta) 10)  ""
68
69 test12 = unT ((sharp f2 . eta) 10) "" == unT (f2 10) ""
70 test12detail = unT ((sharp f2 . eta) 10)  ""
71
72
73 -- Test2 for
74 ---- eta_A^# = id_(TA)
75 -- the theoretical goal in code
76 ---- sharp eta == id
77 -- written extensionally
78 ---- sharp eta (ta::T a) == (ta::T a)
79 -- fully extensionally
80 ---- unT (sharp eta (ta::T a)) (s::String) ==
81 ---- unT (ta::T a) (s::String)
82
83 test21 = unT (sharp eta (f1 10)) "" ==  unT (f1 10) ""
84 test22 = unT (sharp eta (f2 10)) "" ==  unT (f2 10) ""
85
86
87 -- Test3 for
88 ---- g^# o f^# = (g^# o f)^#
89 -- the theoretical goal in code
90 ---- sharp g . sharp f == sharp (sharp g . f)
91 -- written extensionally
92 ---- (sharp g . sharp f) (ta::T a) ==
93 ---- sharp (sharp g . f) (ta::T a)
94 -- fully extensionally
95 ---- unT ((sharp g . sharp f) (ta::T a)) (s::String) ==
96 ---- unT (sharp (sharp g . f) (ta::T a)) (s::String)
97
98 test31 =  unT ((sharp f2 . sharp f1) (eta 10)) "" ==
99           unT (sharp (sharp f2 . f1) (eta 10)) ""
100
101 test3 = [unT ((sharp f2 . sharp f1) (eta v)) s ==
102          unT (sharp (sharp f2 . f1) (eta v)) s |
103          (v,s) <- testArgs ]
104
105 -- other tests
106 (xTest1, stTest1) =
107     (unT $ sharp f2 $ sharp f1 $ eta 10) ""
```

Line 5 defines a functor T. Lines 13-14 implement η as a function eta. Lines 22-28 implement $(-)^\sharp$ as a function sharp. Lines 30 and later prepare test data and code.

Successful construction must satisfy the following conditions:

(1) $f^\sharp \circ \eta_A = f$
(2) $\eta_A^\sharp = 1_{TA}$
(3) $g^\sharp \circ f^\sharp = (g^\sharp \circ f)^\sharp$

Lines 30-53 prepare test data. `test11` and `test12` in lines 56-70, `test21` and `test22` in lines 83-84, and `test3` in line 101 are tests for conditions (1), (2), and (3), respectively. Lines 106-107 give a total test.

Tests are carried out as follows.

```
*CalcWithOutput> test11
True
*CalcWithOutput> test12
True
*CalcWithOutput> test21
True
*CalcWithOutput> test22
True
*CalcWithOutput> test3
[True,True,True,True,True,True]
*CalcWithOutput>
```

Other suggested tests are given as follows.

```
*CalcWithOutput> (unT $ unitT 10) ""
(10,"")
*CalcWithOutput> xTest1
10.0
*CalcWithOutput> putStr stTest1
10^2=100.
sqrt of 100 is 10.0.
```

So much for tests. They are not proofs. Rigorous proofs are given in Appendix C.

6.2.3 Continuation Passing

The idea of continuation passing is related to the concept of dual space. After walking through the sample code, we will reason the design for it.

Listing 6.3 Continuation.hs

```
module Continuation where
-- Building a Kleisli triple for a program of CPS
-- In this version, T takes two arguments.
```

```

newtype T v a = T {unT :: (a -> v) -> v}

-- Definition of eta
---- eta :: A -> TA
-- plays the role of
---- return :: (Monad m) => a -> m a
---- where m = T v.

eta :: a -> T v a
eta a = T (\x -> x a)

-- Definition of sharp
---- (-)^# :: Hom(A,TB) -> Hom(TA,TB)
-- plays the role of
---- (=<<) :: (Monad m) => (a -> m a) -> m a -> m b
---- where m = T v.
-- Types of arguments in the definition below
---- f :: a -> T v b
---- m :: T v a
---- b' :: b -> v

sharp :: (a -> T v b) -> (T v a -> T v b)
sharp f m =
    T (\b' -> unT m (\a -> unT (f a) b'))

-------------
-- Test data
-------------

f1 :: Int -> T Int Int
f1 x = T (\c -> c (x+1))

f2 :: Int -> T Int Int
f2 x = T (\c -> c (x^2))

g1 :: String -> T Int String
g1 x = T (\c -> c (x++"!"))

g2 :: String -> T Int String
g2 x = T (\c -> c (x++x))

-- Test1 for
---- f^# o eta_A = f
-- the theoretical goal in code
---- sharp f . eta == f
-- written extensionally
---- (sharp f . eta) a == f a
-- fully extensionally
---- unT ((sharp f . eta) a) (h::a->v) ==
---- unT (f a) (h::a->v)

test111 = unT ((sharp f1 . eta) 3) (^2) == unT (f1 3) (^2)
test112 = unT ((sharp f2 . eta) 3) (+1) == unT (f2 3) (+1)
test121 = unT ((sharp g1 . eta) "hello") length ==
```

```
59              unT (g1 "hello") length
60 test122 = unT ((sharp g2 . eta) "hello") length ==
61              unT (g2 "hello") length
62 test129 = unT ((sharp g2 . eta) "hello") ((^2) . length) ==
63              unT (g2 "hello") ((^2) . length)
64
65
66 -- Test2 for
67 ---- eta_A^# = id_(TA)
68 -- the theoretical goal in code
69 ---- sharp eta == id
70 -- written extensionally
71 ---- sharp eta (ta::T a) == (ta::T a)
72 -- fully extensionally
73 ---- unT (sharp eta (ta::T a)) (f::a->v) ==
74 ---- unT (ta::T a) (f::a->v)
75
76 test211 = unT (sharp eta (eta 3)) (+1) == unT (eta 3) (+1)
77 test212 = unT (sharp eta (eta 3)) (^2) == unT (eta 3) (^2)
78 test221 = unT (sharp eta (f1 3)) (+1) == unT (f1 3) (+1)
79 test222 = unT (sharp eta (f2 3)) (+1) == unT (f2 3) (+1)
80 test231 = unT (sharp eta (g1 "hello")) length ==
81              unT (g1 "hello") length
82 test232 = unT (sharp eta (g2 "hello")) length ==
83              unT (g2 "hello") length
84
85
86 -- Test3 for
87 ---- g^# o f^# = (g^# o f)^#
88 -- the theoretical goal in code
89 ---- sharp g . sharp f == sharp (sharp g . f)
90 -- written extensionally
91 ---- (sharp g . sharp f) (ta::T a) ==
92 ---- sharp (sharp g . f) (ta::T a)
93 -- fully extensionally
94 ---- unT ((sharp g . sharp f) (ta::T a)) (h::a->v) ==
95 ---- unT (sharp (sharp g . f) (ta::T a)) (h::a->v)
96
97 test311 =
98      unT ((sharp f2 . sharp f1) (eta 3)) (+1) ==
99      unT ((sharp (sharp f2 . f1)) (eta 3)) (+1)
100 test312 =
101     unT ((sharp f2 . sharp f1) (eta 3)) (^2) ==
102     unT ((sharp (sharp f2 . f1)) (eta 3)) (^2)
103 test321 =
104     unT ((sharp g2 . sharp g1) (eta "hello")) length ==
105     unT ((sharp (sharp g2 . g1)) (eta "hello")) length
106 test322 =
107     unT ((sharp g2 . sharp g1) (g1 "hello")) length ==
108     unT ((sharp (sharp g2 . g1)) (g1 "hello")) length
```

Line 5 defines a type constructor `T` that takes two type parameters `v` and `a`. Type constructor T itself in not a functor, but a partially applied `T v` is an instance of the `Functor` type class. We will discuss this in detail later.

Lines 7-28 give the definitions of η and $(-)^\sharp$. Lines 30-44 prepare test data and code. Tests for successful construction follow like those of previous examples.

Let us take a closer look at the definition of η.

```
eta a = T (\x -> x a)
```

T is just a wrapper. For the time being, forget this wrapper. Given a value a of type A, the task of computing `\x -> x a` is to create a function that takes a function x of type $A \to V$ as an argument and returns the result of the function application x a of type V. The role of `eta` is to embed A to $(A \to V) \to V$. The latter is often called the dual space of A.

In what follows, we use uppercase letters for type variables outside the code so that using the same letters for values does not cause confusions. For example, when we want to use a value a of type a (type variable), we write $a \in A$. Recall that we cannot use uppercase letters for type variables in Haskell code.

Lines 26-28 define the $(-)^\sharp$ function.

```
sharp :: (a -> T v b) -> (T v a -> T v b)
sharp f m =
      T (\ b' -> unT m ( \a -> unT (f a) b'))
```

The construction looks a little bit clumsy. However, we can deduce the implementation in a straight forward manner. First, analyze the signature of `sharp`.

- Given `f::a->T v b`, we have `sharp f::T v a->T v b`.
- Given successively `m::T v a`, we have `sharp f m::T v b`.

With all these in mind, let us play with building blocks. At first sight, the only thing we can do with `f::a->T v b` is to apply it to some value $a \in A$ and unbox the result.

```
unT (f a) :: (b->v)->v
```

We can apply it to `b' :: b->v` to obtain the following.

```
unT (f a) b' :: v
```

At this point, a is a free variable. We lambda-bind it to obtain the following.

```
\a -> unT (f a) b' :: a->v
```

Since we have

```
unT m :: (a->v)->v
```

we can match blocks as follows.

```
unT m (\a -> unT (f a) b') :: v
```

At this point, `b'` is a free variable. We lambda-bind it to obtain

```
\b' -> unT m (\a -> unT (f a) b') :: (b->v)->v
```

Finally, we box it with `T` to obtain a value of type `T v b`. Are there any other ways to construct a value of this type?

We summarize the above result in the unboxed(unT) expressions `(a->v)->v` and `(b->v)->v` as in the following diagram.

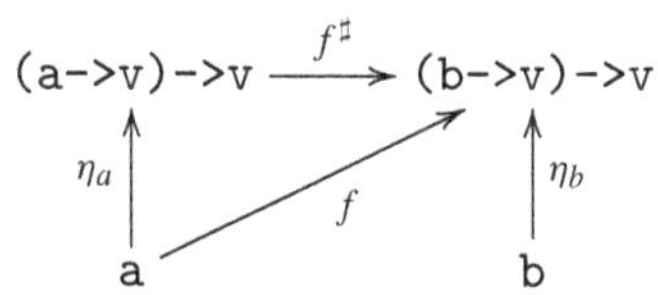

$(-)^\sharp$ makes the above triangle commute.

The conditions for $(-)^\sharp$ are tested as follows.

```
*Continuation> test111
True
...
omitted
...
*Continuation> test322
True
*Continuation>
```

6.2.4 Computation that May Fail and the Maybe Functor

Computation that may fail is provided by the `Maybe` functor. Values from a successful computation are boxed in the `Just` value constructor, while failure yields a polymorphic value `Nothing`. Composition of computation that may fail is performed under the Kleisli triple mechanism.

Listing 6.4 Maybe.hs

```
module Maybe where
-- Building a Kleisli triple for the Maybe functor

type T = Maybe
```

```

-- Definition of
---- eta :: A -> TA
-- plays the role of
---- return :: (Monad m) => a -> m a
---- where m = T.

eta :: a -> T a
eta = Just

-- Definition of
---- (-)^# :: Hom(A,TB) -> Hom(TA,TB)
-- plays the role of
---- (=<<) :: (Monad m) => (a -> m a) -> m a -> m b
---- where m = T.
-- Types of arguments in the definition below
---- f :: a -> T b

sharp :: (a -> T b) -> (T a -> T b)
sharp f Nothing  = Nothing
sharp f (Just x) = f x


-------------
-- Test data
-------------

f1 :: Float -> T Float
f1 0 = Nothing
f1 x = Just (1.0/x)

f2 :: Float -> T Float
f2 x = Just (x+1)

testFuncs  = [f1,f2]
testFloats  = [10.0,100.0,0.1] :: [Float]
testTAs    = [Nothing, Just 0.0, Just 1.0]


-- Test1 for
---- f^# o eta_A = f
-- the theoretical goal in code
---- sharp f . eta == f
-- written extensionally
---- (sharp f . eta) a == f a

test1cases  = [(f, v) | f <- testFuncs, v <- testFloats]
test1 =
    map (\(f,v) -> (sharp f . eta) v ==  f v) test1cases

-- Test2 for
---- eta_A^# = id_(TA)
-- the theoretical goal in code
---- sharp eta == id
-- written extensionally
```

```
---- sharp eta (ta::T a) == (ta::T a)

test2 = map (\ta -> sharp eta ta == ta) testTAs

-- Test3 for
---- g^# o f^# = (g^# o f)^#
-- the theoretical goal in code
---- sharp g . sharp f == sharp (sharp g . f)
-- written extensionally
---- (sharp g . sharp f) (ta::T a) ==
---- sharp (sharp g . f) (ta::T a)

test3cases = [(g,f,ta) |
              g <- testFuncs, f <- testFuncs,
              ta <- testTAs]

test3 = map (\(g,f,ta) ->
              (sharp g . sharp f) ta ==
              sharp (sharp g . f) ta)
            test3cases

test3detail1 =
        map (\(g,f,ta) ->
              (sharp g . sharp f) ta)
              test3cases

test3detail2 =
        map (\(g,f,ta) ->
              sharp (sharp g . f) ta)
            test3cases
```

Test cases are designed in the same way as in the previous examples.

Tests can be done as follows.

```
*Maybe> test1
[True,True,True,True,True,True]
...
omitted
...
*Maybe> test3
[True,True,True,True,True,True,True,True,True,True,True,True]
*Maybe>
```

6.3 Monads

We introduce the concept of Monads and see it is closely related to the concept of Kleisli triples.

Definition 6.2 Let $\mathscr{C}$ be a category. A *Monad* is a triple (T, η, μ), where

- $T : \mathscr{C} \to \mathscr{C}$ is an endofunctor,
- $\eta : \mathrm{Id}_{\mathscr{C}} \to T$ and $\mu : T^2 \to T$ are natural transformations that make the following two diagrams commute for all $A \in \mathrm{Obj}(\mathscr{C})$.

$$\begin{array}{ccccc} TA & \xrightarrow{T(\eta_A)} & T(TA) & \xleftarrow{\eta_{TA}} & TA \\ & {\scriptstyle 1_{TA}}\searrow & \downarrow{\scriptstyle \mu_A} & \swarrow{\scriptstyle 1_{TA}} & \\ & & TA & & \end{array} \qquad \begin{array}{ccc} T(T(TA)) & \xrightarrow{T(\mu_A)} & T(TA) \\ {\scriptstyle \mu_{TA}}\downarrow & & \downarrow{\scriptstyle \mu_A} \\ T(TA) & \xrightarrow[\mu_A]{} & TA \end{array}$$

For later use, we rewrite the above diagrams in a list of equations below. (1) and (2) are derived from the left diagram, and (3) is from the right.

(1) $\mu_A \circ T\eta_A = 1_{TA}$
(2) $\mu_A \circ \eta_{TA} = 1_{TA}$
(3) $\mu_A \circ \mu_{TA} = \mu_A \circ T\mu_A$

These equations are called the *axioms for Monads.*

Remark 6.2 The notation here may confuse the reader. We must first distinguish $T(\eta_A)$ and η_{TA}. Both are morphisms from TA to $T(T(A))$. Let us construct such morphisms from available materials. We have two things at hand: natural transformation η and endofunctor T.

(1) Treat TA as an individual and look for a morphism $\underline{TA} \to T(\underline{TA})$. We notice that η_{TA}, the TA component of η is there to be used.

$$\eta_{TA} : TA \to T(TA)$$

(2) Start with $\eta_A : A \to TA$, an A component of η, take its image by the functor T. Then, we obtain the following.

$$T(\eta_A) : TA \to T(TA)$$

So much for the construction of two morphisms that share their domain and codomain. In general, they are different morphisms. Yet, they are coequated by μ_A as follows.

$$TA \underset{T(\eta_A)}{\overset{\eta_{TA}}{\rightrightarrows}} T(TA) \xrightarrow{\mu_A} TA$$

Further, the left commutative diagram in the above definition requires that the composition yields 1_{TA}.

Remark 6.3 Let us see $T(\mu_A)$ and μ_{TA}. Both are morphisms from $T(T(TA))$ to $T(TA)$. Let us construct such morphisms from available materials. We have two things at hand: natural transformation μ and endofunctor T.

(1) Treat TA as an individual and look for a morphism $T(T(\underline{TA})) \to T(\underline{TA})$. We have the TA component of the natural transformation μ, namely μ_{TA}.

$$\mu_{TA} : T(T(TA)) \to T(TA)$$

(2) Start with $\mu_A : T(TA) \to TA$ and take its image by the functor T. We obtain the following

$$T(\mu_A) : T(T(TA)) \to T(TA)$$

So much for the construction of two morphisms that share their domain and codomain. In general, they are different. Yet, they are coequated by μ_A as follows.

$$T(T(TA)) \underset{T(\mu_A)}{\overset{\mu_{TA}}{\rightrightarrows}} T(TA) \xrightarrow{\mu_A} TA$$

This is the second commutative diagram in Definition 6.2.

Remark 6.4 Required conditions in Definition 6.2 are described with component-wise commutative diagrams. We can restate the conditions with natural transformations as a whole using horizontal composition.

$$\begin{array}{ccccc} T & \xrightarrow{T\circ\eta} & T^2 & \xleftarrow{\eta\circ T} & T \\ & {}_{\mathrm{Id}}\searrow & \downarrow{\scriptstyle\mu} & \swarrow_{\mathrm{Id}} & \\ & & T & & \end{array} \qquad \begin{array}{ccc} T^3 & \xrightarrow{T\circ\mu} & T^2 \\ {\scriptstyle\mu\circ T}\downarrow & & \downarrow{\scriptstyle\mu} \\ T^2 & \xrightarrow[\mu]{} & T \end{array}$$

Given a Kleisli triple, we can construct a Monad.

Theorem 6.1 *Let $\mathscr{C}$ and $T : \mathscr{C} \to \mathscr{C}$ be a category and endofunctor. Given a Kleisli triple $(T, \eta, (-)^\sharp)$, we set*

$$\mu_A = (1_{TA})^\sharp : T(TA) \to TA. \tag{6.1}$$

Then, $\mu : T \circ T \to T$ is a natural transformation, and a triple (T, η, μ) is a Monad.

Remark 6.5 We are naturally led to the definition of μ by Eq. (6.1) if we notice that replacing A with TA, TB with TA, and f with 1_{TA} in the left diagram below gives a specialized diagram at right.

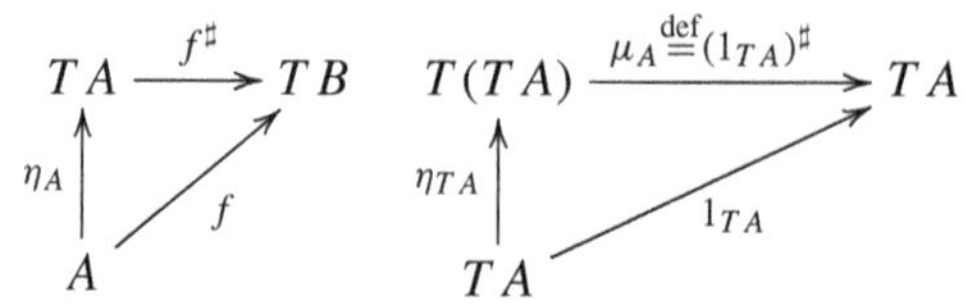

Remember that T and $\sharp$ are related by a commutative diagram appearing in the proof of Proposition 6.1:

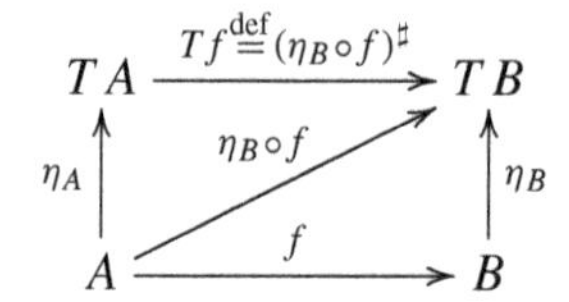

Proof We first show that the family of morphisms $\{\mu_A = 1^\sharp_{TA} \mid A \in \mathrm{Obj}(\mathscr{C})\}$ is a natural transformation. It is sufficient to show that the trapezoid at the right in the following diagram commutes.

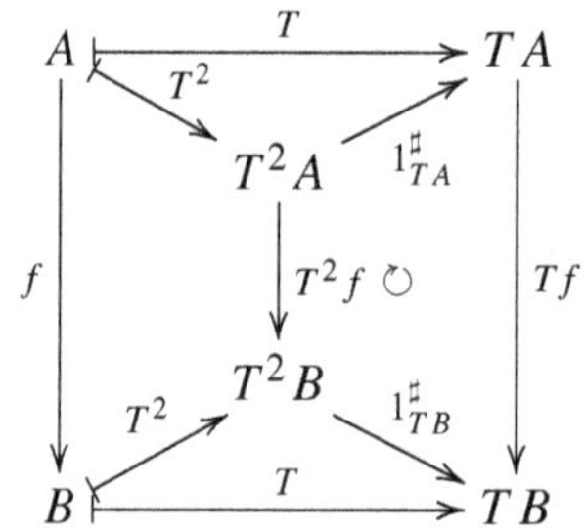

Iterate rewriting as follows. We consider a path that starts with T^2A, goes down, and goes down-right:

$$
\begin{aligned}
&1^\sharp_{TB} \circ (T(Tf)) \\
&= 1^\sharp_{TB} \circ (\eta_{TB} \circ (\eta_B \circ f)^\sharp)^\sharp && \text{twice replacing } T \text{ with } \sharp \\
&= (1^\sharp_{TB} \circ \eta_{TB} \circ (\eta_B \circ f)^\sharp)^\sharp && \text{Condition (3) for } \sharp \\
&= (1_{TB} \circ (\eta_B \circ f)^\sharp)^\sharp && \text{Condition (1) for } \sharp \\
&= (\eta_B \circ f)^{\sharp\sharp} && \text{omitting the identity}
\end{aligned}
$$

Next, we consider a path that starts with T^2A, goes up-right, and goes down:

$$
\begin{aligned}
&Tf \circ 1^\sharp_{TA} \\
&= (\eta_B \circ f)^\sharp \circ 1^\sharp_{TA} && \text{replacing } T \text{ with } \sharp \\
&= ((\eta_B \circ f)^\sharp \circ 1_{TA})^\sharp && \text{Condition (3) for } \sharp \\
&= (\eta_B \circ f)^{\sharp\sharp} && \text{omitting the identity}
\end{aligned}
$$

This establishes the naturality of μ defined via $\mu_A = 1^\sharp_{TA}$.

What remains to show is that $\mu_A = 1^\sharp_{TA}$ satisfies the three equations in Definition 6.2.

(1) We show that $1^\sharp_{TA} \circ T\eta_A = 1_{TA}$, namely the commutativity of the left triangle of the diagram below.

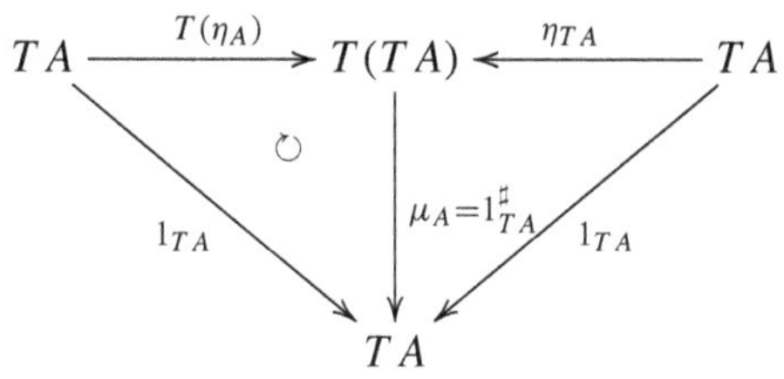

Iterate rewriting as follows.

$$\begin{array}{ll}
1^\sharp_{TA} \circ T\eta_A & \\
= 1^\sharp_{TA} \circ (\eta_{TA} \circ \eta_A)^\sharp & \text{replace } T \text{ with } \sharp\ (*) \\
= (1^\sharp_{TA} \circ \eta_{TA} \circ \eta_A)^\sharp & \text{Condition (3) for } \sharp \\
= (1_{TA} \circ \eta_A)^\sharp & \text{Condition (2) for } \sharp\ (**) \\
= \eta^\sharp_A & \text{omitting the identity} \\
= 1_{TA} & \text{Condition (2) for } \sharp
\end{array}$$

Take a careful look at the diagram

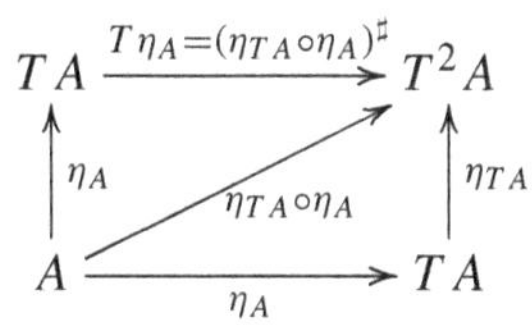

to see $T\eta_A = (\eta_{TA} \circ \eta_A)^\sharp$ in the step (*).
Take a careful look again at the diagram

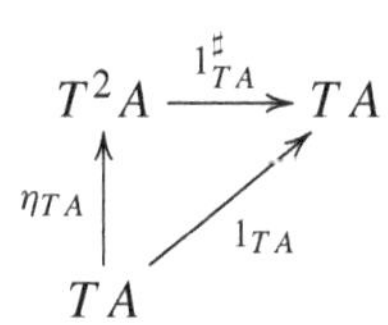

to see $1^\sharp_{TA} \circ \eta_{TA} = 1_{TA}$ in the step (**).

(2) We prove that $1^{\sharp}_{TA} \circ \eta_{TA} = 1_{TA}$ by showing the triangle at right in the following diagram commutes.

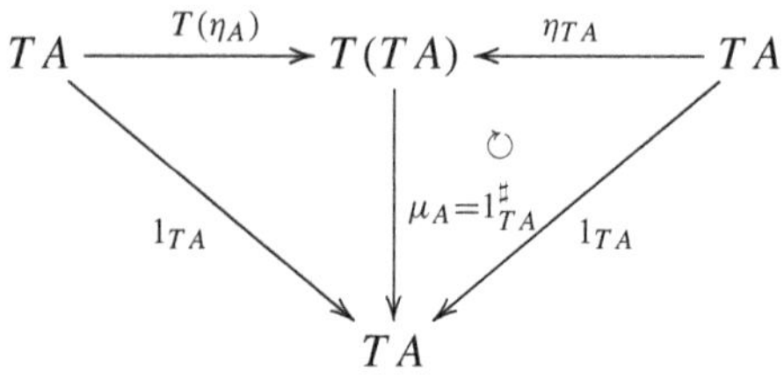

This is a direct consequence of the definition of $1^{\sharp}_{TA}$ and μ. Since the following diagram

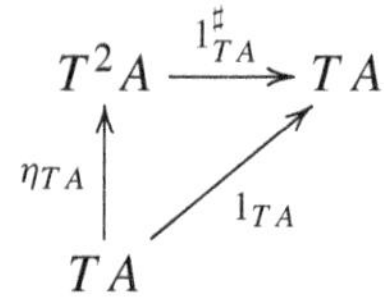

must commute.

(3) Finally, we prove the commutativity of the following diagram.

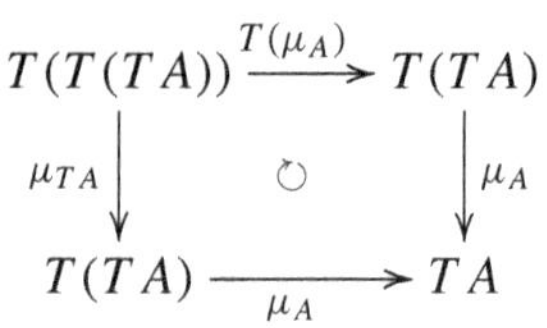

Using the definition of μ, namely $\mu_A = 1^{\sharp}_{TA}$, we will calculate the two paths from the upper-left corner to the lower-right.

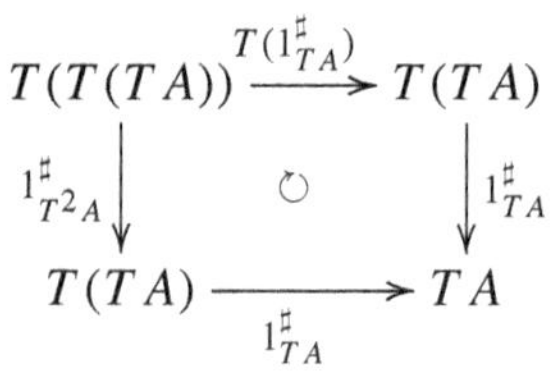

First, we compute the path down-right.

$$\begin{aligned}
&1^{\sharp}_{TA} \circ 1^{\sharp}_{T^2A} \\
&= (1^{\sharp}_{TA} \circ 1_{T^2A})^{\sharp} && \text{formula } (g^{\sharp} \circ f)^{\sharp} = g^{\sharp} \circ f^{\sharp} \\
&= 1^{\sharp\sharp}_{TA} && \text{omitting the identity } 1_{T^2A}
\end{aligned}$$

Next, we compute the path right-down.

$$\begin{aligned}
&1^\sharp_{TA} \circ T(1^\sharp_{TA}) \\
&= 1^\sharp_{TA} \circ (\eta_{TA} \circ 1^\sharp_{TA})^\sharp && T \text{ is replaced by } \sharp \\
&= (1^\sharp_{TA} \circ \eta_{TA} \circ 1^\sharp_{TA})^\sharp && \text{formula } (g^\sharp \circ f)^\sharp = g^\sharp \circ f^\sharp \\
&= (1_{TA} \circ 1^\sharp_{TA})^\sharp && \text{formula } f^\sharp \circ \eta_- = f \\
&= 1^{\sharp\sharp}_{TA} && \text{omitting the identity } 1_{TA}
\end{aligned}$$

Rewriting the first line to the second is inspired by the following commutative diagram, where morphism $T(1^\sharp_{TA})$ appears at the top edge.

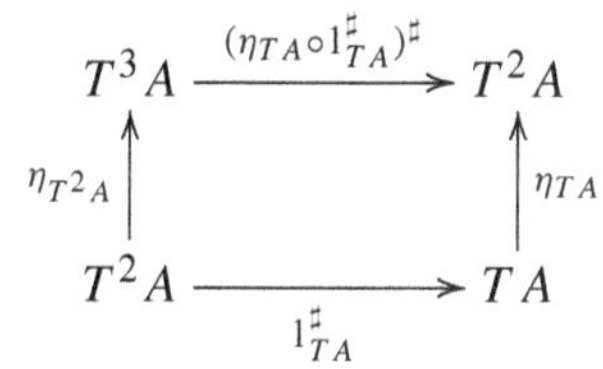

Other rewriting steps are clearly justified by each accompanying comment.

We completed the proof by showing the last commutativity. □

Theorem 6.2 *Given a Monad (T, η, μ) on a category $\mathscr{C}$, we put*

$$\begin{aligned}
(-)^\sharp : \mathscr{C}(A, TB) &\longrightarrow \mathscr{C}(TA, TB) \\
f &\longmapsto \mu_B \circ Tf
\end{aligned}$$

Then, the data $(T, \eta, (-)^\sharp)$ is a Kleisli triple on the category $\mathscr{C}$.

Remark 6.6 We defined the functor T from the $(-)^\sharp$ operation. We are doing the reverse. Compare the two diagrams below.

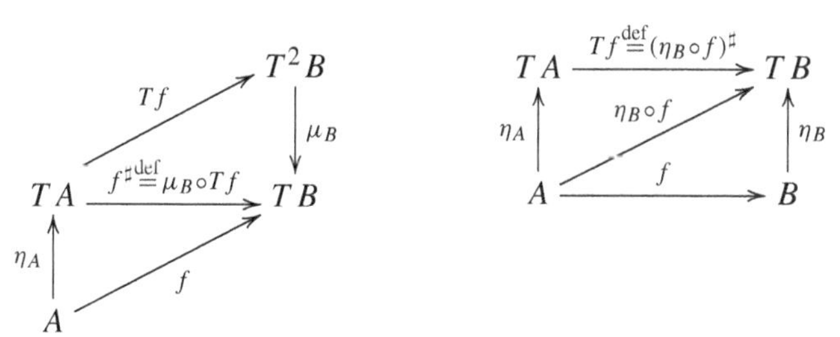

Proof The operation $(-)^\sharp$ defined by $f^\sharp \stackrel{\text{def}}{=} \mu_B \circ Tf$ gives the following equations:

$$f^\sharp \circ \eta_A = f, \quad \eta^\sharp_A = 1_{TA}, \quad g^\sharp \circ f^\sharp = (g^\sharp \circ f)^\sharp,$$

which we will show in order.

1. We prove $f^\sharp \circ \eta_A = f$.
   From the definition of $(-)^\sharp$, we have to show
   $$\mu_B \circ Tf \circ \eta_A = f.$$
   We look for any information available to approach this goal. There is the fact that $\eta : \mathrm{Id} \to T$ is a natural transformation. We put it in a commutative diagram as follows.

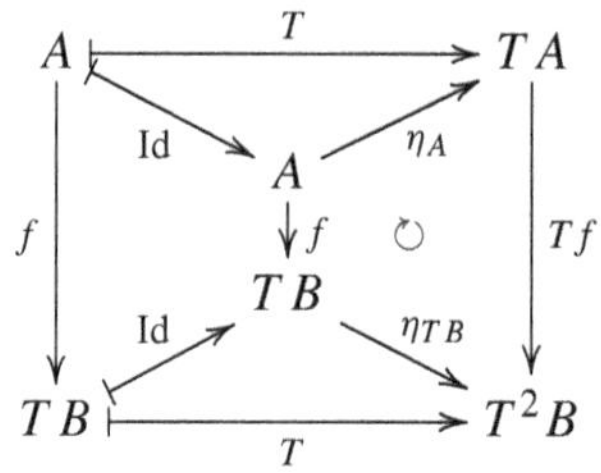

   The trapezoid at right commutes. Namely, we have
   $$Tf \circ \eta_A = \eta_{TB} \circ f.$$
   By composing μ_B from left, we have
   $$\mu_B \circ Tf \circ \eta_A = \mu_B \circ \eta_{TB} \circ f. \tag{6.2}$$
   An axiom of Monads tells that triangles in the diagram

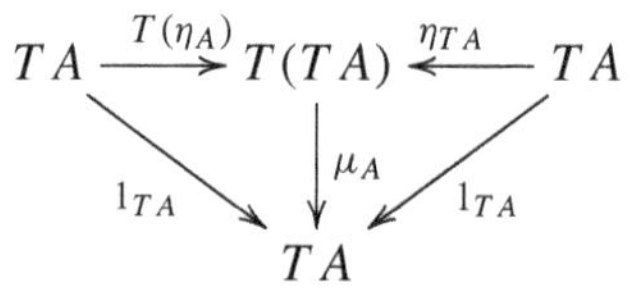

   commute. Especially, the commutative triangle at right gives us $\mu_B \circ \eta_{TB} = 1_{TB}$, which means that the RHS of Equation (6.2) equals to f.
2. We prove $\eta_A^\sharp = 1_{TA}$.
   Recall the definition of $(-)^\sharp$. We only have to show
   $$\mu_A \circ T\eta_A = 1_{TA}.$$
   We have nothing to do since this comes from the commutative triangle at left in the axiom above.
3. We prove $g^\sharp \circ f^\sharp = (g^\sharp \circ f)^\sharp$.
   Recall again the definition of $(-)^\sharp$. We have to show
   $$(\mu_C \circ Tg) \circ (\mu_B \circ Tf) = \mu_C \circ T(\mu_C \circ Tg \circ f).$$
   Since T is a functor, we can rewrite the subexpression of the RHS as follows.
   $$T(\mu_C \circ Tg \circ f) = T\mu_C \circ T^2 g \circ Tf$$

The whole RHS becomes

$$\mu_C \circ T(\mu_C \circ Tg \circ f) = \mu_C \circ T\mu_C \circ T^2 g \circ Tf.$$

We rewrite the RHS further in two steps. Recall that μ_C coequalizes $T\mu_C$ and μ_{TC} by the axiom

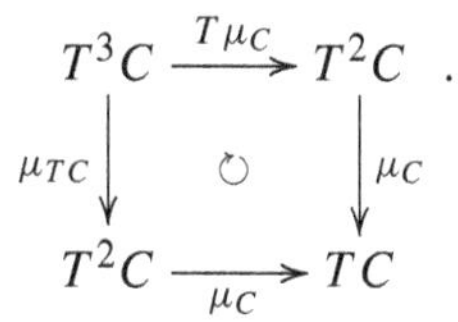

From this, we can rewrite as follows.

$$\underline{\mu_C \circ T\mu_C} \circ T^2 g \circ Tf = \underline{\mu_C \circ \mu_{TC}} \circ T^2 g \circ Tf$$

Next, we use the fact that $\mu : T^2 \to T$ is a natural transformation. We have

$$\mu_C \circ \underline{\mu_{TC} \circ T^2 g} \circ Tf = \mu_C \circ \underline{Tg \circ \mu_B} \circ Tf,$$

which completes the proof.

The bird's eye view of the whole rewriting process is given below.

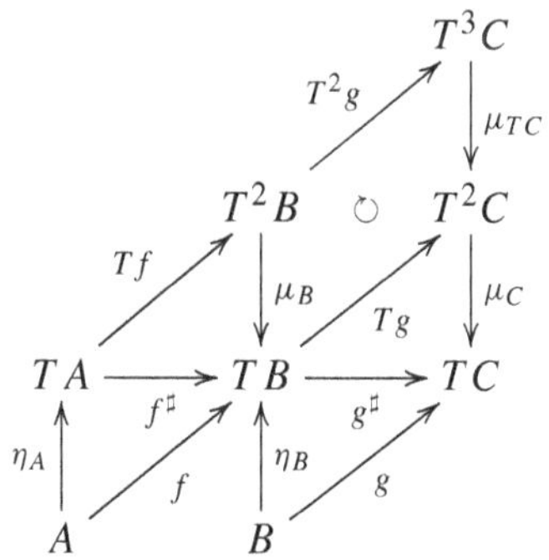

□

Theorems 6.1 and 6.2 established that Monads and Kleisli triples translate themselves into others. The remaining question is whether they have the same data when they go and come back. Fix the endofunctor T and η that are shared between the Monads and the corresponding Kleisli triple. Let φ be the correspondence in Theorem 6.1 that sends $\sharp$ to μ. Namely, we set

$$\varphi(\sharp) = \mu.$$

Likewise, let ψ be the correspondence in Theorem 6.2 that sends μ to $\sharp$. Namely, we set

$$\psi(\mu) = \sharp.$$

Set $\psi(\varphi(\sharp)) = \sharp'$. We do not know if $\sharp' = \sharp$. Set $\varphi(\psi(\mu)) = \mu'$. We do not know if $\mu' = \mu$.

Theorem 6.3 *Assume that a Kleisli triple $(T, \eta, (-)^\sharp)$ and a Monad (T, η, μ) share the functor T and η. Then, we have $\psi(\varphi(\sharp)) = \sharp$ and $\varphi(\psi(\mu)) = \mu$.*

Proof We start with $\sharp$. Let $\varphi(\sharp) = \mu$ and $\psi(\mu) = \sharp'$. We have $\mu_A = 1^\sharp_{TA}$ from the construction in Theorem 6.1. For any $f : A \to TB$ we have

$$\begin{array}{ll} f^{\sharp'} & \\ = \mu_B \circ Tf & \text{definition of } \sharp' \text{ by } \mu \\ = 1^\sharp_{TB} \circ Tf & \text{definition of } \mu \text{ by } \sharp \\ = 1^\sharp_{TB} \circ \eta_{TB} \circ f^\sharp & \text{replacing } T \text{ with } \sharp \\ = 1_{TB} \circ f^\sharp & \text{axiom of } \sharp \\ = f^\sharp & \end{array}$$

Next, we start with μ. Let $\psi(\mu) = \sharp$ and $\varphi(\sharp) = \mu'$. We have $f^\sharp = \mu_B \circ Tf$ for any $f : A \to TB$, recalling the construction in Theorem 6.2. Set $f = 1_{TB}$ and follow the rewriting below.

$$\begin{array}{ll} \mu'_B & \\ = 1^\sharp_{TB} & \text{definition of } \mu' \text{ by } \sharp \\ = \mu_B \circ T1_{TB} & \text{replacing } \sharp \text{ with } T \\ = \mu_B \circ 1_{T^2B} & \text{functoriality of } T \\ = \mu_B & \end{array}$$

We completed the proof that φ and ψ are mutual inverses. □

6.4 Monad Instances in Haskell

Examples presented in Sect. 6.2 are revisited as instances of type classes Applicative and Monad in the Haskell standard library.

6.4.1 Computation with Output(revisited)

We extract the necessary part of the sample code presented earlier, aiming to recall the definitions of

`eta :: a -> T a` and `sharp :: (a -> T b) -> (T a -> T b)`.

Listing 6.5 CalcWithOutput.hs(excerpt)

```
newtype T a = T {unT :: String -> (a,String)}

-- Definition of
---- (-)^# :: Hom(A,TB) -> Hom(TA,TB)
-- plays the role of
---- return :: (Monad m) => a -> m a
---- where m = T.

eta :: a -> T a
eta x = T (\s -> (x, s))

-- Definition of
----    (-)^# :: Hom(A,TB) -> Hom(TA,TB)
-- plays the role of
---- (=<<) :: (Monad m) => (a -> m a) -> m a -> m b
---- where m = T.

sharp :: (a -> T b) -> (T a -> T b)
sharp f h =
    T (\st -> let
                   (y,st2) = unT h st
                   (z,st3) = unT (f y) st2
              in
                   (z,st3))
```

f1 and f2 are test cases.

Listing 6.6 CalcWithOutput.hs(excerpt)

```
f1 :: Integer -> T Integer
f1 x = T (\s -> (x^2,
                 s ++ show x ++ "^2=" ++ show (x^2) ++
                 ".\n"))

f2 :: Integer -> T Float
f2 x = T (\s -> (sqrt(fromInteger x),
                 s ++
                 "sqrt of " ++ show x ++
                 " is " ++ show (sqrt(fromInteger x)) ++
                 ".\n"))
```

We import this module and make the type constructor T an instance of type classes Functor, Applicative, and Monad.

Listing 6.7 CalcWithOutput2.hs

```
module CalcWithOutput2 where
import Control.Monad
import CalcWithOutput

-- After building a Kleisli triple
-- for a program with simple output,
-- we declare instances of
```

```
-- Functor, Applicative, and Monad

-----------
-- Functor
-----------

instance Functor T where
-- fmap :: (a -> b) -> (T a -> T b)
  fmap f ta =
    T (\s ->  let (a,s2) = (unT ta) s
              in  (f a, s2)
      )

{- suggested tests
-- f1 and f2 are imported from CalcWithOutput

unT (fmap (+1) (f1 10)) "hello\n"
-}

----------------
-- Applicative
----------------

instance Applicative T where
  pure x = T (\s -> (x,s))
  tf <*> ta =
     T (\s ->
              let (h, s2) = unT tf s
                  (a, s3) = unT ta s2
              in  (h a, s3))
  -- pure :: a -> T a
  -- (<*>) :: (T (a -> b)) -> T a -> T b

-------------
-- Test data
-------------

appTest1 :: T (Integer -> Integer)
appTest1 =
    T (\s -> ((\x -> x * 3),
              s ++ "multiplied by 3 <= ")
      )

{- suggested tests
:t appTest1 <*> f1 10
:t unT $ appTest1 <*> f1 10
unT (appTest1 <*> f1 10) ""
-}

----------
-- Monad
----------

instance Monad T where
  return = pure
```

```
63   h >>= f = sharp f h
64   --  return :: a -> T a
65   --  (>>=) :: T a -> (a -> T b) -> (T b)
66
67 {- suggested tests
68 (unT $ return 10) ""
69
70 let (xTest1, stTest1) = (unT $ sharp f2 $ sharp f1 $ return
71      10) ""
72 putStr stTest1
73
74 let (xTest2, stTest2) = unT (return 10 >>= f1 >>= f2) ""
75 putStr stTest2
76
77 let (xTest3, stTest3) = unT (do  x <- return 10; y <- f1 x; f2
       y) ""
78 putStr stTest3
79 -}
```

Lines 14-19 declare T as an instance of the type class `Functor` implementing the `fmap` function. A suggested test described in line 24 is carried out as follows.

```
*CalcWithOutput2> unT (fmap (+1) (f1 10)) "hello\n"
(101,"hello\n10^2=100.\n")
```

The proofs for two equations that `fmap` must satisfy are given in Appendix C.

Lines 27-55 give an instance of the type class `Applicative`. The real meaning of this implementation will be made clear with the `Monad` instance.

A test described in line 54 yields the following.

```
*CalcWithOutput2> unT (appTest1 <*> f1 10) ""
(300,"multiplied by 3 <= 10^2=100.\n")
*CalcWithOutput2>
```

Refer to Appendix C for proofs concerning various equations required to be an instance of `Applicative`.

Lines 61-63 implement the `Monad` instance. Suggested tests can be carried out as follows.

```
*CalcWithOutput2> (unT $ return 10) ""
(10,"")
*CalcWithOutput2> (xTest2,stTest2)=unT (return 10>>=f1>>=f2) ""
*CalcWithOutput2> putStr stTest2
10^2=100.
sqrt of 100 is 10.0.
*CalcWithOutput2> (xTest2,stTest2)=unT (return 10>>=f1>>=f2) ""
*CalcWithOutput2> putStr stTest2
10^2=100.
sqrt of 100 is 10.0.
*CalcWithOutput2> let (xTest3, stTest3) =
          unT (do  x <- return 10; y <- f1 x; f2 y) ""
*CalcWithOutput2> putStr stTest3
10^2=100.
sqrt of 100 is 10.0.
*CalcWithOutput2>
```

Refer to Appendix C for proofs concerning various equations required to be an instance of `Monad`.

6.4.2 Computation with Continuation(revisited)

We extract the necessary part of the sample code presented earlier, aiming to recall the definitions of
`eta :: a -> T a` and `sharp :: (a -> T b) -> (T a -> T b)`.

Listing 6.8 Continuation.hs(excerpt)

```
newtype T v a = T {unT :: (a -> v) -> v}

-- Definition of eta
---- eta :: A -> TA
-- plays the role of
---- return :: (Monad m) => a -> m a
---- where m = T v.

eta :: a -> T v a
eta a = T (\x -> x a)

-- Definition of sharp
---- (-)^# :: Hom(A,TB) -> Hom(TA,TB)
-- plays the role of
---- (=<<) :: (Monad m) => (a -> m a) -> m a -> m b
---- where m = T v.
-- Types of arguments in the definition below
---- f :: a -> T v b
---- m :: T v a
---- b' :: b -> v

sharp :: (a -> T v b) -> (T v a -> T v b)
sharp f m =
    T (\b' -> unT m (\a -> unT (f a) b'))
```

Test data have been given as follows.

Listing 6.9 Continuation.hs(excerpt)

```
f1 :: Int -> T Int Int
f1 x = T (\c -> c (x+1))

f2 :: Int -> T Int Int
f2 x = T (\c -> c (x^2))

g1 :: String -> T Int String
g1 x = T (\c -> c (x++"!"))

g2 :: String -> T Int String
g2 x = T (\c -> c (x++x))
```

We import this module and instantiate Functor, Applicative, and Monad type classes for the continuation functor.

Listing 6.10 Continuation2.hs

```
module Continuation2 where
import Control.Monad
import Continuation

-- After building a Kleisli triple
-- for a continuation,
-- we declare instances of
-- Functor, Applicative, and Monad

-----------
-- Functor
-----------

---- instance declarations

-- T v a is a bifunctor.
-- However, we declare a Fuctor instance only for (T v).

instance Functor (T v) where
  fmap f h =
    T (\b' -> unT h (b'.f))

  -- where
  -- fmap  :: (a -> b) -> (T v a -> T v b)
  -- f     ::  a -> b
  -- unT h :: (a -> v) -> v
  -- b' ::  b -> v
  -- result::  v


-- Note that T itself is not an Applicative, but (T v) is.

instance (Applicative (T v)) where
  pure = eta
  -- pure :: a -> T v a
```

```
  tf <*> ta =
     T (\b' -> unT tf $ \f -> unT ta (b'. f))
  -- where
  -- (<*>)    :: (T v (a -> b)) -> T v a -> T v b
  -- unT tf :: ((a->b)->v) -> v
  -- unT ta :: (  a    ->v) -> v
  -- b'       ::    b   ->v
  -- f        ::    a   ->b
  -- b'.f     ::    a   ->v


-- Note that T itself is not a Monad, but (T v) is.
instance (Monad (T v)) where
  return = pure
  h >>= f = sharp f h
  --  return :: a -> T v a
  --  (>>=) :: T v a -> (a -> T v b) -> (T v b)

--- for tests

{- defined in Continuation.hs
f2 :: Int -> T Int Int
f2 x = T (\c -> c (x^2))

g2 :: String -> T Int String
g2 x = T (\c -> c (x++x))
-}

test9 =
  do x <- return "hello"
     y <- g2 x
     g2 y

-- unT test9 length

{-
unT (f2 3) (+1) ---> 10
unT (return 3 >>= f2) (+1)
unT (sharp f2 (eta 3)) (+1)

unT (sharp f2 $ sharp f2 $ eta 3) id   -----> 81
unT (return 3 >>= f2 >>= f2) id

unT (g2 "hello") length -----> 10
unT (sharp g2 (eta "hello") ) length
unT (return "hello" >>= g2) length

(unT $ sharp g2 $ sharp g2 $ eta "hello") length -------> 20
unT (return "hello" >>= g2 >>= g2) length
-}

-------------
-- Pythagoras
-------------
```

```
sq' :: (Num a) => a -> T v a
sq' x = return (x*x)
-- sq' x = T (\f -> f (x*x))
-- unT (sq' 3) id ----------> 9

add' :: (Num a) => a -> a -> T v a
add' x y = return (x+y)
-- add' x y = T (\f -> f (x+y))
-- unT (add' 11 19) id ----------> 30

sqrt' :: (Floating a) => a -> T v a
sqrt' x = return (sqrt x)
-- sqrt' x = T (\f -> f (sqrt x))
-- unT (sqrt' 10) (\x -> (x*x))

pyth x y = do
  x2 <- sq' x
  y2 <- sq' y
  z  <- add' x2 y2
  sqrt' z

-- unT (pyth 3 4) id

pyth2 x y =
  sq' x >>= \x2 -> sq' y >>= \y2 -> add' x2 y2 >>= sqrt'

-- unT (pyth2 3 4) id

----------------
-- factorial
----------------
factorial' :: Integer -> T v Integer
factorial' 0 = return 1
factorial' n = do
  fmn1 <- factorial' (n-1)
  return (n*fmn1)

-- unT (factorial' 5) id

factorial2' :: Integer -> T v Integer
factorial2' 0 = return 1
factorial2' n = factorial2' (n-1) >>= \fmn1 -> return (n*fmn1)

-- unT (factorial2' 5) id

----------------
-- factorial''
-- tail recursive
----------------
factorial'' :: Integer -> T v Integer
factorial'' n = faux' n 1

faux' :: Integer -> Integer -> T v Integer
faux' 0 k = return k
faux' n k = faux' (n-1) (k*n)
```

```
146
147 -- unT (factorial'' 5) id
```

Lines 19-21 implement an instance of `Functor`. Lines 33-37 implement an instance of `Applicative`. Lines 48-50 implement an instance of `Monad`.

So defined Monad instance makes all the following

```
unT (f2 3) (+1)
unT (return 3 >>= f2) (+1)
unT (sharp f2 (eta 3)) (+1)
```

evaluate to 10.

All the following

```
unT (sharp f2 $ sharp f2 $ eta 3) id
unT (return 3 >>= f2 >>= f2) id
```

evaluate to 81.

All the following

```
unT (g2 "hello") length
unT (sharp g2 (eta "hello") ) length
unT (return "hello" >>= g2) length
```

evaluate to 10.

All the following

```
(unT $ sharp g2 $ sharp g2 $ eta "hello") length
unT (return "hello" >>= g2 >>= g2) length
```

evaluate to 20.

Lines from 87 to the end include familiar materials that demonstrate the power of Monads.

The `pyth` function calculates the sum of squares and passes the result to a given continuation. The `pyth2` function is an equivalent version of `pyth`, replacing the do-notation with the sequence of bind(>>=) operations.

A suggested test is carried out as follows.

```
*Continuation2> unT (pyth 3 4) id
5.0
*Continuation2>
```

6.5 Functor, Applicative, and Monad

In the previous section, we mentioned applicative functors without clearly defining them. Let us see briefly here.

Given an endofunctor T : Hask $\to$ Hask, objects `a` and `b` of Hask, then we know that `a->b` is also an object of Hask. These objects are mapped to `T a`, `T b`, `T(a->b)`, respectively.

One may naturally be led to the question if there are any relations between `T (a->b)` and `T a->T b`. In some cases, we can define an operator

```
<*> :: T (a->b) -> (T a->T b) .
```

If a functor T satisfies some conditions related to this question, we say that T is an *Applicative* functor. Monad functors are known to be applicative.

In this section, we will discuss the relations among the concepts of functor, applicative, and Monad.

The functor class defines the following interface.

```
class Functor f where
fmap :: (a -> b) -> f a -> f b
```

A function of a signature is further required to satisfy constraints, namely the functor rules below.

```
fmap id = id
fmap (g . f) = fmap g . fmap f
```

The Applicative class defines the following interface.

```
class Functor f => Applicative f where
    pure :: a -> f a
    (<*>) :: f (a -> b) -> f a -> f b
```

The functions of these signatures are required to satisfy the constraints, namely the Applicative rules below.

```
pure id <*> v                 = v
pure (.) <*> u <*> v <*> w = u <*> (v <*> w)
f <*> pure x                  = pure (f x)
u <*> pure y                  = pure ($ y) <*> u
```

The Monad class defines the following interface.

```
class Monad f where
   (>>=) :: m a -> (a -> m b) -> m b
   return :: a -> m a
```

The functions of these signatures are required to satisfy the constraints, namely the Monad rules below.

```
return a >>= f = f a
m >>= return = m
\( m >>= f) >>= g = m >>= (\x -> f x >>= g)
```

We try to demonstrate the derivation (1), (2), and (3) from rules to rules.

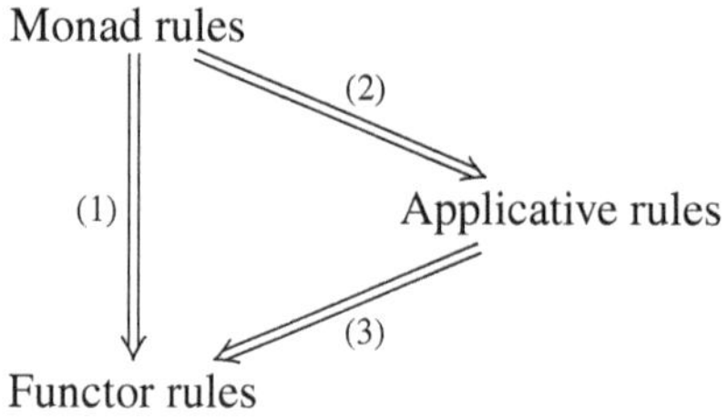

In the derivation (1), we implement `fmap` by `return`.

```
fmap f m = m >>= \a -> return (f a)
```

In the derivation (2) we implement `pure` and `<*>` by the bind operator `>>=` and `return`.

```
pure = return
 mf <*> mx =
    mf >>= \f ->
    mx >>= \x ->
    return (f x)
```

In the derivation (3), we implement `fmap` by `pure` and `<*>`.

```
fmap f x = pure f <*> x
```

Proposition 6.3 *In the rewriting of (1),* `fmap` *defined by* `>>=` *and* `return` *satisfies the functor rules.*

Proof We use Monad rules.

```
fmap id m
        -- rewrite with (>>=) and (return)
  = m >>= \a -> return (id a)
        -- eliminate id
  = m >>= \a -> return a
        -- eta transformation
  = m >>= return
        -- the 2nd Monad rule
  = m
```

```
fmap g . fmap f $ m
  = fmap g (fmap f m)
       -- rewrite (fmap f) by (>>=) and (return)
  = fmap g (m >>= return . f)
       -- rewrite (fmap g) by (>>=) and (return)
  = (m >>= (return . f)) >>= (return . g)
       -- use the 3rd Monad rule
  = m >>= (\x -> (return . f) x >>= (return . g) )
       -- rewrite function composition
  = m >>= (\x -> return (f x) >>= (return . g) )
       -- use the 1st Monad rule
  = m >>= (\x -> (return . g) (f x) )
       -- rewrite function composition
  = m >>= (\x -> (return . g . f) x)
       -- use Eta transformation
  = m >>= return . g . f
       -- (.) is associative. Insert a redundant parentheses.
  = m >>= return . (g . f)
       -- definition of fmap by Monad
  = fmap (g . f) m
```

□

Proposition 6.4 *In the rewriting of (2),* `pure` *and* `<*>` *defined by* `>>=` *satisfy all the Applicative rules.*

Proof We first show `pure id <*> = id`.

```
pure id <*> v
```

```
      -- rewrite pure and <*>
  = return id >>= \f -> (v >>= \x -> return (f x))
      -- return is a left unit
  = v >>= \x -> return (id x)
      -- omit id
  = v >>= \x -> return x
      -- Eta transformation
  = v >>= return
      -- return is a right unit
  = v
```

Next, we show the following.

```
pure (.) <*> u <*> v <*> w = u <*> (v <*> w)
```

We rewrite the LHS.

```
pure (.) <*> u
      -- replace (pure) with (return)
  = return (.) <*> u
      -- replace <*> with (>>=)
  = return (.) >>= \f -> (u >>= \x -> return (f x))
      -- return is a left unit
  = u >>= \x -> return ((.) x)
      -- Eta transformation
  = u >>= (return . (.))

pure (.) <*> u <*> v
  = (u >>= (return . (.))) <*> v
      -- rewrite <*>
  = (u >>= (return . (.))) >>=
    \f -> (v >>= \x -> return (f x))
      -- (>>=) is associative
  = u >>= (\y -> (return . (.)) y >>=
                 \f -> (v >>= \x -> return (f x)) )
      -- rewrite the first instance of (return)
  = u >>= (\y -> (return ((.) y) >>=
                 \f -> (v >>= \x -> return (f x)) )
      -- the 1st instance of (return) is a left unit
  = u >>= (\y -> v >>= \x -> return ((.) y x)) )
      -- streamline the subexpression
  = u >>= (\y -> v >>= \x -> return (y . x) )

The scope of lambda binding extends to as right as possible.
So, the parentheses at the top level are redundant. Just for readability.

pure (.) <*> u <*> v <*> w
  = (u >>= (\y -> v >>= \x -> return (y . x)) ) <*> w
      -- rewrite <*>
  = (u >>= (\y -> v >>= \x -> return (y . x)) ) >>=
```

```
    \f -> w >>= \z -> return (f z)
      -- (>>=) is associative
  = u >>= (\g -> (\y -> v >>= \x -> return (y . x)) g >>=
                 \f -> w >>= \z -> return (f z) )
      -- Beta reduction
  = u >>= (\g -> (v >>= \x -> return (g . x)) >>=
                 \f -> w >>= \z -> return (f z) )
      -- (>>=) is associative
  = u >>= \g -> v >>=
                (\h -> (\x -> return (g . x)) h >>=
                             \f -> w >>= \z -> return (f z) )
      -- Beta reduction
  = u >>= \g -> v >>=
                (\h -> return (g . h) >>=
                       \f -> w >>= \z -> return (f z) )
      -- the 1st instance of (return) is a left unit
  = u >>= \g -> v >>= \h -> w >>= \z -> return (g (h z))
```

So much for the LHS. Next, we focus on the RHS.

```
v <*> w
  = v >>= \f -> w >>= \x -> return (f x)

u <*> (v <*> w)
  = u <*> (v >>= \f -> w >>= \x -> return (f x))
      -- rewrite <*>
  = u >>= \g -> (v >>= \f -> w >>= \x -> return (f x)) >>=
                \h -> return (g h)
      -- (>>=) is associative
  = u >>= \g -> v >>=
                \j -> (\f -> w >>= \x -> return (f x)) j >>=
                      \h -> return (g h)
      -- Beta reduction
  = u >>= \g -> v >>=
                \j -> (w >>= \x -> return (j x)) >>=
                      \h -> return (g h)
      -- (>>=) is associative
  = u >>= \g -> v >>=
                \j -> w >>=
                      \f -> (\x -> return (j x)) f >>=
                            \h -> return (g h)
      -- Beta reduction
  = u >>= \g -> v >>=
                \j -> w >>=
                      \f -> return (j f) >>=
                            \h -> return (g h)
      -- The 1st instance of (return) is a left unit
  = u >>= \g -> v >>=
                \j -> w >>=
                      \f -> return (g (j f))
```

Changing local variables' names, we see that the LHS and RHS are identical. Next, we prove `pure f <*> pure x = pure (f x)`. We calculate the LHS.

```
pure f <*> pure x
      -- rewrite (pure) and <*>
  = return f >>= \g -> (return x >>= return (g x))
      -- the 2nd instance of (return) is a left unit
  = return f >>= \g -> return (g x)
      -- the 1st instance of (return) is a left unit
  = return (f x)
  = pure (f x)
```

Finally, we show u <*> `pure y = pure ($ y) <*>` u. We calculate the LHS.

```
u <*> pure y
      -- rewrite <*>
  = u >>= \f -> pure y >>= \x -> return (f x)
      -- rewrite (pure)
  = u >>= \f -> return y >>= \x -> return (f x)
      -- the 1st instance of r(eturn) is a left unit
  = u >>= \f -> return (f y)
```

We calculate the RHS.

```
pure ($ y) <*> u
      -- rewrite (pure) and <*>
  = return ($ y) >>= \f -> (u >>= \x -> return (f x))
      -- the 1st instance of (return) is a left unit
  = u >>= \x -> return (($ y) x)
  = u >>= \x -> return (x y)
```

Changing local variables' names, we see that the LHS and RHS are identical. □

Proposition 6.5 *In the derivation (3),* `fmap` *defined by* `pure` *satisfies all the functor rules.*

Proof We show `fmap id = id`.

```
fmap id x
      -- rewrite (fmap)
  = pure id <*> x
      -- (pure id) is a unit
  = x
```

Next, we show `fmap (g . f) = fmap g . fmap f`.

```
fmap (g . f) x
```

```
        -- rewrite fmap
    = pure (g . f) <*> x
    = pure ((.) g f) <*> x

  (fmap g . fmap f) x
    = fmap g (fmap f x)
        -- rewrite fmap
    = pure g <*> pure f <*> x
        -- composition rule for (.) and <*>
    = pure (.) <*> pure g <*> pure f <*> x
        -- homomorphism
    = pure ((.) g) <*> pure f <*> x
        -- homomorphism
    = pure ((.) g f) <*> x
```

This concludes that `fmap (g . f) = fmap g . fmap f`. □

We have established implications among the concepts of Functor, Applicative, and Monad.

6.6 Monads and Adjoints

Monads and adjoints are tightly related. An adjoint gives a Monad. A Monad gives an adjoint. In this section, we will show that we can construct a Monad for a given adjoint. We leave the question if there always is an adjoint for a given Monad. This problem has been completely solved, but we only give an example of the list Monad case.

6.6.1 Adjoints to Monads

Theorem 6.4 *Let $F : \mathscr{X} \to \mathscr{A}$ be a left adjoint of $U : \mathscr{A} \to \mathscr{X}$ with unit η and counit ε. Let T be an endofunctor defined by $T = U \circ F : \mathscr{X} \to \mathscr{X}$. Let μ be a natural transformation defined by $\mu = U\varepsilon F$. Then, the triple (T, η, μ) is a Monad on $\mathscr{X}$.*

Proof We show that the triple (T, η, μ) satisfies the Monad axioms:

(1) $\mu_A \circ T\eta_A = 1_{TA}$,
(2) $\mu_A \circ \eta_{TA} = 1_{TA}$,
(3) $\mu_A \circ \mu_{TA} = \mu_A \circ T\mu_A$.

We use the second alternative definition of adjoint justified by Theorem 5.1: there exist natural transformations $\eta : \mathrm{Id}_{\mathscr{X}} \to UF$ and $\varepsilon : FU \to \mathrm{Id}_{\mathscr{A}}$ that make the following two diagrams commute.

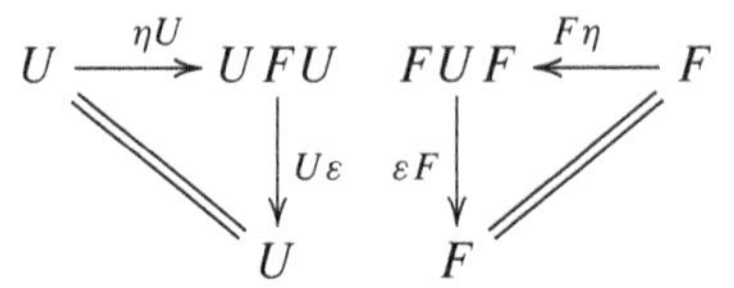

(6.3)

1. We prove $\mu T\eta = T$.
   Substituting T with UF and μ with $U\varepsilon F$ in, we can rewrite the goal as follows.

$$U\varepsilon FUF\eta = UF$$

   We calculate horizontal and vertical compositions with functors and natural transformations intermixed. Note that horizontal arrows are drawn from left to right in the diagrams, which may confuse the reader when one compares a formula to its corresponding diagram.
   We start with a subexpression $UF\eta$ of the LHS. It is a horizontal composition of the natural transformation η and the functor UF. We can draw a diagram:

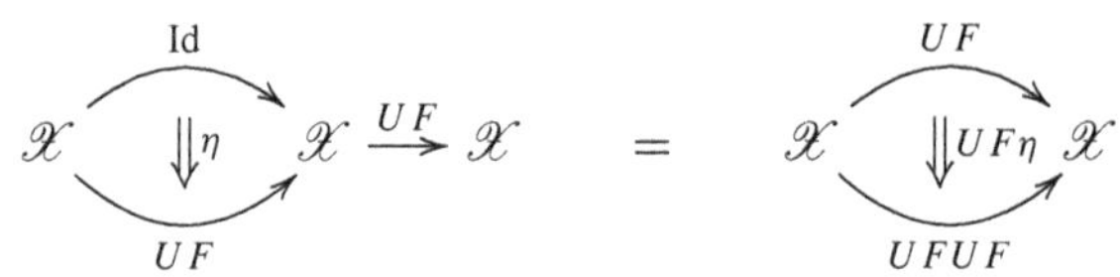

   We vertically compose this result with a natural transformation $U\varepsilon F : UFUF \to UF$ to obtain the whole LHS. Note that functor U can be horizontally factored out:

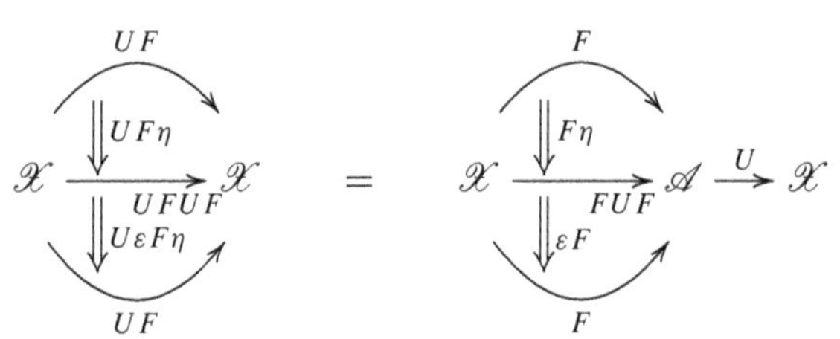

   The triangle at right in the diagram (6.3) tells us that the vertical composition of $F\eta$ and εF is $\mathrm{Id} : F \to F$. Thus, the result is Id_{UF}, which is an alternative expression for UF. This completes the first goal.
2. We prove $\mu\eta T = T$.
   Substituting T with UF and μ with $U\varepsilon F$ in, we can rewrite the goal as follows.

$$U\varepsilon F\eta UF = UF$$

We factor the LHS as the horizontal composition of $U\varepsilon F$ and ηUF.

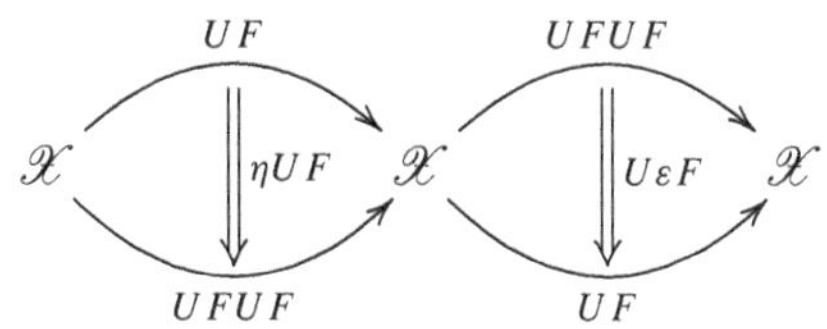

We immediately find that this can be rewritten as a vertical composition:

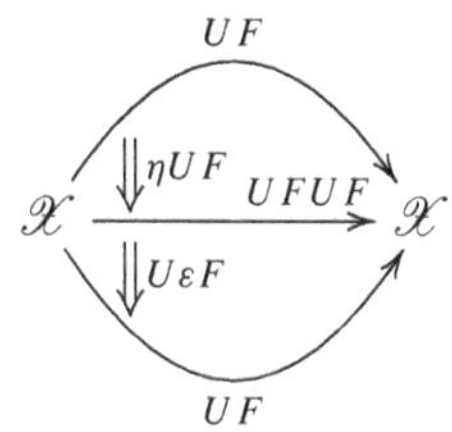

We can factor out F.

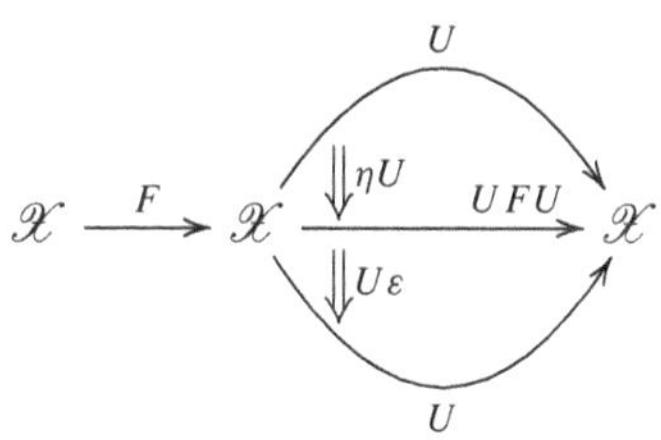

The triangle at left in the diagram (6.3) tells us that the vertical composition is $\mathrm{Id} : U \to U$. Thus, the result is UF. This completes the second goal.

3. We prove $\mu\mu T = \mu T\mu$.
Substituting T with UF and μ with $U\varepsilon F$ in, we can rewrite the goal as follows.

$$U\varepsilon FU\varepsilon FUF = U\varepsilon FUFU\varepsilon F$$

The naturality of $\varepsilon : FU \to \mathrm{Id}$ is expressed as $\varepsilon FU = FU\varepsilon$. We have

$$U\varepsilon FU\underline{\varepsilon FU}F = U\varepsilon FU\underline{FU\varepsilon}F.$$

The linerlined subexpressions are identical. This completes the third goal. □

6.6.2 List Monad as Adjoint

The list Monad can be decomposed as a pair of adjoint functors.

Definition 6.3 Given a set A and a binary operation $\bullet : A \times A \to A$, we say the pair $(A, \bullet_A)$ is a *monoid* if it satisfies the following conditions:

(1) $\bullet_A$ is associative.
(2) $\bullet_A$ has left and right unit.

Having left and right units means that there is $1_A \in A$ such that $1_A \bullet a = a \bullet 1_A = a$ holds for any $a \in A$. Set A is called an *underlying set*. We say simply that A is a monoid if it is clear from the context.

Given two monoids $(A, \bullet_A)$, $(B, \bullet_B)$, a mapping $f : A \to B$ is said to be a *monoid homomorphism* if the following conditions are satisfied.

(1) $f(x \bullet_A y) = f(x) \bullet_B f(y)$
(2) $f(1_A) = 1_B$

Definition 6.4 Given an alphabet Σ, namely a set of letters to be used. The set of all finite strings consisting of elements of Σ is a monoid with string concatenation as the binary operation $\bullet$, and the empty string ε as a unit. There are no relations other than those appearing in the axioms. We call this monoid a *free monoid* over Σ and write Σ^*. When we work with a programming language, we often adopt a list representation. For example, a finite string $x_1 x_2 \cdots x_n$ is represented by a list $[x_1, x_2, \ldots, x_n]$. The empty string is represented by the empty list [ ]. The concatenation of strings is translated into a concatenation of lists.

Remark 6.7 We observe the terminology in formal language theory. An alphabet is the set of letters that are used to construct a language.

Definition 6.5 Category **Mon** is called the *category of all monoids*. Its objects are all monoids. Its morphisms are all homomorphisms between the monoids.

We consider two functors between categories **Set** and **Mon**.

- A *forgetful functor* U : **Mon** $\to$ **Set** with function on objects:

$$\begin{aligned} U_0 &: \mathrm{Obj}(\mathbf{Mon}) \to \mathrm{Obj}(\mathbf{Set}) \\ &(A, \bullet_A) \mapsto A \end{aligned}$$

  For any monoid $M = (A, \bullet_A)$, we write $UM = U_0(M) = A = \underline{M}$. Function on objects is defined as follows. For any pair of monoids M and N, we set

$$\begin{aligned} U_1(M, N) &: \mathbf{Mon}(M, N) \to \mathbf{Set}(\underline{M}, \underline{N}) \\ &f \mapsto \underline{f} \end{aligned}$$

  We write $\underline{f}$ to indicate it is obtained from f forgetting the monoid structures.
- A *free functor* F with function on objects:

$$\begin{aligned} F_0 &: \mathrm{Obj}(\mathbf{Set}) \to \mathrm{Obj}(\mathbf{Mon}) \\ &X \mapsto (\mathrm{List}(X), \mathrm{++}) \end{aligned}$$

It maps a set X to the free monoid on X. The underlying set of $F(X)$ is List(X), consisting of all the finite lists $[x_1, \cdots, x_n]$. Function on morphisms is defined as follows. For any pair of sets X and Y, we set

$$\begin{aligned} F_1(X, Y) :&\mathbf{Set}(X, Y) \to \mathbf{Mon}(\mathrm{List}(X), \mathrm{List}(Y)) \\ &f \mapsto \texttt{map}\ f, \end{aligned}$$

where `map` is the same as those that have the same name in various programming languages.

We consider the composite $FU : \mathbf{Mon} \to \mathbf{Mon}$. Starting with a monoid $(A, \bullet_A, 1_A)$, the object function sends it to UA, the underlying set of the monoid A. We further apply F to UA to get FUA which is a free monoid on UA. An element of FUA is a list of elements in A. The binary operation is the concatenation of lists. Let $[a_1, \cdots, c_m]$ and $[b_1, \cdots, b_n]$ be a list of elements of A, which is, as stated above, elements of the monoid FUA. We have

$$[a_1, \cdots, a_m] \bullet_{FUA} [b_1, \cdots, b_n] = [a_1, \cdots, c_m, b_1, \cdots, b_n]$$

and

$$[\ ] \bullet_{FUA} [a_1, \cdots, a_m] = [a_1, \cdots, a_m] \bullet_{FUA} [\ ] = [a_1, \cdots, a_m].$$

We construct a natural transformation ε from the functor $FU : \mathbf{Mon} \to \mathbf{Mon}$ to $\mathrm{Id}_{\mathbf{Mon}} : \mathbf{Mon} \to \mathbf{Mon}$. For any monoid $(A, \bullet_A, 1_A)$, as we have just seen, FUA is a free monoid. We define a homomorphism from FUA to A as follows. Only we have to do is determine the image of each generator of the free monoid FUA.

$$[a_1, \cdots, a_n] \mapsto a_1 \bullet_A a_2 \bullet_A \cdots \bullet_A a_n,$$

which extends straightforwardly to the whole domain, namely the free monoid. The homomorphism turns out to be the A component of a natural transformation. We leave the proof of naturality to the reader.

We consider the composite $UF : \mathbf{Set} \to \mathbf{Set}$. Starting with a set X, the object function sends it to FX, a free monoid on X. The underlying set of this monoid FX, that is UFX, is the set of all lists that consist of elements of X allowing repetition.

We construct a natural transformation η from $\mathrm{Id}_{\mathbf{Set}} : \mathbf{Set} \to \mathbf{Set}$ to $UF : \mathbf{Set} \to \mathbf{Set}$. For any set X, we define

$$\eta_X : x \mapsto [x].$$

This map turns out to be the X component of a natural transformation. We leave the proof of naturality to the reader.

We can claim the following proposition concerning U, F, η, and ε so constructed.

Proposition 6.6 *F : **Set** $\to$ **Mon** is a left adjoint of U : **Mon** $\to$ **Set**, with a unit η and a counit ε. (List $= UF, \eta, \mu = U\varepsilon F$) is a Monad obtained from this adjunction.*

Proof We repeat diagrams (6.3) below.

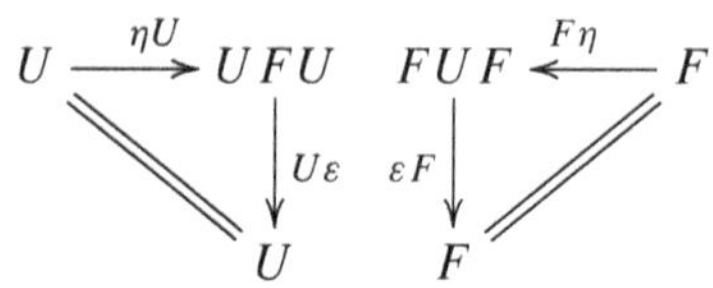

It is sufficient to show that the two triangles commute.

1. We calculate the composition of two edges in the triangle at left. Let A be a monoid. UA is the underlying set of A. η_{UA} is a mapping defined by

$$\begin{aligned} \eta_{UA} :& UA \to UFUA \\ & \underline{x} \mapsto \underline{[x]}, \end{aligned}$$

   where we used underlines to indicate that we forgot the algebraic structure. Before proceeding to the calculation of $U\varepsilon_A$, let us recall ε_A itself.

$$\begin{aligned} \varepsilon_A :& FUA \to A \\ & [x_1, \cdots, x_n] \mapsto x_1 \cdots x_n \end{aligned}$$

   We forget the monoid structure by operating U. We have

$$\begin{aligned} U\varepsilon_A :& UFUA \to UA \\ & \underline{[x_1, \cdots, x_n]} \mapsto \underline{x_1 \cdots x_n} \end{aligned}$$

   Compose them to obtain

$$\begin{aligned} U\varepsilon_A \circ \eta_{UA} :& UA \to UA \\ & \underline{x} \mapsto \underline{x}, \end{aligned}$$

   which shows that $U\varepsilon_A \circ \eta_{UA} = 1_{UA}$ for all A. Thus, we conclude $U\eta \circ \varepsilon U = \mathrm{Id}_U$.
2. We calculate the composition of two edges in the triangle at right. Let X be a set. We have

$$\begin{aligned} \eta_X :& X \to UFX \\ & x \mapsto [x]. \end{aligned}$$

   If we further apply F to it, we obtain the following:

$$\begin{aligned} F\eta_X :& FX \to FUFX \\ & \hat{x} \mapsto [\hat{x}], \end{aligned}$$

where $\hat{x}$ is x regarded as an element of the free monoid.

$$\begin{aligned}\varepsilon_{FX} :&FUFX \to FX\\ &[\hat{x_1}, \hat{x_2}, \cdots, \hat{x_n}] \mapsto \hat{x_1}\hat{x_2}\cdots\hat{x_n}\end{aligned}$$

We compose these to obtain

$$\begin{aligned}\varepsilon_{FX} \circ F\eta_X :&FX \to FX\\ &\hat{x} \mapsto \hat{x},\end{aligned}$$

which shows $\varepsilon_{FX} \circ F\eta_X = 1_{FX}$. Thus, we conclude that $\varepsilon F \circ F\eta = \mathrm{Id}_{FX}$.

□

Example 6.1 Let U : **Mon** $\to$ **Set** be a forgetful functor, F : **Set** $\to$ **Mon** a free functor. We define a monoid $(A, \bullet_A)$ as follows.

(1) $A = \{1, a, a^2\}$ is the underlying set.
(2) The binary operation is given in the table below:

	1	a	a^2
1	1	a	a^2
a	a	a^2	1
a^2	a^2	1	a

To sum up, this is a monoid with a generator a and a sole relation $a^3 = 1$. The forgetful functor U : **Mon** $\to$ **Set** gives $UA = \{1, a, a^2\}$. Function on morphisms sends a monoid homomorphism

$$\begin{aligned}f :&A \to A\\ 1 &\mapsto 1\\ a &\mapsto a^2\\ a^2 &\mapsto a\end{aligned}$$

to a plain mapping $Uf : A \to A$ forgetting its being a homomorphism. For examples of an element of $FU(A)$, we may list

$$[a, a^2], \quad [1], \quad [1, a^2], \quad [a, a, a, a].$$

We can generate other elements of $FU(A)$ by the free functor.

$FU(A)$ is a monoid as well as A. We can define a monoid homomorphism $\varepsilon_A : FU(A) \to A$ as follows.

ε_A sends a given list to the result of all possible monoid calculations. This setting makes ε_A a monoid homomorphism. For example, the calculation goes like this:

$$\varepsilon_A : [a] + +[a^2, 1] + +[a, a] \mapsto \varepsilon_A([a, a^2, 1, a, a]) = a^2$$

Reference

1. Moggi E, Notions of computation and monads. https://person.dibris.unige.it/moggi-eugenio/ftp/ic91.pdf

7 Representable Functors

In this chapter, aiming at Yoneda's lemma, we introduce the notion of representable functors. Roughly speaking, Yoneda's lemma shows that natural transformations are determined by local information. It can also be described as coherence or resonance. In the section on "reverse engineering," the resonance phenomenon is verified by Yoneda's lemma.

Closely related to the notions of limit and adjoint, it is possible to paraphrase the same mathematical phenomenon in different terms. Some examples are introduced at the end of the chapter.

It is cumbersome to properly write the naturality associated with various concepts, and so with the naturality that appears in Yoneda's lemma. The reader is requested to trace all the diagrams to get a feel of what naturality is all about.

7.1 Representation of Covariant Functors

Definition 7.1 Let $\mathscr{A}$ be a locally small category and A a fixed object. For every object A', we consider the set of morphisms $\mathscr{A}(A, A')$. We write this correspondence as follows:

$$H^A = \mathscr{A}(A, -) : \mathrm{Obj}(\mathscr{A}) \to \mathbf{Set}$$
$$A' \mapsto H^A(A') = \mathscr{A}(A, A')$$

We extend this definition to morphisms. For any morphism $f : A' \to A''$ of $\mathscr{A}$, we define a mapping $H^A(f)$ as follows:

$$H^A(f) = \mathscr{A}(A, f) : \quad \mathscr{A}(A, A') \longrightarrow \mathscr{A}(A, A'')$$
$$g \longmapsto f \circ g$$

The situation is depicted in Fig. 7.1.

S. Yukita, *Category Theory Using Haskell*, Computer Science Foundations and Applied Logic, https://doi.org/10.1007/978-3-031-68538-5_7

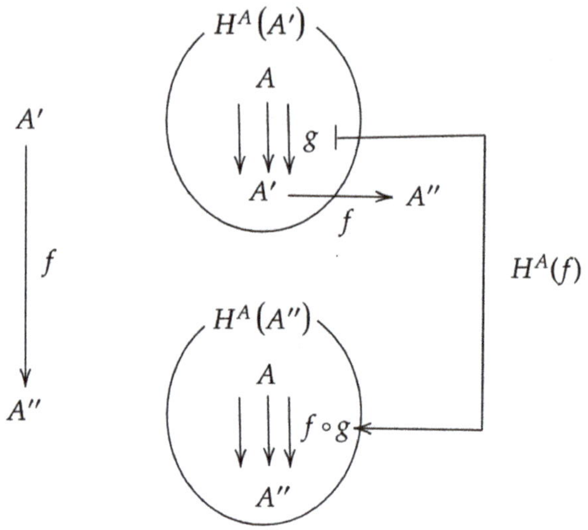

Fig. 7.1 Function on morphisms of the functor H^A

Remark 7.1 We sometimes write $f \circ -$ instead of $H^A(f)$. In a lambda expression, we may write as follows:

$$H^A(f) = \lambda g \in \mathscr{A}(A, A').f \circ g$$

Definition 7.2 Let $\mathscr{A}$ be a locally small category. Let $X : \mathscr{A} \to \mathbf{Set}$ be a functor. If the functor X is naturally isomorphic to H^A for some $A \in \mathrm{Obj}(\mathscr{A})$, we say X is *representable*. A *representation* of X is a pair (A, ϕ) where ϕ in

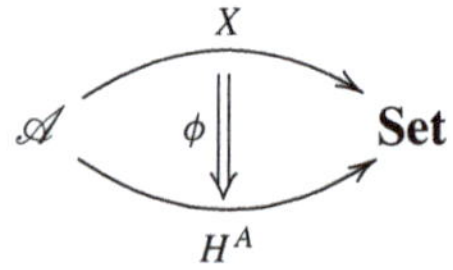

is a natural isomorphism.

Functor $H^A : \mathscr{A} \to \mathbf{Set}$ is automatically a representable functor.

Example 7.1 A forgetful functor $U : \mathbf{Grp} \to \mathbf{Set}$ is naturally isomorphic to $\mathbf{Grp}(Z, -)$. Given a group G, if we want a homomorphism φ from an infinite cyclic group φ to G, only we have to do is to determine the image $\varphi(a) \in G$ of the generator $a \in Z$. In fact, we have no option but to set $\varphi(a^n) = (\varphi(a))^n$. Let us consider the other way around. Given any $x \in U(G)$, if we set $\varphi(a^n) = x^n$, we get a homomorphism $\varphi : Z \to G$.

We construct a natural transformation $\alpha : \mathbf{Grp}(Z, -) \to U$, which turns out to be a natural isomorphism.

We define α as follows:

$$\begin{aligned}\alpha_G : \quad \mathbf{Grp}(Z, G) &\longrightarrow U(G)\\ \varphi &\longmapsto \varphi(a)\end{aligned}$$

Being a natural transformation requires the trapezoid at right in Fig. 7.2

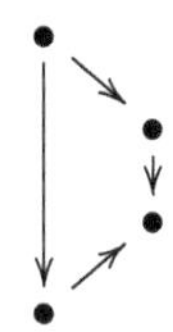

to commute.

To see this, we will show that the two paths of length two that originate at φ terminate at the same $\bullet$ in Fig. 7.3. The path passing $\varphi(a)$ terminates at $(Uf)(\varphi(a)) = f(\varphi(a))$. The path passing $f \circ \varphi$ terminates at $(f \circ \varphi)(a)$. Notice that these terminating points are identical since both are the result of applying f and φ to $a \in Z$.

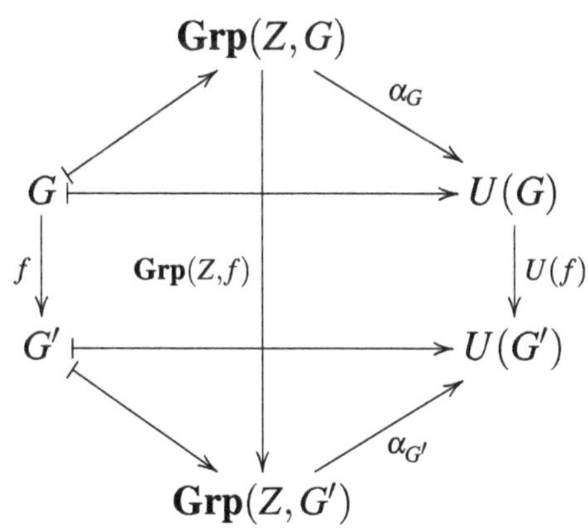

Fig. 7.2 Natural transformation $\alpha : \mathbf{Grp}(Z, -) \to U$

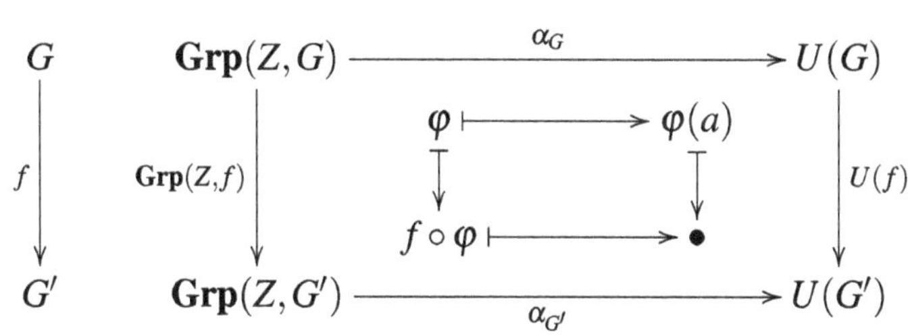

Fig. 7.3 Naturality of α

We have already known that for each G we have the inverse of α_G as follows:

$$\begin{aligned}\alpha_G^{-1}: \quad U(G) &\longrightarrow \mathbf{Grp}(Z, G)\\ x &\longmapsto (a^n \mapsto x^n)\end{aligned}$$

Lemma 2.1 established that if all the components of a natural transformation are isomorphisms then the natural transformation itself is an isomorphism.

We can argue that representability is closely related to adjoints.

Theorem 7.1 *Let $\mathscr{A}$ and $\mathscr{X}$ be locally small categories with adjunction:*

$$\phi_{X,A} : \mathscr{A}(FX, A) \cong \mathscr{X}(X, UA).$$

*Then, the functor $\mathscr{X}(X, U(-)) : \mathscr{A} \to$ **Set** is representable.*

Proof We have to show that

$$\phi_{X,-} : \mathscr{A}(FX, -) \cong \mathscr{X}(X, U(-))$$

is a natural transformation (consequently a natural isomorphism).

Recall the correspondence of an adjunction introduced in Sect. 5.1

$$\frac{FX \xrightarrow{h} A \xrightarrow{a} A'}{X \underset{f}{\rightarrow} UA \underset{Ua}{\longrightarrow} UA'}$$

This correspondence claims that the square in Fig. 7.4 commutes, which means $\phi_{X,-}$ is natural. We know that for each A, $\phi_{X,A}$ is an isomorphism. Therefore, we can conclude that $\phi_{X,-}$ is a natural isomorphism by Lemma 2.1.

Rewrite $\mathscr{A}(FX, -)$ as H^{FX}. We have the following natural isomorphism:

$$\phi_{X,-} : H^{FX} \to \mathscr{X}(X, U(-)),$$

which shows that $\mathscr{X}(X, U(-))$ is representable. □

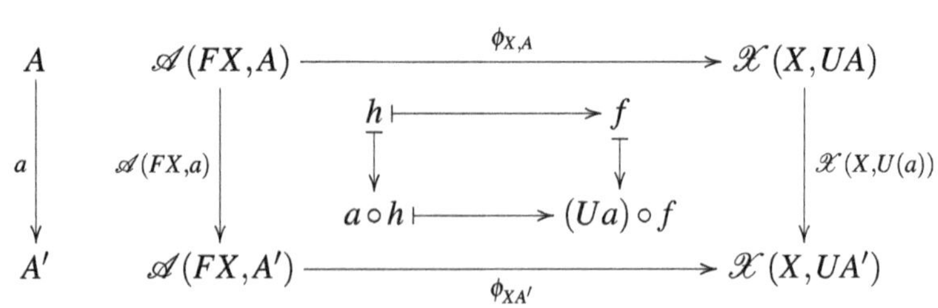

Fig. 7.4 Naturality of $\phi_{X,-}$

We established that the composite functor $\mathscr{X}(X, U(-))$ is representable. We can also show the right adjoint U is itself representable when it is a **Set**-valued functor.

Theorem 7.2 *If there is an adjunction ϕ between locally small categories $\mathscr{A}$ and* **Set***:*

$$\phi_{X,A} : \mathscr{A}(FX, A) \cong \mathbf{Set}(X, UA),$$

then the functor $U : \mathscr{A} \to$ **Set** *is representable.*

Proof Let 1 be a singleton set. We have

$$U(A) \cong \mathbf{Set}(1, U(A)).$$

This gives $U \cong \mathbf{Set}(1, U(-))$. Since we have already shown the RHS is representable, so is the LHS. □

We consider the correspondence $A \mapsto H^A$, the function on objects that sends an object to a functor. We would like to augment it a function on morphisms that will make $A \mapsto H^A$ a functor from the category $\mathscr{A}$ to the functor category $[\mathscr{A}, \mathbf{Set}]$.

Recall that Obj($[\mathscr{A}, \mathbf{Set}]$) consists of all the functors from $\mathscr{A}$ to **Set**, and that morphisms in $[\mathscr{A}, \mathbf{Set}]$ are natural transformations. If we only focus on function on objects, we might take

$$\mathscr{A} \to [\mathscr{A}, \mathbf{Set}].$$

But, on morphisms, we should be careful about the directions of morphisms. The solution is that we should construct the following contravariant functor:

$$\mathscr{A}^{\mathrm{op}} \to [\mathscr{A}, \mathbf{Set}].$$

Definition 7.3 Let $\mathscr{A}$ be a locally small category. We define a functor

$$H^- : \mathscr{A}^{\mathrm{op}} \to [\mathscr{A}, \mathbf{Set}]$$

as follows. The function on objects is given by $A \mapsto H^-(A) = H^A$. In lambda expression, we write

$$H^- = \lambda A.\lambda B.\mathscr{A}(A, B)$$

Given a morphism $f : A' \to A$. A natural transformation

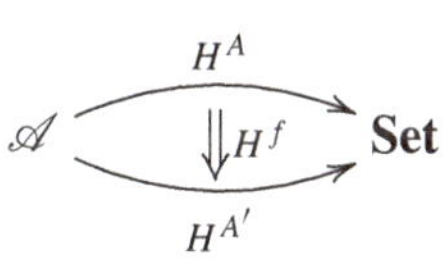

is defined by taking the B-component H^f_B as

$$H^-(f)_B = H^f_B : \quad H^A(B) \longrightarrow H^{A'}(B)$$
$$g \longmapsto g \circ f$$

In the λ expression we write

$$H^f = \lambda B.\lambda g \in \mathscr{A}(A, B).g \circ f$$

7.2 Representation of Contravariant Functors

Discussions proceed almost parallelly with the previous section. However, examples in category **Set**, the most important ones, need some attention to asymmetry in arrow directions. Therefore, we repeat the whole arguments along with concrete examples.

Definition 7.4 Let $\mathscr{A}$ be a locally small category and A a fixed object. For every object A', we consider the set of morphisms $\mathscr{A}(A', A) \in \text{Obj}(\mathbf{Set})$. We write this correspondence as follows:

$$H_A = \mathscr{A}(-, A) : \text{Obj}(\mathscr{A}) \to \text{Obj}(\mathbf{Set})$$
$$A' \mapsto \mathscr{A}(A', A).$$

We extend this correspondence to a function on morphisms in a unique way. For any morphism $f : A'' \to A'$ of $\mathscr{A}$, we define a mapping $H_A(f)$ as follows:

$$H_A(f) = \mathscr{A}(f, A) : \quad \mathscr{A}(A', A) \longrightarrow \mathscr{A}(A'', A) \ .$$
$$g \longmapsto g \circ f$$

Thus, we have the functor

$$H_A = \mathscr{A}(-, A) : \mathscr{A}^{\text{op}} \to \mathbf{Set}.$$

The situation is depicted in Fig. 7.5.

Remark 7.2 We sometimes write $- \circ f$ instead of $H_A(f)$. In a lambda expression, we may write as follows:

$$H_A(f) = \lambda g \in \mathscr{A}(A', A).g \circ f$$

Definition 7.5 Let $\mathscr{A}$ be a locally small category. Let $X : \mathscr{A}^{\text{op}} \to \mathbf{Set}$ be a contravariant functor. If the functor X is naturally isomorphic to H_A for some $A \in \text{Obj}(\mathscr{A})$, we say X is *representable*. A *representation* of X is a pair (A, ϕ)

Fig. 7.5 Contravariant functor H_A acts on morphisms

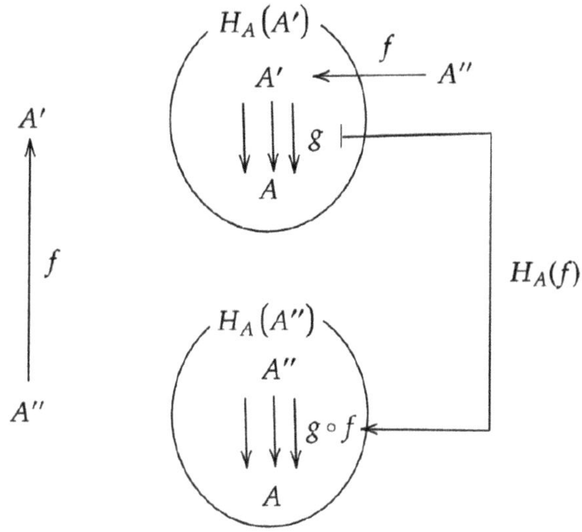

where ϕ in

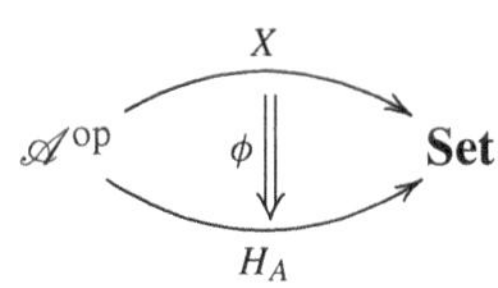

is a natural isomorphism.

Contravariant functor $H_A : \mathscr{A}^{\mathrm{op}} \to \mathbf{Set}$ is automatically a representable contravariant functor.

Example 7.2 We consider a contravariant functor $\mathscr{P} : \mathbf{Set}^{\mathrm{op}} \to \mathbf{Set}$. Function on objects sends a set X to its power set $\mathscr{P}(X)$. Function on morphisms sends a morphism $f : X' \to X$ to

$$\begin{aligned} \mathscr{P}(f) : \quad \mathscr{P}(X) &\longrightarrow \mathscr{P}(X') \\ U &\longmapsto f^{-1}(U), \end{aligned}$$

where

$$f^{-1}(U) = \{x' \in X' \mid f(x') \in U\}.$$

The characteristic function of a subset $U \subset X$ is a function $X \to \{0, 1\}$. We write $\mathbf{2} = \{0, 1\}$. Then we have $\mathscr{P} \cong H_{\mathbf{2}}$, which means that $\mathscr{P}$ is representable.

We now study the correspondence $A \mapsto H_A$. Its variance will turn out to be covariant:

$$\mathscr{A} \to [\mathscr{A}^{\mathrm{op}}, \mathbf{Set}].$$

Definition 7.6 Let $\mathscr{A}$ be a locally small category. We aim to define a functor

$$H_{-} : \mathscr{A} \to [\mathscr{A}^{\mathrm{op}}, \mathbf{Set}].$$

Function on objects is given by $A \mapsto H_A$. We can write it in the lambda expression like

$$H_{-} = \lambda A.\lambda B.\mathscr{A}(B, A) \quad .$$

Function on morphisms is defined as follows. Given a morphism $f : A \to A'$, we have to construct the corresponding natural transformation

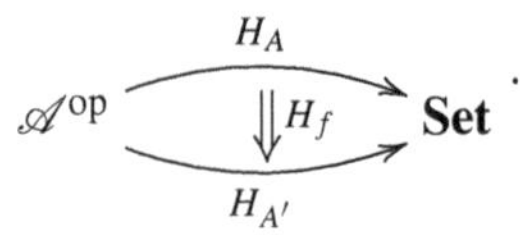

The B component $(H_f)_B$ is given by

$$\begin{aligned} H_{-}(f)_B = (H_f)_B : \quad H_A(B) &\longrightarrow H_{A'}(B) \ . \\ g &\longmapsto f \circ g \end{aligned}$$

We can write it in the lambda expression like

$$H_f = \lambda B.\lambda g \in \mathscr{A}(B, A).f \circ g \quad .$$

Definition 7.7 Let $\mathscr{A}$ be a locally small category. The functor

$$H_{-} : \mathscr{A} \to [\mathscr{A}^{\mathrm{op}}, \mathbf{Set}]$$

is called the Yoneda embedding.

7.3 The Yoneda Lemma

The Yoneda lemma may be called a *resonance theorem*, which the reader will agree with after going through the whole discussion in this section. We will study the cases of covariant functors, contravariant functors (presheaves), and embeddings, in order.

7.3.1 Covariant Case

We fixed an object A of a category $\mathscr{A}$ and defined a functor $H^A = \mathscr{A}(A, -) : \mathscr{A} \to \mathbf{Set}$ by $A' \mapsto \mathscr{A}(A, A')$. In what follows, we assume that $\mathscr{A}$ is a small category. We consider a natural transformation $\alpha : H^A \to X$ from the functor $H^A : \mathscr{A} \to \mathbf{Set}$ to $X : \mathscr{A} \to \mathbf{Set}$.

Remark 7.3 We want to work within the functor category. It is possible for a locally small category $\mathscr{A}$ to give the functor category $[\mathscr{A}, \mathbf{Set}]$ which is not locally small. To avoid this, we restrict $\mathscr{A}$ to a small category or introduce another universe. We take the former option.

Let us observe a resonance caused by a natural transformation $\alpha : H^A \to X$ in Fig. 7.6.

- First, focus on the A component of α, namely, a mapping α_A from the set $H^A(A)$ to the set $X(A)$. The set $H^A(A)$ has the identity morphism 1_A as an element.
- This identity morphism is sent to $\alpha_A(1_A) \in X(A)$.
- Let A' be any object of $\mathscr{A}$. The A' component of α is a mapping from the set $H^A(A')$ to the set $X(A')$.
- Let f be any element of $H^A(A')$. This f is, at the same time, a morphism $f : A \to A'$. Therefore, functor H^A sends f to

$$H^A(f) : H^A(A) \to H^A(A'),$$

  which is a mapping between sets.
- Functor X sends morphism f to $X(f)$. By naturality of α, the two paths in Fig. 7.6 originating from 1_A must be identical as morphisms.
- To sum up, $\alpha_A(1_A)$ determines all the components of α. This is the resonance we are talking about.

We have just observed that there is a bijection between the set $X(A)$ and the set of all natural transformations from the functor H^A to X, namely, $[\mathscr{A}, \mathbf{Set}](H^A, X)$. We still want to know how the bijection depends on A and X. To this end we need lots of preparation.

Our goal is given below.

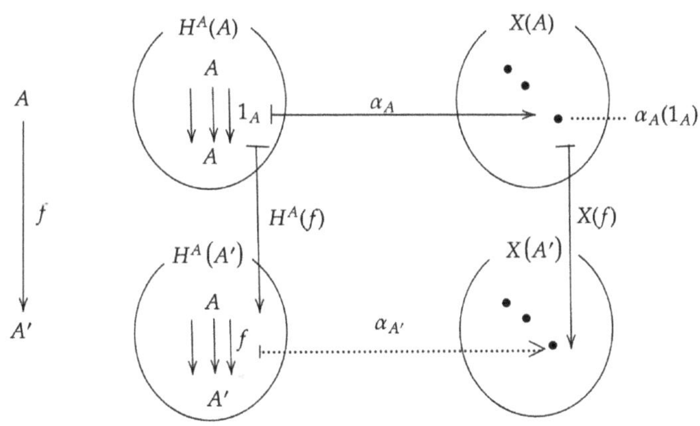

Fig. 7.6 Resonance between a covariant functor and a natural transformation

Theorem 7.3 (The Yoneda Lemma) *Let $\mathscr{A}$ be a small category. Then, the following isomorphism*

$$[\mathscr{A}, \mathbf{Set}](H^A, X) \simeq X(A)$$

is natural in A and X.

Naturality, in this case, means that

- The LHS and RHS are functors, say $L : \mathscr{A} \times [\mathscr{A}, \mathbf{Set}] \to \mathbf{Set}$ and $R : \mathscr{A} \times [\mathscr{A}, \mathbf{Set}] \to \mathbf{Set}$, respectively.
- There is a natural isomorphism $z : L \to R$.

At first sight, the LHS of the equation in Theorem 7.3 looks like a functor of two arguments A and X. We give a rigorous description as follows. We use the construction of a product of categories. Review Definition 1.5.

The function on objects $\mathrm{Obj}(A, X) \to \mathrm{Obj}(\mathbf{Set})$ is given as follows:

$$\begin{aligned} \mathscr{A} \times [\mathscr{A}, \mathbf{Set}] &\to \mathbf{Set} \\ (A, X) &\mapsto [\mathscr{A}, \mathbf{Set}](H^A, X) \end{aligned}$$

To be precise, this correspondence is the composition of two operations.

$$\begin{array}{ccccc} \mathscr{A} \times [\mathscr{A}, \mathbf{Set}] & \xrightarrow{(H^-)^{\mathrm{op}} \times 1} & [\mathscr{A}, \mathbf{Set}]^{\mathrm{op}} \times [\mathscr{A}, \mathbf{Set}] & \xrightarrow{[\mathscr{A}, \mathbf{Set}](-,-)} & \mathbf{Set} \\ (A, X) & \longmapsto & (H^A, X) & \longmapsto & [\mathscr{A}, \mathbf{Set}](H^A, X) \end{array} .$$

Let us take a closer look at this composition. In the first stage, $A \mapsto H^A$ is given by the contravariant functor $H^- : \mathscr{A}^{\mathrm{op}} \to [\mathscr{A}, \mathbf{Set}]$. In the second stage, the Hom functor $[\mathscr{A}, \mathbf{Set}](-, -)$ expects its first argument H^A contravariantly. In this way, the composition observes the covariant-contravariant restriction.

Let us consider the functor $\mathscr{A} \times [\mathscr{A}, \mathbf{Set}] \to \mathbf{Set} : (A, X) \mapsto [\mathscr{A}, \mathbf{Set}](H^A, X)$ with its second argument X fixed. We observe how its image depends on A. The image of $f : A \to A'$ is found in the following diagram:

$$\begin{array}{ccccccc} \mathscr{A} & \xrightarrow{(H^-)^{\mathrm{op}}} & [\mathscr{A}, \mathbf{Set}]^{\mathrm{op}} & \longrightarrow & [\mathscr{A}, \mathbf{Set}](H^-, X) & & \\ A & \longmapsto & H^A & \longmapsto & [\mathscr{A}, \mathbf{Set}](H^A, X) & & \ni \alpha \\ {\scriptstyle f}\downarrow & & \uparrow{\scriptstyle H_f} & & \downarrow{\scriptstyle -\circ H^f} & & \downarrow \\ A' & \longmapsto & H^{A'} & \longmapsto & [\mathscr{A}^{\mathrm{op}}, \mathbf{Set}](H^{A'}, X) & & \ni \alpha \circ H^f \end{array}$$

Next, we fix A and observe the dependency on X. Let us consider a morphism $\theta : X \to X'$ in $[\mathscr{A}, \mathbf{Set}]$

$$\mathscr{A} \underset{X'}{\overset{X}{\rightrightarrows}} \mathbf{Set} \quad (\Downarrow \theta)$$

which can be regarded as a natural transformation:

$$\begin{array}{ccccc} X & & [\mathscr{A}, \mathbf{Set}](H^A, X) & \ni & \alpha \\ {\scriptstyle\theta}\downarrow & & \downarrow{\scriptstyle\theta\circ-} & & \downarrow \\ X' & & [\mathscr{A}, \mathbf{Set}](H^A, X') & \ni & \theta \circ \alpha \end{array}$$

namely, $\theta \mapsto (\theta \circ -)$. We have to check if the functor really preserves the composition of morphisms:

$$(\theta' \circ \theta) \circ - = (\theta' \circ -) \circ (\theta \circ -),$$

which can be easily seen by the following diagram:

$$\begin{array}{ccccc} X & & [\mathscr{A}, \mathbf{Set}](H^A, X) & & \alpha \\ {\scriptstyle\theta}\downarrow & & \downarrow{\scriptstyle\theta\circ-} & & \downarrow \\ X' & {\scriptstyle\theta'\circ\theta} & [\mathscr{A}, \mathbf{Set}](H^A, X') & {\scriptstyle(\theta'\circ\theta)\circ-} & \theta \circ \alpha \\ {\scriptstyle\theta'}\downarrow & & \downarrow{\scriptstyle\theta'\circ-} & & \downarrow \\ X'' & & [\mathscr{A}, \mathbf{Set}](H^A, X'') & & \theta' \circ \theta \circ \alpha \end{array}$$

Next, let us define the right-hand side of the equation in Theorem 7.3 as a functor in two variables A and X.

$$\begin{array}{ccc} \mathscr{A} \times [\mathscr{A}, \mathbf{Set}] & \longrightarrow & \mathbf{Set} \\ (A, X) & \longmapsto & X(A) \end{array}$$

This is called an *evaluation functor*. We fix X as a constant. For any morphism $f : A \to A'$ we have

$$\begin{array}{ccc} A & \longmapsto & X(A) \\ {\scriptstyle f}\downarrow & & \downarrow{\scriptstyle X(f)} \\ A' & \longmapsto & X(A') \end{array},$$

which is a consequence of X being a functor $\mathscr{A} \to \mathbf{Set}$.

Fixing A, we want to define a functor $[\mathscr{A}, \mathbf{Set}]$ with a function on objects $X \mapsto X(A)$ and an appropriate function on morphisms as follows. For any morphism θ in $[\mathscr{A}, \mathbf{Set}]$

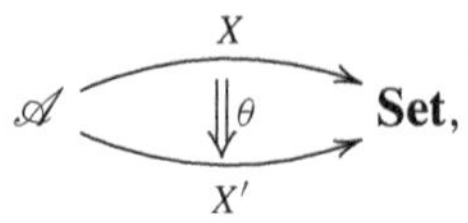

which is actually a natural transformation between functors X and X', we can simply take $\theta \mapsto \theta_A$. Composition of morphisms $\theta : X \to X'$ and $\theta' : X' \to X''$ is given by

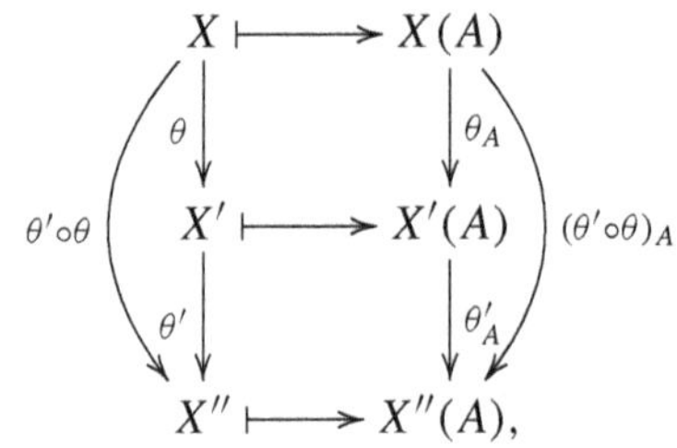

where we have commutative triangles at the left and right sides. This comes from the vertical composition of natural transformations θ and θ'. Combining the discussion before, we now have

$$\mathscr{A} \times [\mathscr{A}, \mathbf{Set}] \to \mathbf{Set} : (A, X) \mapsto X(A)$$

is a functor of two variables.

Proof (The Yoneda Lemma—Covariant case—Naturality in A)
The preparation above lets us proceed to prove that the following correspondence:

$$y : [\mathscr{A}, \mathbf{Set}](H^{-}, X) \to X(-),$$

whose component at A is given by

$$y_A = [\mathscr{A}, \mathbf{Set}](H^{A}, X) \to X(A) : \alpha \mapsto \alpha_A(1_A)$$

can be made into a natural transformation. For each morphism $f : A \to A'$, we have to show that the two paths from α to $X(A')$ in Fig. 7.7 are identical. In other words, we have to show the following equation holds:

$$(\alpha \circ H^{f})_{A'}(1_{A'}) = (Xf)(\alpha_A(1_A)).$$

We transform the LHS step by step as follows:

$$\begin{aligned} &(\alpha \circ H^{f})_{A'}(1_{A'}) \\ &= \alpha_{A'}((H^{f})_{A'}(1_{A'})) \\ &= \alpha_{A'}(1_{A'} \circ f) \\ &= \alpha_{A'}(f). \end{aligned}$$

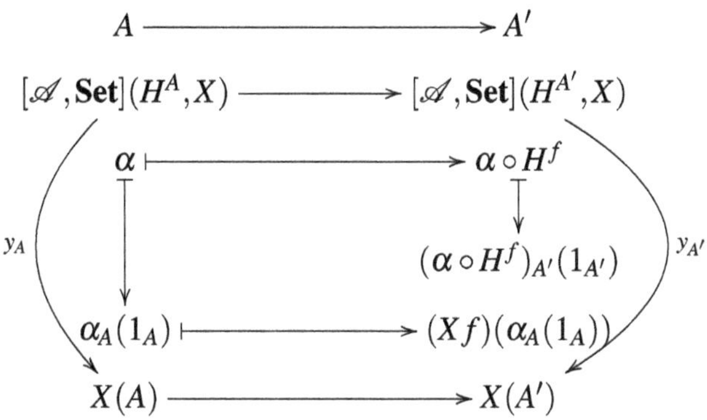

Fig. 7.7 Naturality with A (covariant version)

The last expression above is equal to the RHS since α is a natural transformation that makes the following square commute:

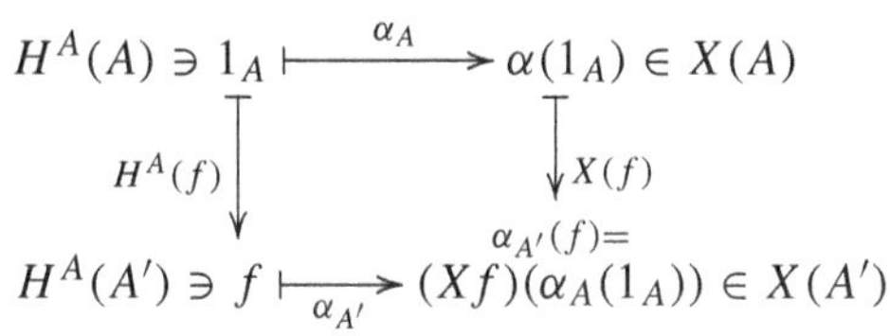

which lives in Fig. 7.6. This completes the proof. □

Proof (The Yoneda Lemma—Covariant case—Naturality in X)
We fix A and show that the following correspondence:

$$z : [\mathscr{A}, \mathbf{Set}](H^A, -) \to (-)(A),$$

whose component at X is given by

$$z_X = [\mathscr{A}, \mathbf{Set}](H^A, X) \to X(A) : \alpha \mapsto \alpha_A(1_A)$$

can be made into a natural transformation.

Let θ be a morphism in $[\mathscr{A}, \mathbf{Set}]$.

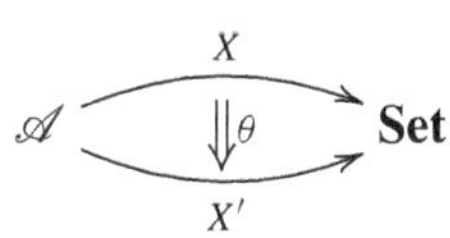

Only we have to show is that the two paths in Fig. 7.8 starting from α coincide, which is clear from the definition of function composition. □

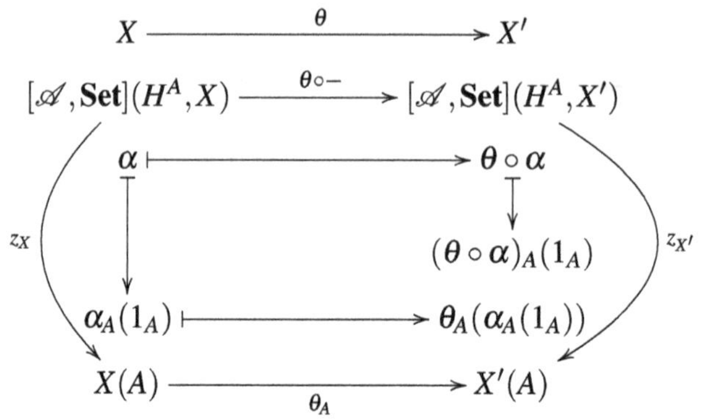

Fig. 7.8 Naturality in X (covariant version)

7.3.2 Contravariant Case (Presheaves)

We fixed an object A of a category $\mathscr{A}$ and defined a contravariant functor $H_A = \mathscr{A}(-, A) : \mathscr{A}^{\mathrm{op}} \to \mathbf{Set}$ by $A' \mapsto \mathscr{A}(A', A)$. Let us consider a natural transformation $\alpha : H_A \to X$ between functors $H_A : \mathscr{A}^{\mathrm{op}} \to \mathbf{Set}$ and $X : \mathscr{A}^{\mathrm{op}} \to \mathbf{Set}$. Reversing all arrows in the covariant case gives us the result for the contravariant case. We repeat the whole parallel argument in the following. We suppose that $\mathscr{A}$ is a small category from the same reason described in Remark 7.3. Let us observe a resonance caused by a natural transformation $\alpha : H_A \to X$ in Fig. 7.9.

- First, focus on the A component of α, namely, a mapping α_A from the set $H_A(A)$ to the set $X(A)$. The set $H_A(A)$ has the identity morphism 1_A as an element.
- This identity morphism is sent to $\alpha_A(1_A) \in X(A)$.
- Let A' be any object of $\mathscr{A}$. The A' component of α is a mapping from the set $H_A(A')$ to the set $X(A')$.
- Let f be any element of $H_A(A')$. This f is, at the same time, a morphism $f : A' \to A$. Therefore, contravariant functor H_A sends f to

$$H_A(f) : H_A(A) \to H_A(A'),$$

  which is a mapping between sets.
- Contravariant functor X sends morphism f to $X(f)$. By naturality of α, the two paths in Fig. 7.9 originating from 1_A must be identical as morphisms.
- To sum up, $\alpha_A(1_A)$ determines all the components of α. This is the resonance we are talking about.

We have just observed that there is a bijection between the set $X(A)$ and the set of all natural transformations from the functor H_A to X, namely, $[\mathscr{A}^{\mathrm{op}}, \mathbf{Set}](H_A, X)$.We

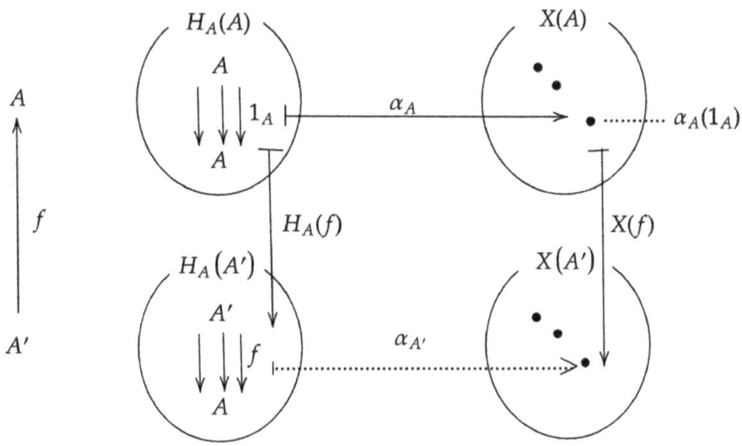

Fig. 7.9 Contravariant functors and the resonance of a natural transformation

still want to know how the bijection depends on A and X. To this end we need lots of preparation.

Our goal is given below.

Theorem 7.4 (The Yoneda Lemma (contravariant case)) *Let $\mathscr{A}$ be a small category. Then, the following isomorphism*

$$[\mathscr{A}^{\mathrm{op}}, \mathbf{Set}](H_A, X) \simeq X(A)$$

is natural in A and X.

We need preparations to clearly state the naturality in A and X. We expect the LHS of the equation in Theorem 7.4 to be a functor in two variables A and X. Let us define the function on objects as follows:

$$\begin{aligned} \mathscr{A}^{\mathrm{op}} \times [\mathscr{A}^{\mathrm{op}}, \mathbf{Set}] &\to \mathbf{Set} \\ (A, X) &\mapsto [\mathscr{A}^{\mathrm{op}}, \mathbf{Set}](H_A, X). \end{aligned}$$

To be precise, this correspondence is the composition of two operations.

$$\begin{array}{ccccc} \mathscr{A}^{\mathrm{op}} \times [\mathscr{A}^{\mathrm{op}}, \mathbf{Set}] & \xrightarrow{H_-^{\mathrm{op}} \times 1} & [\mathscr{A}^{\mathrm{op}}, \mathbf{Set}]^{\mathrm{op}} \times [\mathscr{A}^{\mathrm{op}}, \mathbf{Set}] & \xrightarrow{[\mathscr{A}^{\mathrm{op}}, \mathbf{Set}](-,-)} & \mathbf{Set} \\ (A, X) & \longmapsto & (H_A, X) & \longmapsto & [\mathscr{A}^{\mathrm{op}}, \mathbf{Set}](H_A, X) \end{array}.$$

Let us take a closer look at this composition. In the first stage, $A \mapsto H_A$ is given by the covariant functor $H_-^{\mathrm{op}} : \mathscr{A}^{\mathrm{op}} \to [\mathscr{A}^{\mathrm{op}}, \mathbf{Set}]^{\mathrm{op}}$, which can be simply rewritten as $H_- : \mathscr{A} \to [\mathscr{A}, \mathbf{Set}]$ as far as only objects are concerned. However, in the second stage, the Hom functor $[\mathscr{A}^{\mathrm{op}}, \mathbf{Set}](-,-)$ expects its first argument H_A contravariantly.

Therefore, in the first stage we write $\mathscr{A}^{\mathrm{op}} \to [\mathscr{A}^{\mathrm{op}}, \mathbf{Set}]^{\mathrm{op}}$ for the left component. This way, the composition observes the covariant-contravariant restriction.

Let us consider the functor $\mathscr{A}^{\mathrm{op}} \times [\mathscr{A}^{\mathrm{op}}, \mathbf{Set}] \to \mathbf{Set} : (A, X) \mapsto [\mathscr{A}^{\mathrm{op}}, \mathbf{Set}](H_A, X)$ with its second argument X fixed. We observe how its image depends on A. The image of $f : A' \to A$ is found to be $- \circ H_f$ in the following diagram:

$$\begin{array}{ccccccc}
\mathscr{A}^{\mathrm{op}} & \xrightarrow{H_-^{\mathrm{op}}} & [\mathscr{A}^{\mathrm{op}}, \mathbf{Set}]^{\mathrm{op}} & \longrightarrow & [\mathscr{A}^{\mathrm{op}}, \mathbf{Set}](H_-, X) & & \\
A & \longmapsto & H_A & \longmapsto & [\mathscr{A}^{\mathrm{op}}, \mathbf{Set}](H_A, X) & \ni & \alpha \\
{\scriptstyle f}\uparrow & & \uparrow{\scriptstyle H_f} & & \downarrow{\scriptstyle -\circ H_f} & & \downarrow \\
A' & \longmapsto & H_{A'} & \longmapsto & [\mathscr{A}^{\mathrm{op}}, \mathbf{Set}](H_{A'}, X) & \ni & \alpha \circ H_f
\end{array}$$

Next, we fix A, the left argument of the functor $\mathscr{A}^{\mathrm{op}} \times [\mathscr{A}^{\mathrm{op}}, \mathbf{Set}] \to \mathbf{Set} : (A, X) \mapsto [\mathscr{A}^{\mathrm{op}}, \mathbf{Set}](H_A, X)$, and observe the dependency on X. Let us consider a morphism $\theta : X \to X'$ in $[\mathscr{A}, \mathbf{Set}]$

$$\mathscr{A}^{\mathrm{op}} \underset{X'}{\overset{X}{\rightrightarrows}} \mathbf{Set} \qquad \Downarrow \theta$$

which can be regarded as a natural transformation:

$$\begin{array}{cccccc}
X & \qquad & [\mathscr{A}^{\mathrm{op}}, \mathbf{Set}](H_A, X) & \ni & \alpha \\
{\scriptstyle \theta}\downarrow & & \downarrow{\scriptstyle \theta\circ -} & & \downarrow \\
X' & & [\mathscr{A}^{\mathrm{op}}, \mathbf{Set}](H_A, X') & \ni & \theta \circ \alpha
\end{array}$$

namely, $\theta \mapsto (\theta \circ -)$.

We have to check if the functor really preserves the composition of morphisms:

$$(\theta' \circ \theta) \circ - = (\theta' \circ -) \circ (\theta \circ -)$$

which can be easily seen by the following diagram:

$$\begin{array}{ccccc}
X & [\mathscr{A}^{\mathrm{op}}, \mathbf{Set}](H_A, X) & & \alpha \\
{\scriptstyle \theta}\downarrow & \downarrow{\scriptstyle \theta\circ -} & & \downarrow \\
X' \;\; {\scriptstyle (\theta'\circ\theta)} & [\mathscr{A}^{\mathrm{op}}, \mathbf{Set}](H_A, X') & {\scriptstyle (\theta'\circ\theta)\circ -} & \theta \circ \alpha \\
{\scriptstyle \theta'}\downarrow & \downarrow{\scriptstyle \theta'\circ -} & & \downarrow \\
X'' & [\mathscr{A}^{\mathrm{op}}, \mathbf{Set}](H_A, X'') & & \theta' \circ \theta \circ \alpha
\end{array}$$

Next, let us define the right-hand side of the equation in Theorem 7.4 as a functor in two variables A and X.

$$\begin{array}{ccc} \mathscr{A}^{\mathrm{op}} \times [\mathscr{A}^{\mathrm{op}}, \mathbf{Set}] & \longrightarrow & \mathbf{Set} \\ (A, X) & \longmapsto & X(A) \end{array} .$$

This is called an *evaluation functor* which has been introduced already. We fix X as a constant. For any morphism $f : A' \to A$ we have

$$\begin{array}{ccc} A & \longmapsto & X(A) \\ {\scriptstyle f}\uparrow & & \downarrow{\scriptstyle X(f)} \\ A' & \longmapsto & X(A') \end{array} ,$$

which is a consequence of X being a functor $\mathscr{A} \to \mathbf{Set}$.

Fixing A, we want to define a functor $[\mathscr{A}^{\mathrm{op}}, \mathbf{Set}]$ with a function on objects $X \mapsto X(A)$ and an appropriate function on morphisms as follows. For any morphism θ in $[\mathscr{A}^{\mathrm{op}}, \mathbf{Set}]$

$$\mathscr{A}^{\mathrm{op}} \underset{X'}{\overset{X}{\rightrightarrows}} \mathbf{Set} \qquad (\Downarrow \theta)$$

which is actually a natural transformation between functors X and X', we can simply take $\theta \mapsto \theta_A$. Composition of morphisms $\theta : X \to X'$ and $\theta' : X' \to X''$ is given by

$$\begin{array}{ccccc} & X & \longmapsto & X(A) & \\ & {\scriptstyle \theta}\downarrow & & \downarrow{\scriptstyle \theta_A} & \\ \theta' \circ \theta & X' & \longmapsto & X'(A) & (\theta' \circ \theta)_A \\ & {\scriptstyle \theta'}\downarrow & & \downarrow{\scriptstyle \theta'_A} & \\ & X'' & \longmapsto & X''(A), & \end{array}$$

where we have commutative triangles at the left and right sides. This comes from the vertical composition of natural transformations θ and θ'. Combining the discussion before, we now have

$$\mathscr{A}^{\mathrm{op}} \times [\mathscr{A}^{\mathrm{op}}, \mathbf{Set}] \to \mathbf{Set} : (A, X) \mapsto X(A)$$

is a functor of two variables.

Proof (The Yoneda Lemma—Cotravariant case—Naturality in A)
The preparation above lets us proceed to prove that the following correspondence:

$$y : [\mathscr{A}^{\mathrm{op}}, \mathbf{Set}](H_{-}, X) \to X(-),$$

whose component at A is given by

$$y_A = [\mathscr{A}^{\mathrm{op}}, \mathbf{Set}](H_A, X) \to X(A) : \alpha \mapsto \alpha_A(1_A)$$

can be made into a natural transformation. For each morphism $f : A \to A'$, we have to show that the two paths from α to $X(A')$ in Fig. 7.10 are identical. In other words, we have to show the following equation holds:

$$(\alpha \circ H_f)_{A'}(1_{A'}) = (Xf)(\alpha_A(1_A)). \tag{7.1}$$

We transform the LHS step by step as follows:

$$\begin{aligned} &(\alpha \circ H_f)_{A'}(1_{A'}) \\ &= \alpha_{A'}((H_f)_{A'}(1_{A'})) \\ &= \alpha_{A'}(f \circ 1_{A'}) \\ &= \alpha_{A'}(f). \end{aligned}$$

The last expression above is equal to the RHS in Eq. (7.1) since α is a natural transformation that makes the following square commute:

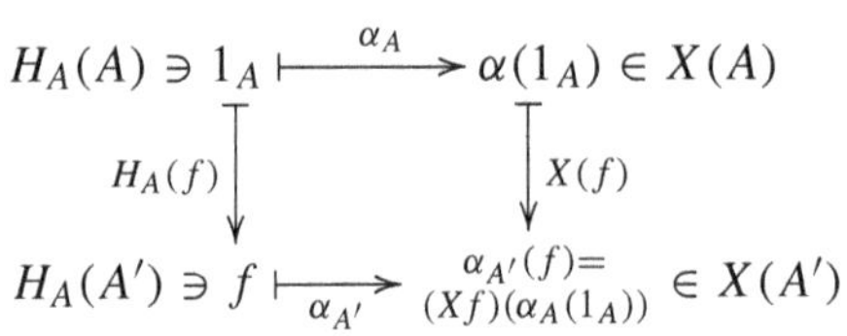

which lives in Fig. 7.9. This completes the proof. □

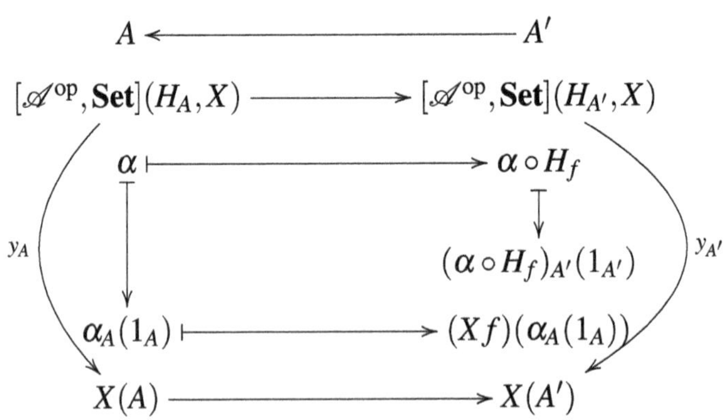

Fig. 7.10 Naturality in A (contravariant case)

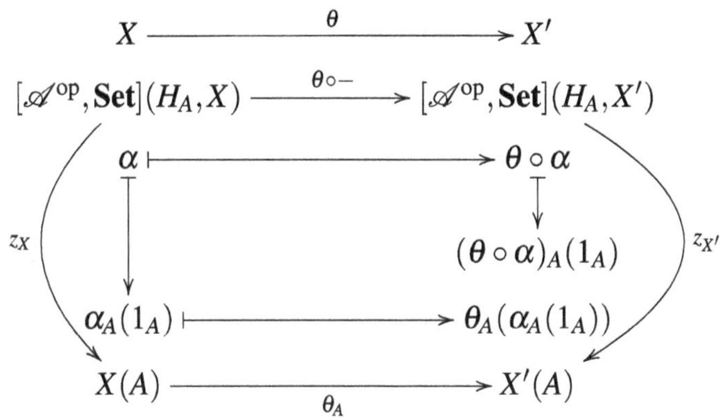

Fig. 7.11 Naturality in X (contravariant case)

Proof (The Yoneda Lemma—Contravariant version—Naturality in X)
We fix A and show that the following correspondence

$$z : [\mathscr{A}^{\mathrm{op}}, \mathbf{Set}](H_A, -) \to (-)(A),$$

whose component at X is given by

$$z_X = [\mathscr{A}^{\mathrm{op}}, \mathbf{Set}](H_A, X) \to X(A) : \alpha \mapsto \alpha_A(1_A)$$

can be made into a natural transformation.

Let θ be a morphism in $[\mathscr{A}^{\mathrm{op}}, \mathbf{Set}]$.

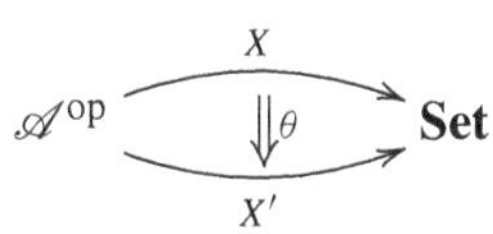

Only we have to show is that the two paths in Fig. 7.11 starting from α coincide, which is clear from the definition of function composition. □

7.4 Milewski's "Understanding Yoneda"

In this and the next section, we will see that it is often the case that when creating a polymorphic function in a functional programming language, if a signature is given, the implementation may be almost determined by Yoneda's lemma.

7.4.1 Specification Determines Implementation

We will see an example of natural transformation `imager` and a polymorphic function `iffie` in the article of Milewski [1] (partially omitted and modified). The author himself calls it a teaser problem. Even if you don't understand it at first glance, you are requested to be patient.

Listing 7.1 yoneda.hs

```
1  {-# LANGUAGE ExplicitForAll #-}
2  imager :: forall r . ((Bool -> r) -> [r])
3  imager iffie = fmap iffie [True, False, True, True]
4
5  data Color = Red | Green | Blue        deriving Show
6  data Note  = C | D | E | F | G | A | B deriving Show
7
8  colorMap x = if x then Blue else Red
9  heatMap  x = if x then 32   else 212
10 soundMap x = if x then C    else G
11
12 idBool :: Bool -> Bool
13 idBool x = x
14
15 {- suggested tests
16 imager colorMap
17 imager heatMap
18 imager soundMap
19 -}
```

Try running this program to see what happens.

```
*Main> imager colorMap
[Blue,Red,Blue,Blue]
*Main> imager heatMap
[32,212,32,32]
*Main> imager soundMap
[C,G,C,C]
```

The signature of `imager` in Line 2 says that it is a natural transformation from the Hom functor H^{Bool} to List functor `[ ]`. According to Yoneda's lemma, this natural transformation has a one-to-one correspondence with an element of the "set" $\text{List Bool} = [\text{Bool}]$. Figure 7.12 is a redrawn version of the diagram used in the general theory of the Yoneda lemma to suit this example. According to Yoneda's lemma, we have

$$[\text{Hask}, \mathbf{Set}](H^{\text{Bool}}, \text{List}) \cong \text{List}(\text{Bool}).$$

In this formula, **Set** is written in, but informal alternative Hask might be replaced with. As mentioned above, it is not mathematically rigorous. Nevertheless, there is no practical problem, so let's continue the discussion. In Fig. 7.12, `iffie` is a representative in general of `colorMap`, `heatMap`, and `soundMap`. These are the

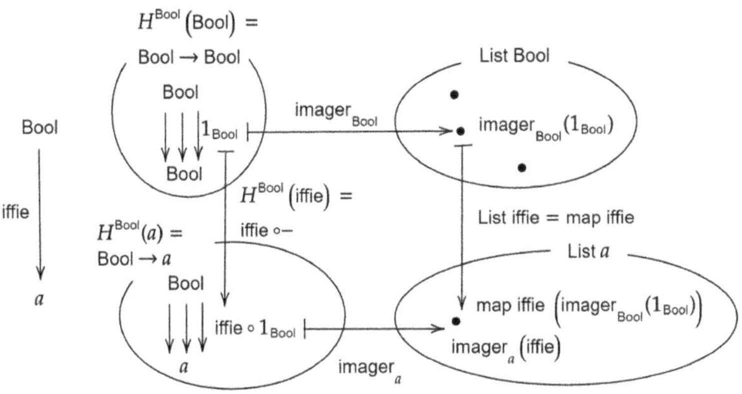

Fig. 7.12 Natural transformation `imager`

arbitrarily chosen morphisms Bool $\rightarrow a$. For example, when a is taken as `Color`, there are $2^3 = 8$ possible morphisms (ignoring `undefined` is one of the reasons for the lack of rigorousness mentioned above). Here are some examples.

```
colorMap1 _ = Red
colorMap2 x = if x then Red else Blue
colorMap3 x = if x then Red else Green
...
```

If we want to know the element of List Bool that corresponds the natural transformation `imager`, what should we do? The Yoneda lemma says that it is enough to find the image of 1_{Bool} by $\text{imager}_{\text{Bool}}$. We can do it as follows:

```
*Main> imager idBool
[True,False,True,True]
```

The obtained list of `Bool` is an internal information about the implementation of `imager`. This example gives us a lesson that an implementation details can be disclosed by the Yoneda lemma.

The information on the component $\text{imager}_{\text{Bool}}$ can be automatically inferred by the Haskell-type inference functionality. So, it needn't specify the component explicitly.

7.4.2 A Function That Has the Same Signature as the Identity Function

Let us consider to what extent we can limit the possible implementation of a function that has the signature `forall a . ((() -> a) -> a)`. The *unit type* is introduced as `()`, which has the only element (value) `()`. Since Haskell's name spaces for types and values are separated, we can use the same identifier for a type and value. To avoid confusion, we use $\star$ for a value. A function that has this signature can be regarded

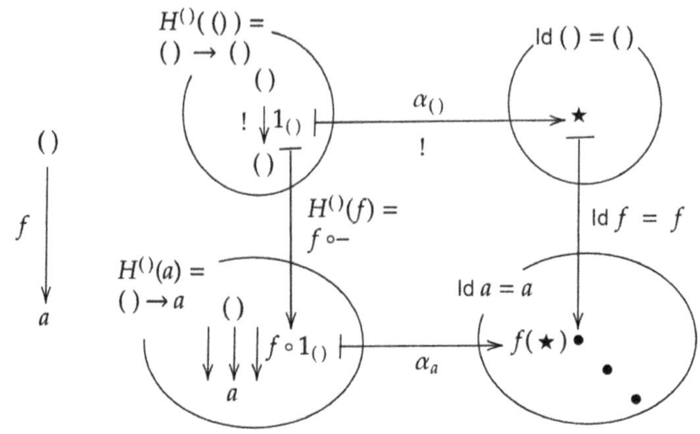

Fig. 7.13 Natural transformation α

as a natural transformation $\alpha : H^{()} \to \mathrm{Id}$. The type `()->()` has only one element $1_{()}$. Thus, α is a natural isomorphism (Fig. 7.13).

The Yoneda lemma says that such a natural transformation α corresponds to an element of $\mathrm{Id}() = ()$, which is the only one possible. Since `()->a` and `a` are isomorphic, there is exactly one function that has signature `forall a . a -> a`. We conclude that there is exactly one Haskell polymorphic function with signature `id::a->a`.

7.5 Reverse Engineering by the Yoneda Lemma

We will see examples that have origins in an excellent article [2].

7.5.1 A Machine That Has a Hidden Parameter

Listing 7.2 unbox1.hs

```
{-# LANGUAGE RankNTypes #-}

factory1 :: a -> (forall b . (a -> b) -> b)
factory1 a f = f a

unbox1 :: (forall b . (a -> b) -> b) -> a
unbox1 t = t id

-- testdata
machine1 = factory1 10

{- suggested tests
```

```
13 machine1 (\x -> x*x)
14 unbox1 machine1
15 -}
```

In Line 3, the signature of `factory1` can be interpreted as follows: Type `(a -> b)` is a functor H^a with a being fixed. The rightmost `b` in `(a -> b) -> b` is an image of `b` by the identity functor Id. We can explain roughly that type `(forall b . (a -> b) -> b)` corresponds to the natural transformation $H^a \to \mathrm{Id}$. To sum up, given a value of type `a`, we get a natural transformation parameterized by `a`. It is kind of a factory of natural transformation.

The function `unbox1` in lines 6–7 reveals the secrecy of implementation. Using the Yoneda lemma

$$\alpha : [\mathrm{Hask}, \mathrm{Hask}]((a \to), \mathrm{Id}) \simeq \mathrm{Id}\, a,$$

it seeks the image $\alpha_a(1_a)$ of $1_a \in H^a(a)$.

Line 10 gives a value 10 to `factory1`.

A test can be carried out as follows:

```
*Main> machine1 (\x -> x*x)
100
*Main> unbox1 machine1
10
```

Figure 7.14 explains the naturality.

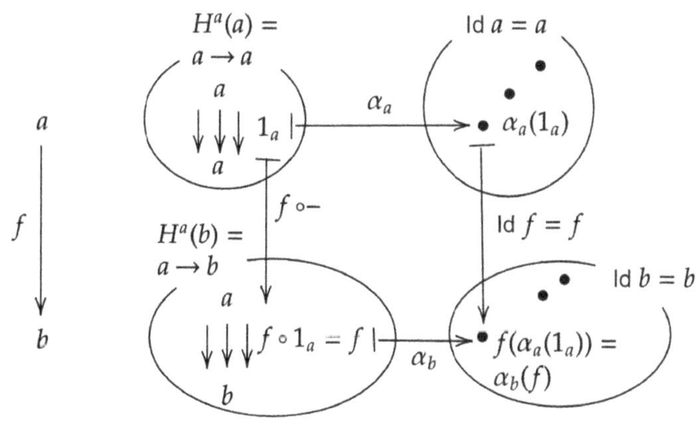

Fig. 7.14 Machine1

7.5.2 A Machine That Hides a List Inside

We consider a machine that hides implementation detail which is actually a list.

Listing 7.3 unbox2.hs

```
1  {-# LANGUAGE RankNTypes #-}
2
3  factory2 :: [a] -> (forall b . (a -> b) -> [b])
4  factory2 a f = map f a
5
6  unbox2 :: (forall b . (a -> b) -> [b]) -> [a]
7  unbox2 t = t id
8
9  -- test data
10 machine2 = factory2 [1,2,3,4]
11
12 {- tests
13 machine2 (\x -> x*x)
14 unbox2 machine2
15 -}
16
17 --------------------
18 factory2' :: [a] -> (forall b . (a -> b) -> [b])
19 factory2' a f = reverse $ map f a
20
21 --------------------
22 rotateL :: forall a . [a] -> [a]
23 rotateL []     = []
24 rotateL (x:xs) = xs ++ [x]
25
26 factory2'' :: [a] -> (forall b . (a -> b) -> [b])
27 factory2'' a f = rotateL $ map f a
```

Let us give an interpretation to Line 3. We interpret the signature of `factory2` in Line 3 from a category theoretical point of view. Let us focus on the subexpression `forall b . (a->b) -> [b]`, where `a` is free. We may regard free variables in a `forall` expression as a fixed object. Therefore, `(a -> b)` is a functor H^a with a being fixed, and `[b]` is an image of the List functor. So, `forall b . (a->b) -> [b]` is a natural transformation $H^a \to [\ ]$. Figure 7.15 depicts this situation. Haskell function `factory2` is a factory that produces a natural transformation, given a value of type `[a]`.

Function `unbox2` reverse-engineers the `machine2`. It simply uses the Yoneda lemma.

$$[\,\mathrm{Hask}, \mathrm{Hask}]((a \to), \mathrm{List}] \simeq \mathrm{List}\ a$$

See Fig. 7.15.

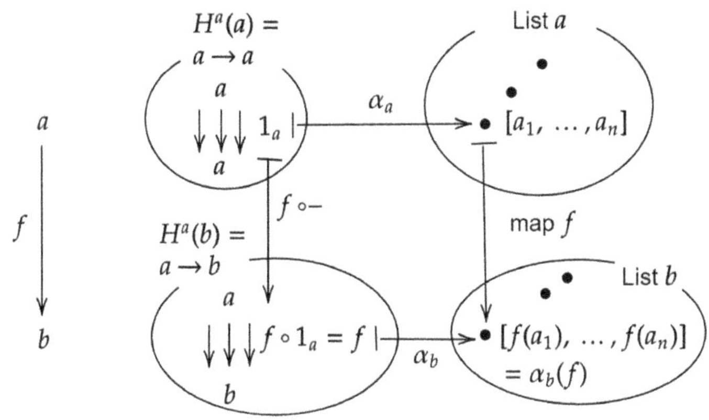

Fig. 7.15 Machine2

Suggested tests are as follows:

```
*Main> machine2 (\x -> x*x)
[1,4,9,16]
*Main> unbox2 machine2
[1,2,3,4]
```

Function `factory2'` in lines 18–19 has the same signature with `factory2`. So, it is also a factory of natural transformations. See how it behaves.

```
*Main> factory2' [1,2,3,4] (\x -> x*x)
[16,9,4,1]
*Main> unbox2 $ factory2' [1,2,3,4]
[4,3,2,1]
```

The result of reverse engineering is slightly different. Remember this. We add another factory `factory2''` in lines 26–27 of natural transformations.

```
*Main> factory2'' [1,2,3,4] (\x -> x*x)
[4,9,16,1]
*Main> unbox2 $ factory2'' [1,2,3,4]
[2,3,4,1]
```

The observed phenomena of getting different results from reverse engineering are all related to the signature `forall a . [a] -> [a]`. Functions of this signature use `reverse` and `rotateL` internally, which are both natural transformations

List → List. These internal natural transformations are used in the horizontal composition. As an example, we depict the case of `reverse` as follows:

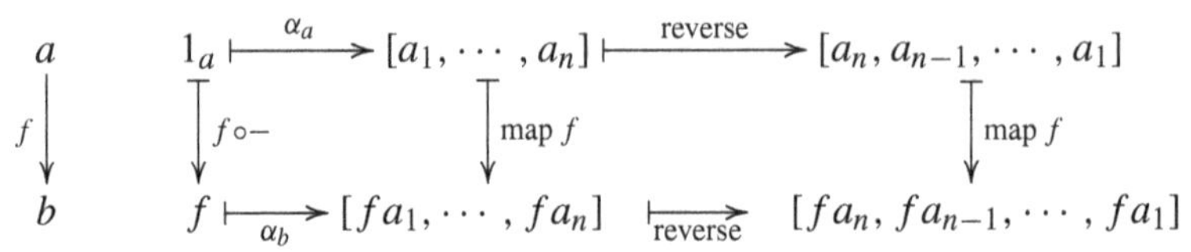

The two squares are commutative. Therefore, the whole rectangle is also commutative. The commutativity of the square at right can be written in pseudocode as follows:

```
reverse . map f == map f . reverse
```

We have the following:

$$[\text{Hask}, \text{Hask}]((a \rightarrow), \text{reverse} \circ \text{List}] \simeq (\text{reverse} \circ \text{List})\ (a)$$

This isomorphism tells us that we had to apply the Yoneda lemma with respect to a natural transformation from the functor $(a \rightarrow)$ to functor reverse ∘ List. Writing `unbox2'` is a trivial task.

The same is true for all functions with signature `forall a . [a] -> [a]`.

7.5.3 A Machine That Has a Secret Morphism

Listing 7.4 unbox3.hs

```
1  {-# LANGUAGE RankNTypes #-}
2
3  factory3 :: (c -> a) -> (forall b . (a -> b) -> (c -> b))
4  factory3 a f = f . a
5
6  unbox3 :: (forall b . (a -> b) -> (c -> b)) -> (c -> a)
7  unbox3 t = t id
8
9  -- test data
10 machine3 = factory3 (\x -> x + 1)
11
12 {- tests
13 map (machine3 (\x -> x*x)) [1..10]
14
15 map (unbox3 machine3) [1..10]
16 -}
```

Function `factory3` in lines 3–4 is a factory of machines. Sources are morphisms. Products (machines) are natural transformations with signature `(forall b . (a -> b) -> (c -> b))`, so to say, $\alpha : (a \rightarrow) \rightarrow (c \rightarrow)$. See Fig. 7.16. Function `unbox3` in lines 6–7 reverse-engineers the machine. The machine in line 10 is a test machine.

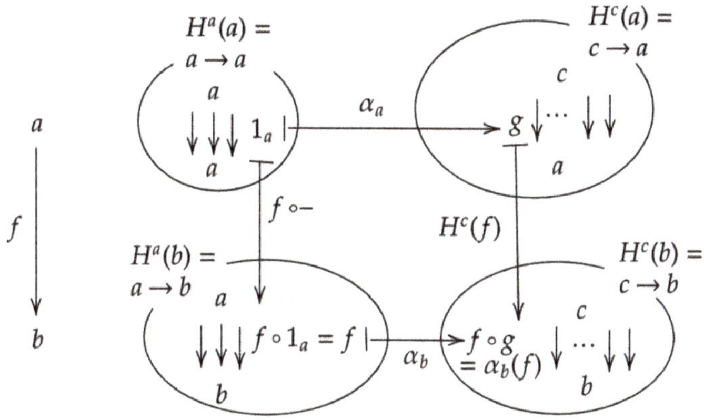

Fig. 7.16 Machine3

Tests can be carried out as follows:

```
*Main> map (machine3 (\x -> x*x)) [1..10]
[4,9,16,25,36,49,64,81,100,121]
*Main> map (unbox3 machine3) [1..10]
[2,3,4,5,6,7,8,9,10,11]
```

We can infer that the machine is equipped with a function that adds one to the argument.

7.5.4 The Design Pattern with Yoneda

We can abstract a design pattern for reverse engineering machines.

Listing 7.5 unbox.hs

```
{-# LANGUAGE RankNTypes #-}

data I a = I a deriving Show
instance Functor I where
  fmap f (I a) = I (f a)

{- Defined in GHC.Base
instance Functor ((-> a) where
  fmap f = (.) f
-}

factory :: Functor f => f a -> (forall b . (a -> b) -> f b)
factory a f = fmap f a

unbox :: (forall b . (a -> b) -> f b) -> f a
unbox t = t id

-- test data
machine41 = factory (I 10)
```

```
20 machine42 = factory [1,2,3,4]
21 machine43 = factory (\x -> x + 1)
22
23 {- tests
24 map machine41 [(\x -> x*2), (\x -> x*x), (\x -> x*x*x)]
25 unbox machine41
26
27 machine42 (\x -> x*x)
28 unbox machine42
29
30 map (machine43 (\x -> x*x)) [1..10]
31 map (unbox machine43) [1..10]
32 -}
```

Identity functor `I` is introduced in lines 3–5. An instance of Functor `((->) a)` is shown in lines 8–9 as an excerpt from a basic library.

A frequently observed pattern is abstracted in lines 12–13, where `factory` is a factory of natural transformations. The corresponding reverse engineering is defined in lines 15–16.

Lines 19–21 provide with Yoneda machines to test the identity, List, Hom functors. Tests can be carried out as follows:

```
*Main> map machine41 [(\x -> x*2), (\x -> x*x), (\x -> x*x*x)]
[I 20,I 100,I 1000]
*Main> unbox machine41
I 10
*Main> machine42 (\x -> x*x)
[1,4,9,16]
*Main> unbox machine42
[1,2,3,4]
*Main> map (machine43 (\x -> x*x)) [1..10]
[4,9,16,25,36,49,64,81,100,121]
*Main> map (unbox machine43) [1..10]
[2,3,4,5,6,7,8,9,10,11]
```

7.6 Adjoints Preserve Limits

Let $\mathscr{A}$ be a small category. If we specialize $X : \mathscr{A} \to \mathbf{Set}$ to $H^{A'}$ in the Yoneda Lemma 7.3, we get the following:

$$[\mathscr{A}, \mathbf{Set}](H^A, H^{A'}) \simeq \mathscr{A}(A', A).$$

The Yoneda Lemma 7.4 for contravariant functors gives, by specializing $X : \mathscr{A}^{\mathrm{op}} \to \mathbf{Set}$ to $H_{A'}$, the following equivalence:

$$[\mathscr{A}^{\mathrm{op}}, \mathbf{Set}](H_A, H_{A'}) \simeq \mathscr{A}(A, A')$$

These observations lead us to the following theorem.

Theorem 7.5 *For any small category $\mathscr{A}$, we have the following equivalences:*

$$H^A \simeq H^{A'} \Leftrightarrow A \simeq A'$$
$$H_A \simeq H_{A'} \Leftrightarrow A \simeq A'.$$

We show that adjoints preserve limits.

Theorem 7.6 *Let functors $F : \mathscr{X} \to \mathscr{A}$ and $U : \mathscr{A} \to \mathscr{X}$ be adjoints; $F \dashv U$. Namely,*

$$\mathscr{A}(FX, A) \simeq \mathscr{X}(X, UA).$$

Then, U preserves limits, and F preserves colimits.

Proof We begin with the right adjoint U. Let J be an index category, $D : J \to \mathscr{A}$ be a J-shaped diagram.

For any object X we have the following isomorphisms:

$$\begin{aligned}
&\mathscr{X}(X, U(\varprojlim D)) && \\
&\simeq \mathscr{A}(FX, \varprojlim D) && \text{adjointness} \\
&\simeq \varprojlim \mathscr{A}(FX, D) && \text{exchage of Hom and limits} \\
&\simeq \varprojlim \mathscr{X}(X, U \circ D) && \text{adjointness} \\
&\simeq \mathrm{Cones}(X, U \circ D) && \text{limiting cones} \\
&\simeq \mathscr{X}(X, \varprojlim(U \circ D)) && \text{limiting cones}
\end{aligned}$$

Then, we have

$$\mathscr{X}(-, U(\varprojlim D)) \simeq \mathscr{X}(-, \varprojlim(U \circ D)).$$

Theorem 7.5 guarantees the following:

$$U(\varprojlim D) \simeq \varprojlim(U \circ D).$$

It says that right adjoint functors preserve limits.

Next, we study the left adjoint functor F.

For any object A, we have the following isomorphisms:

$$\begin{aligned}
&\mathscr{A}(F(\varinjlim D), A) && \\
&\simeq \mathscr{X}(\varinjlim D, UA) && \text{adjointness} \\
&\simeq \varprojlim \mathscr{X}(D, UA) && \text{exchange of Hom and colimits} \\
&\simeq \varprojlim \mathscr{A}(F \circ D, A) && \text{adjointness} \\
&\simeq \mathrm{Cones}(F \circ D, A) && \text{limiting cocones} \\
&\simeq \mathscr{A}(\varinjlim(F \circ D), A) && \text{limiting cocones}
\end{aligned}$$

Then, we have

$$\mathscr{A}(F(\varinjlim D), -) \simeq \mathscr{A}(\varinjlim(F \circ D), -).$$

Theorem 7.5 guarantees the following:

$$F(\varinjlim D) \simeq \varinjlim(F \circ D)$$

It says that left adjoint functors preserve colimits. □

References

1. Milewski B, Understanding yoneda. https://www.schoolofhaskell.com/user/bartosz/understanding-yoneda
2. Piponi D (2006) Reverse engineering machine with the yoneda lemma. Weblog entry, A Neighborhood of Infinity. http://blog.sigfpe.com/2006/11/yoneda-lemma.html

Monoidal Categories and Coherence 8

Monoidal categories allow a binary operation, often called a *tensor product*, among objects. Monoidal categories have been used in many areas of mathematics. Today, monoidal categories and their impressive pictorial representations are extensively used in research and application areas such as quantum computing [1], topological field theory [2], natural language processing [3], and other emerging areas. Many introductory articles and textbooks on monoidal categories are available today. So, learners have greater choice.

However, most textbooks do not spend enough lines and pages to concretely illustrate naturality every time it matters. The goal of this chapter is twofolds. The first is to introduce the reader to coherence problems with special emphasis on basic handling of functors of bracketed products and naturality of associators and unitors. The second is to provide the reader with lots of illustrating materials to digest basic concepts of functors and natural transformations through coherence problems.

8.1 Categories with Tensor Products

We first see a formal definition and then walk through concrete examples. So, please be patient for a moment.

Definition 8.1 A category $\mathscr{C}$ is called a *monoidal category* if it is equipped with

- a functor $\otimes : \mathscr{C} \times \mathscr{C} \to \mathscr{C}$ called the *tensor product* or the *monoidal product*,
- an object $\mathbb{1} \in \operatorname{Obj} \mathscr{C}$ called the *unit object*,
- a family of isomorphisms called the *associator*

$$\alpha = \left\{ \alpha_{X,Y,Z} : X \otimes (Y \otimes Z) \to (X \otimes Y) \otimes Z \;\middle|\; X, Y, Z \in \operatorname{Obj}(\mathscr{C}) \right\} ,$$

S. Yukita, *Category Theory Using Haskell*, Computer Science Foundations and Applied Logic, https://doi.org/10.1007/978-3-031-68538-5_8

- a family of isomorphisms called the *left unitor*

$$\lambda = \{\lambda_X : \mathbb{1} \otimes X \to X \mid X \in \mathrm{Obj}(\mathscr{C})\},$$

and
- a family of isomorphisms called the *right unitor*

$$\rho = \{\rho_X : X \otimes \mathbb{1} \to X \mid X \in \mathrm{Obj}(\mathscr{C})\} ,$$

satisfying the following conditions:

(i) For all $X, Y, Z, W \in \mathrm{Obj}(\mathscr{C})$, the following *pentagon diagram* commutes:

$$\begin{array}{ccccc} & & (X \otimes Y) \otimes (Z \otimes W) & & \\ & \overset{\alpha_{X,Y,Z\otimes W}}{\nearrow} & & \overset{\alpha_{X\otimes Y,Z,W}}{\searrow} & \\ X \otimes (Y \otimes (Z \otimes W)) & & & & ((X \otimes Y) \otimes Z) \otimes W \\ \downarrow{\scriptstyle 1_X \otimes \alpha_{Y,Z,W}} & & & & \uparrow{\scriptstyle \alpha_{X,Y,Z} \otimes 1_W} \\ X \otimes ((Y \otimes Z) \otimes W) & & \xrightarrow[\alpha_{X,Y\otimes Z,W}]{} & & (X \otimes (Y \otimes Z)) \otimes W \end{array} \tag{8.1}$$

This condition is called the *pentagon coherence*.

(ii) For all $X, Y \in \mathrm{Obj}(\mathscr{C})$, the following triangle diagram commutes:

$$\begin{array}{ccc} X \otimes (\mathbb{1} \otimes Y) & \xrightarrow{\alpha_{X,\mathbb{1},Y}} & (X \otimes \mathbb{1}) \otimes Y \\ & {\scriptstyle 1_X \otimes \lambda_Y} \searrow \quad \swarrow {\scriptstyle \rho_X \otimes 1_Y} & \\ & X \otimes Y & \end{array} \tag{8.2}$$

This condition is called the *triangle coherence*.

(iii) For all triplets of morphisms $f : X \to X'$, $g : Y \to Y'$, and $h : Z \to Z' \in \mathrm{Mor}(\mathscr{C})$, the following diagram commutes:

$$\begin{array}{ccc} X \otimes (Y \otimes Z) & \xrightarrow[f\otimes(g\otimes h)]{} & X' \otimes (Y' \otimes Z') \\ \downarrow{\scriptstyle \alpha_{X,Y,Z}} & & \downarrow{\scriptstyle \alpha_{X',Y',Z'}} \\ (X \otimes Y) \otimes Z & \xrightarrow{(f\otimes g)\otimes h} & (X' \otimes Y') \otimes Z' \end{array}$$

(iv) For all morphisms $f : X \to X' \in \mathrm{Mor}(\mathscr{C})$, the following diagram commutes:

$$\begin{array}{ccc} \mathbb{1} \otimes X & \xrightarrow{1_{\mathbb{1}}\otimes f} & \mathbb{1} \otimes X' \\ \downarrow{\scriptstyle \lambda_X} & & \downarrow{\scriptstyle \lambda_{X'}} \\ X & \xrightarrow[f]{} & X' \end{array}$$

(v) For all morphisms $f : X \to X' \in \mathrm{Mor}(\mathscr{C})$, the following diagram commutes:

$$\begin{array}{ccc} X \otimes \mathbb{1} & \xrightarrow{f \otimes 1_{\mathbb{1}}} & X' \times \mathbb{1} \\ {\scriptstyle \rho_X}\downarrow & & \downarrow{\scriptstyle \rho_{X'}} \\ X & \xrightarrow[f]{} & X' \end{array}$$

A monoidal category is often presented as a six-tuple $(\mathscr{C}, \otimes, \mathbb{1}, \alpha, \lambda, \rho)$.

Relaxed versions can be defined and often used. However, we will not need those for our purpose.

The functor $\otimes : \mathscr{C} \times \mathscr{C} \to \mathscr{C}$ is often called a *bifunctor*. A morphism in $\mathscr{C} \times \mathscr{C}$ is given by a pair of morphisms $f : X \to Y$ and $g : X' \to Y'$. Morphism (f, g) in $\mathscr{C} \times \mathscr{C}$ is sent to a morphism $\otimes(f, g)$, which if often written as $f \otimes g$.

Let us see the above axioms can be realized in concrete examples.

Example 8.1 The category **Set** is a monoidal category. The tensor product in this case is the Cartesian product functor $\times$: **Set** $\times$ **Set** $\to$ **Set** whose function on objects is defined as

$$\times : (X, Y) \mapsto X \times Y \ ,$$

and whose function on morphisms is defined as

$$\times : (f, g) \mapsto f \times g \ .$$

Recall that for $f : X \to X'$ and $g : Y \to Y'$

$$\begin{aligned} f \times g : & X \times Y \to X' \times Y' \\ & (x, y) \mapsto (f(x), g(y)) \ . \end{aligned}$$

Any singleton set plays the role of the unit object. We pick one $\{*\}$. The associativity constraint is defined by

$$\begin{aligned} \alpha_{X,Y,Z} : & X \times (Y \times Z) \to (X \times Y) \times Z \\ & (x, (y, z)) \mapsto ((x, y), z) \end{aligned}$$

for all triplets of sets X, Y, and Z. Note that each $\alpha_{X,Y,Z}$ is an isomorphism, in **Set**, a bijection. The left unitor is defined by

$$\begin{aligned} \lambda_X : & \{*\} \times X \to X \\ & (*, x) \mapsto x \ . \end{aligned}$$

The right unitor is defined by

$$\begin{aligned} \rho_X :& X \times \{*\} \to X \\ &(x, *) \mapsto x \ . \end{aligned}$$

Let us check the five conditions described in Definition 8.1 one by one.

(i) We calculate pointwise or elementwise, which is possible since we are working with sets, the two paths from $X \otimes (Y \otimes (Z \otimes W))$ to $((X \otimes Y) \otimes Z) \otimes W$ in the pentagon. Let x, y, z, and w be any elements of X, Y, Z, and W, respectively. The lower path goes like this.

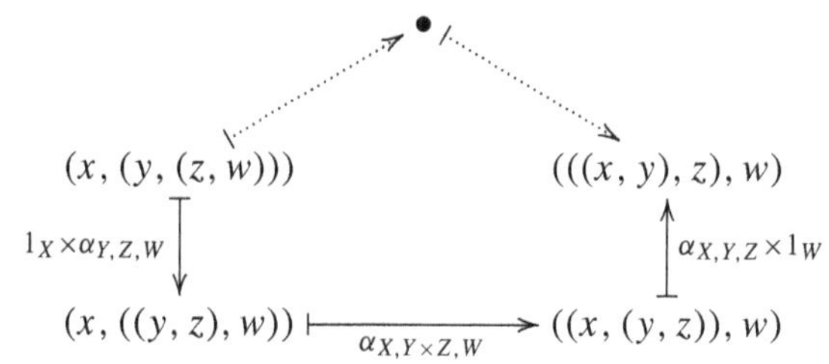

The upper path goes like this.

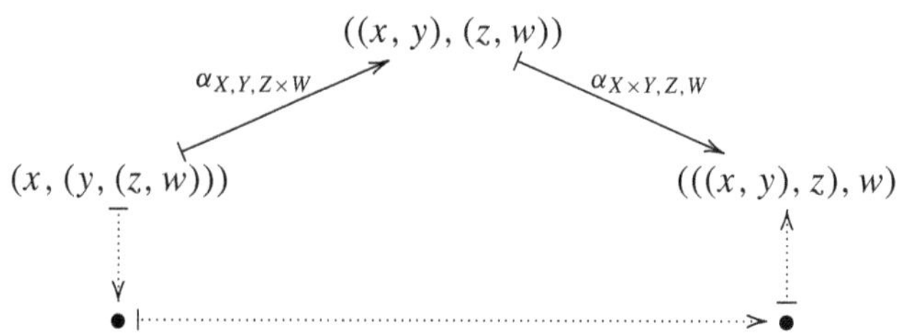

We see that the two paths yield the same result.

(ii) We calculate pointwise the two paths from $X \times (\{*\} \times Y)$ to $X \times Y$ in the triangle. Let x and y be any elements of X and Y, respectively. The upper path goes as follows:

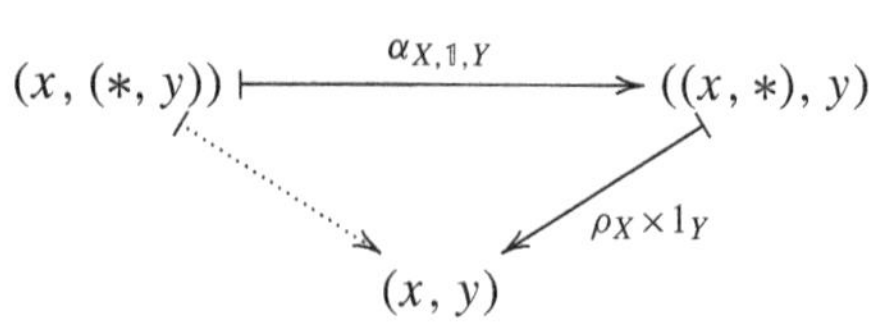

The lower one step path goes as follows:

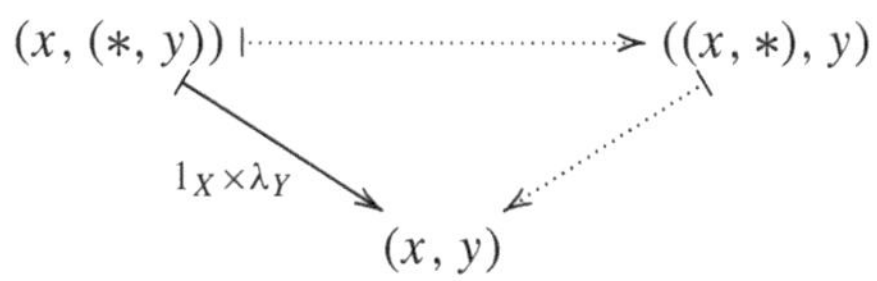

The two paths yield the same result, which proves the commutativity of the triangle.

(iii-v) Easy exercises.

Conditions (iii), (iv), and (v) in Definition 8.1 for the associator and unitors say that these collections of isomorphisms yield natural isomorphisms between some functors. It is worthwhile to clearly state this fact since naturality of an associator and unitors plays the central role in the discussion of coherence later.

We need the notion of a finite product of categories. Recall the concept of product categories introduced in Definition 1.5.

Definition 8.2 Let $\mathscr{C}_1, \mathscr{C}_2, \ldots, \mathscr{C}_n$ be categories. The product of these categories denoted by

$$\mathscr{C}_1 \times \mathscr{C}_2 \times \cdots \times \mathscr{C}_n$$

is a category with the following sets of objects and morphisms. The objects are all n-tuples

$$(X_1, X_2, \ldots, X_n) \text{ or often written as } X_1 \times X_2 \times \cdots X_n$$

for $X_i \in \mathrm{Obj}(\mathscr{C})$. The morphisms are all n-tuples

$$(f_1, f_2, \ldots, f_n) \text{ or often written as } f_1 \times f_2 \times \cdots f_n$$

for $f_i \in \mathrm{Mor}(\mathscr{C}_i)$. Composition of morphisms is performed componentwise. Componentwise functors can be introduced in an obvious way.

Let us begin with the associator α. Consider a functor

$$\begin{aligned} 1_{\mathscr{C}} \times \otimes : \mathscr{C} \times \underline{\mathscr{C} \times \mathscr{C}} &\to \mathscr{C} \times \mathscr{C} \\ (X, Y, Z) &\mapsto (X, Y \otimes Z) \end{aligned}$$

and its composition with $\otimes$

$$\begin{aligned} \otimes(1_{\mathscr{C}} \times \otimes) : \mathscr{C} \times \mathscr{C} \times \mathscr{C} &\to \mathscr{C} \\ (X, Y, Z) &\mapsto X \otimes (Y \otimes Z) \ . \end{aligned}$$

Consider another functor

$$\begin{aligned} \otimes \times 1_{\mathscr{C}} : \underline{\mathscr{C} \times \mathscr{C}} \times \mathscr{C} &\to \mathscr{C} \times \mathscr{C} \\ (X, Y, Z) &\mapsto (X \otimes Y, Z) \end{aligned}$$

and its composition with $\otimes$

$$\begin{aligned}\otimes(1_{\mathscr{C}} \times \otimes) :&\mathscr{C} \times \mathscr{C} \times \mathscr{C} \to \mathscr{C} \\ &(X, Y, Z) \mapsto (X \otimes Y) \otimes Z \ .\end{aligned}$$

Now, we can consider a natural transformation between the two functors

$$\begin{aligned}\alpha :&\otimes (1_{\mathscr{C}} \times \otimes) \to \otimes(\otimes \times 1_{\mathscr{C}}) \\ \alpha_{X,Y,Z} :&X \otimes (Y \otimes Z) \to (X \otimes Y) \otimes Z \ ,\end{aligned}$$

which is required to be an isomorphism by the axiom in Definition 8.1. We often use more intuitive notation, where functors are presented with place holders. For example, α can be written as

$$\alpha : - \otimes (- \otimes -) \to (- \otimes -) \otimes - \ .$$

The following diagram may help understand the situation. The trapezoid at right commutes for all triplets (f, g, h).

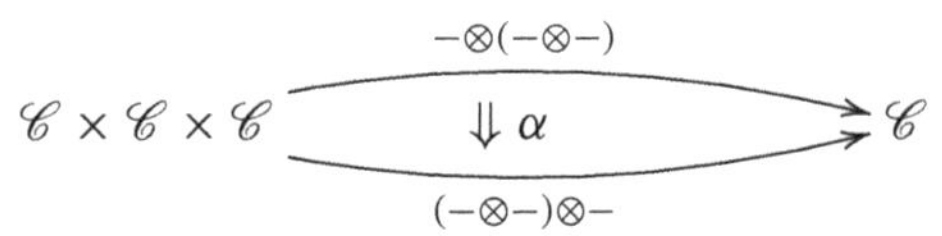

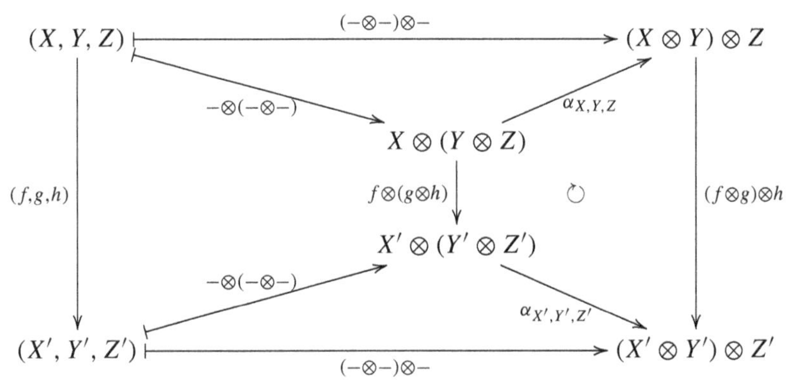

The natural transformation λ relates the functor $\mathbb{1} \otimes -$ and the identity functor $1_{\mathscr{C}}$ just denoted by $-$. See the following diagram. The trapezoid at right commutes for all $f \in \mathrm{Mor}(\mathscr{C})$.

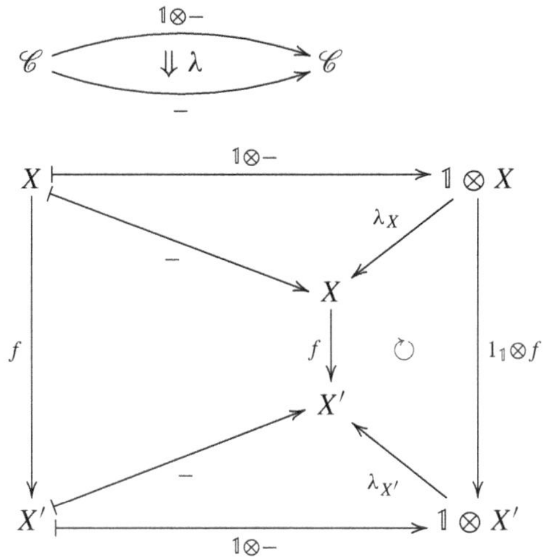

The natural transformation ρ relates the functor $- \otimes \mathbb{1}$ and the identity functor $-$. See the following diagram. The trapezoid at right commutes for all $f \in \mathrm{Mor}(\mathscr{C})$.

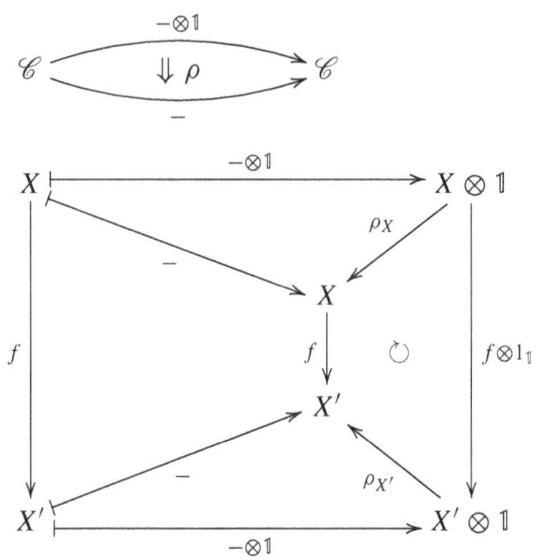

Example 8.2 The category **Set** in Example 8.1 can be treated without referring to elements. We will work within a category with finite products and a terminal object. So, the following argument applies to various categories other than **Set**.

We first show that $X_1 \times (X_2 \times X_3)$ is isomorphic to $X_1 \times X_2 \times X_3$. Let Z be any object with three morphisms as follows.

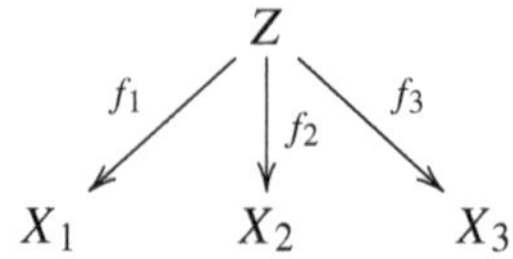

Then, we have the following diagram:

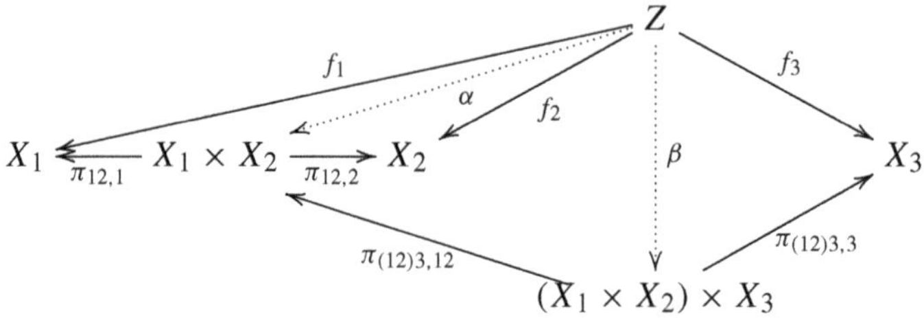

By the universality of $X_1 \times X_2$, a unique morphism α makes the adjacent triangles commute. Then, by the universality of $(X_1 \times X_2) \times X_3$, a unique morphism β makes the two triangles in the following diagram commute:

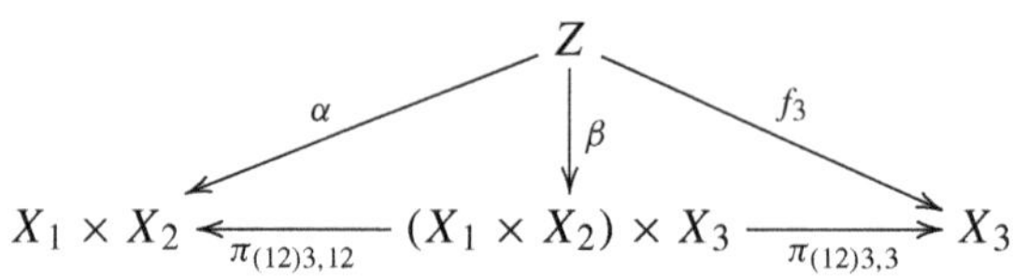

This shows that the morphism β is a unique mediating morphism from the span (Z, f_1, f_2, f_3) to the universal span with apex $(X_1 \times X_2) \times X_3$ and projections

$$\begin{aligned} p_1 &= \pi_{12,1} \circ \pi_{(12)3,12} : (X_1 \times X_2) \times X_3 \to X_1 \ , \\ p_2 &= \pi_{12,2} \circ \pi_{(12)3,12} : (X_1 \times X_2) \times X_3 \to X_2 \ , \text{and} \\ p_3 &= \pi_{(12)3,3} : (X_1 \times X_2) \times X_3 \to X_3 \ . \end{aligned}$$

Likewise, we can show that $(X_1 \times (X_2 \times X_3), q_1, q_2, q_3)$ is a universal span, where

$$\begin{aligned} q_1 &= \pi_{1(23),1} : X_1 \times (X_2 \times X_3) \to X_1 \ , \\ q_2 &= \pi_{23,2} \circ \pi_{1(23),23} : X_1 \times (X_2 \times X_3) \to X_2 \ , \text{and} \\ q_3 &= \pi_{23,3} \circ \pi_{1(23),23} : X_1 \times (X_2 \times X_3) \to X_3 \ . \end{aligned}$$

Since universal objects are isomorphic, there is a unique isomorphism

$$\alpha_{X_1,X_2,X_3} : X_1 \times (X_2 \times X_3) \to (X_1 \times X_2) \times X_3 \ .$$

We are going to show that the collection $\{\alpha_{X_1,X_2,X_3} \mid X_1, X_2, X_3 \in \mathrm{Obj}(\mathscr{C})\}$ forms a natural transformation

$$\alpha : - \times (- \times -) \quad \to \quad (- \times -) \times - \ .$$

For any morphisms $f_i : X_i \to Y_i$ for $(i = 1, 2, 3)$, we want to show the trapezoid at left is commutative in the following diagram:

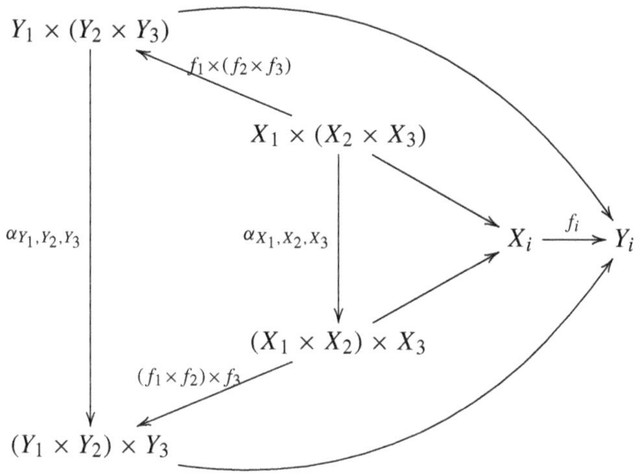

where unlabeled arrows are obvious projections depending on i's. The outermost and innermost triangles commute by the definition of α's. The quadrilaterals formed by projections and f_i's commute. See below:

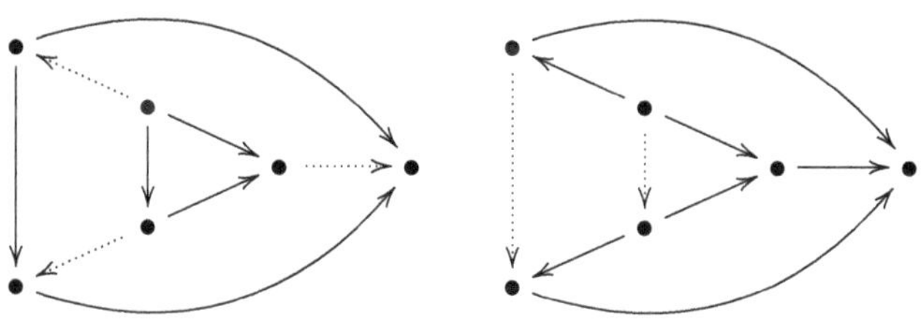

Combining all these commutative pieces, we see the following two paths from ◯ to $\triangle$ coincide for all i's:

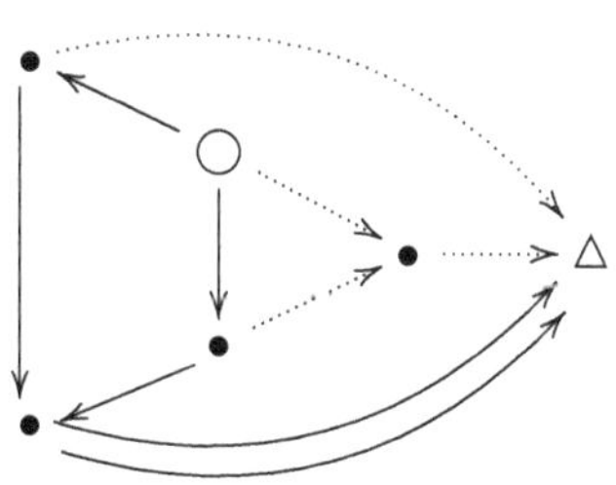

Now, we can consider spans to $\{Y_1, Y_2, Y_3\}$ as follows:

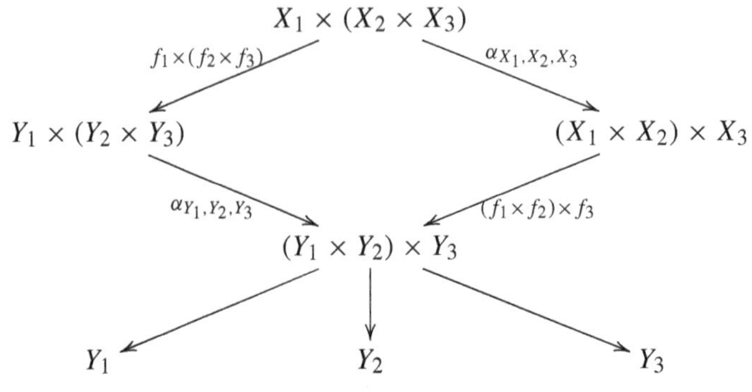

In the above diagram, from two apexes $X_1 \times (X_2 \times X_3)$ and $(Y_1 \times Y_2) \times Y_3$ we find two spans to $\{Y_1, Y_2, Y_3\}$. The object $(Y_1 \times Y_2) \times Y_3$ with its projections is a universal span. So, the two mediating morphisms appearing in the diamond must be identical. We conclude that α is a natural transformation, in our case especially a natural isomorphism.

We have to show that the two paths from $X \times (Y \times (Z \times W))$ to $((X \times Y) \times Z) \times W$ in the pentagon diagram below coincide.

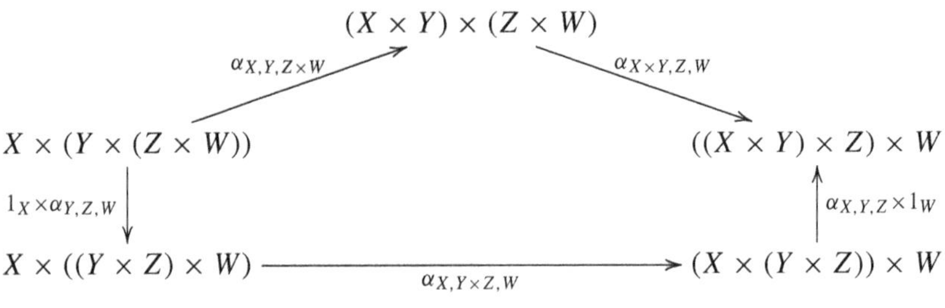

Let us consider spans to $\{X, Y, Z, W\}$ from each vertex of the pentagon. Take two vertices at left, for example, to get the following morphism between the spans:

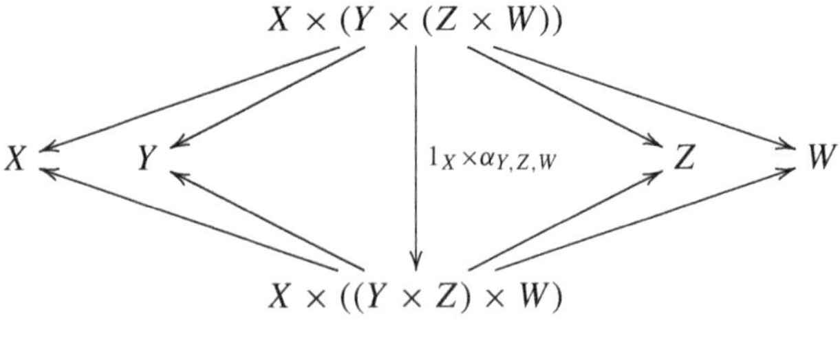

where unlabeled arrows are obvious projections. The morphism $1_X \times \alpha_{Y,Z,W}$ is a unique mediating morphism between the two spans. Likewise, all the morphisms appearing in the pentagon diagram above are mediating morphisms between universal spans to $\{X, Y, Z, W\}$. Any compositions of the morphisms are also a mediating morphism. Then, we use the universality of $((X \times Y) \times Z) \times W$. A mediating morphism from $X \times (Y \times (Z \times W))$ to $((X \times Y) \times Z) \times W$ must be unique. So, the two paths must be identical. We conclude that the natural isomorphism α makes the pentagon diagram commute.

Let T be a terminal object. Any one will do. We will soon see how terminal objects do great jobs. We define $\lambda_X : T \times X \to X$ and $\rho_X : X \times T \to X$ as projections. We first show that λ_X is an isomorphism.

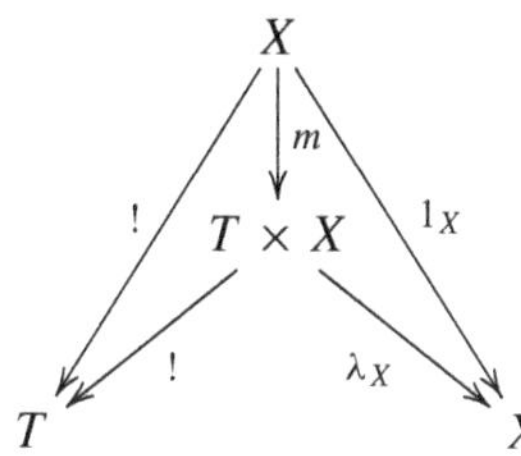

(8.3)

The symbol "!" in this diagram denotes the unique morphisms to the terminal object. The object $T \times X$ with its projections is a universal span to $\{T, X\}$. Thus, the mediating morphism m is unique and $\lambda_X \circ m = 1_X$. Notice also that X with projections ! and 1_X is another universal span to $\{T, X\}$. In fact, let $f : Z \to X$ be any morphism. We have the diagram below:

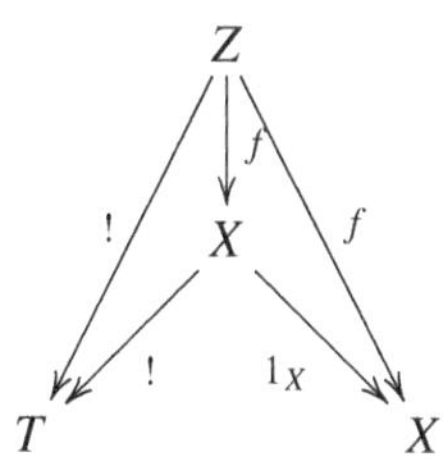

where the two triangles obviously commute. This means $(X, !, 1_X)$ is a universal span. Let us replace Z with $T \times X$. We have the following:

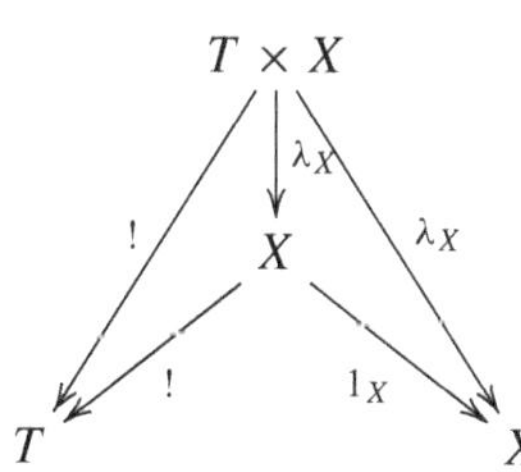

(8.4)

where λ_X plays the role of a mediating morphism. Connecting two diagrams (8.3) and (8.4), we have the following:

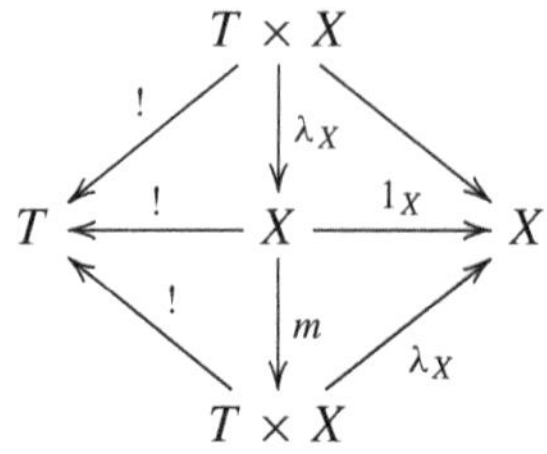

Since $T \times X$ with its projections is a universal span to $\{T, X\}$, we have $m \circ \lambda_X = 1_{T \times X}$ in addition to $\lambda_X \circ m = 1_X$. Thus, the morphism λ_X is an isomorphism.

Likewise, ρ_X is shown to be an isomorphism.

Next, we see if the triangle coherence holds. Let Z be any object with morphisms to X and Y.

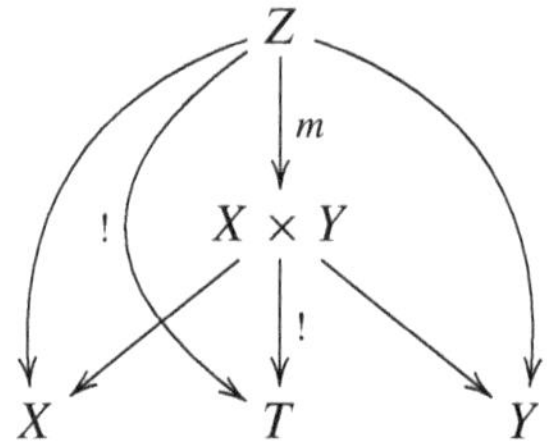

We see that $X \times Y$ with its projections is not only a universal span to $\{X, Y\}$ but also a universal span from $X \times Y$ to $\{X, T, Y\}$. The mediator m is shared by both universal spans. Now, consider the following diagram:

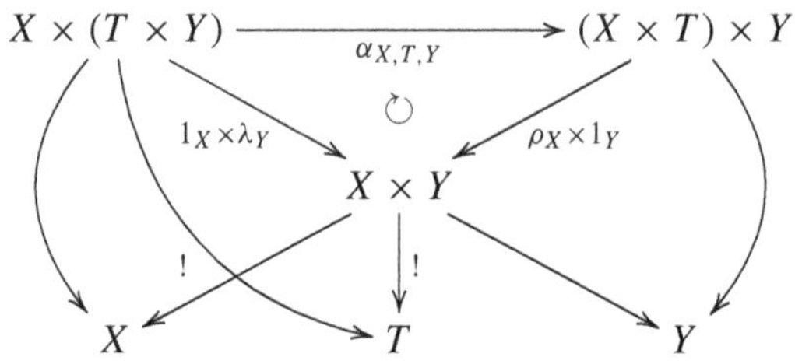

We find a span from $X \times (T \times Y)$ to $\{X, T, Y\}$. Two mediating morphisms to a universal span $(X \times Y, \{X, T, Y\})$ must be identical. Thus, the triangle coherence holds.

8.2 Coherence—Part One

We often treat tensor products in monoidal categories as though they are associative. In general they are only associative up to isomorphisms. Coherence prevents such an attitude from causing troubles.

The pentagon diagram (8.1) is the simplest instance of coherence. We will show that it is powerful enough to derive all other instances of coherence with the help of the triangle diagram (8.2). The commutativity of the pentagon diagram says that the two ways of using associators to get to the same object yield the same morphism.

The goal of this section is twofold.

- To streamline the discussion on coherence we add a working lemma that deals with two more instances of unitor coherence.
- The lemma and its proof is designed to provide the reader with lots of concrete examples of functoriality and naturality.

The following lemma appeared as one of the exercise problems in [4, p. 165].

Lemma 8.1 *Let $\mathscr{C}$ be a monoidal category. For all X, Y, Z, and $W \in \mathrm{Obj}(\mathscr{C})$, the following two triangles commute:*

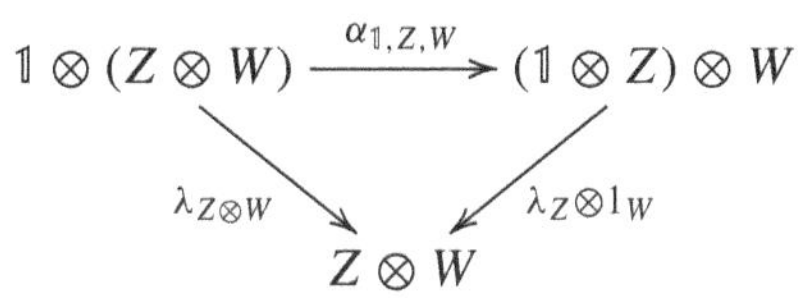

(8.5)

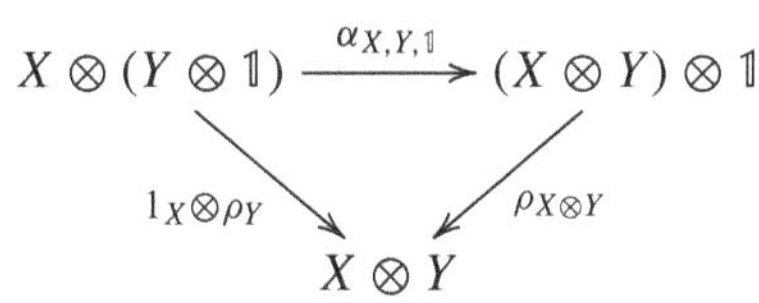

(8.6)

Proof (First half) We prove the triangle (8.5) commutes.

If we substitute X and Y by $\mathbb{1}$ in the pentagon diagram adding two instances ① and ③ of the unitor coherence triangle, we obtain the following diagram:

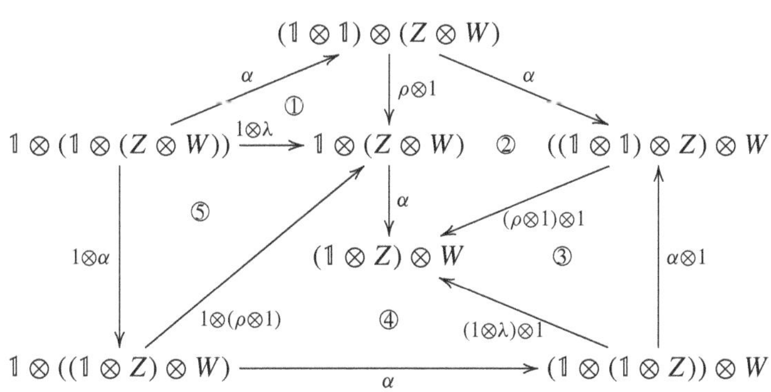

where component subscripts of associators, unitors, and identities are omitted for readability.

We will show triangles ①, ③, and quadrilaterals ②, ④ all commute. Then, using the pentagon coherence, we will show the triangle ⑤ commutes, which implies then the triangle (8.5) commutes.

We can confirm that the triangle ① is a true instance of the unitor coherence as follows:

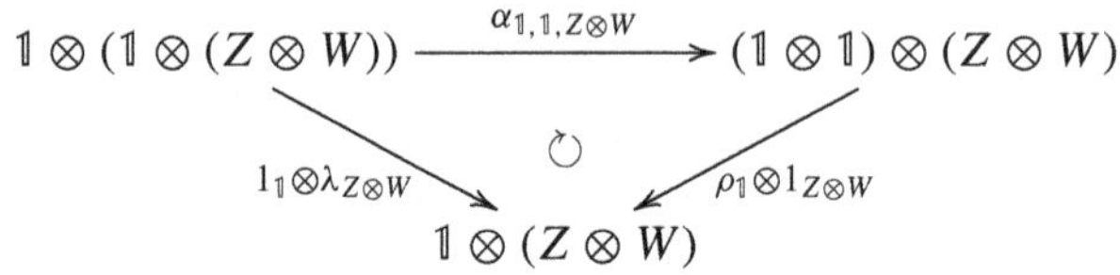

The quadrilateral ② can be seen as a naturality diagram for the natural transformation between the two functors

$$\alpha_{-,Z,W} : - \otimes (Z \otimes W) \to (- \otimes Z) \otimes W$$

at the morphism $\rho_{\mathbb{1}} : \mathbb{1} \otimes \mathbb{1} \to \mathbb{1}$ as follows:

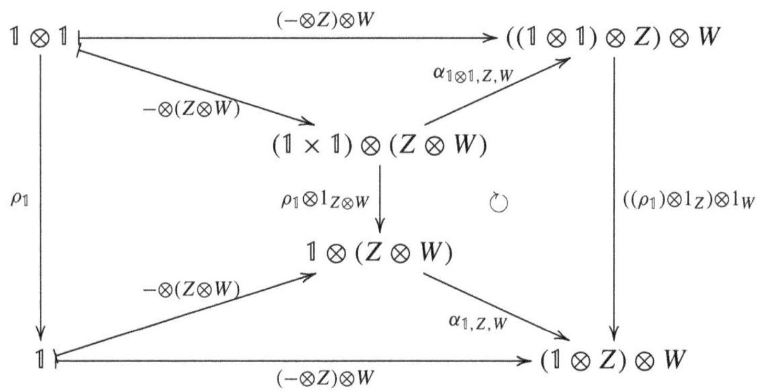

The trapezoid at right is the quadrilateral ②.

The triangle ③ is the image of an instance of the unitor coherence

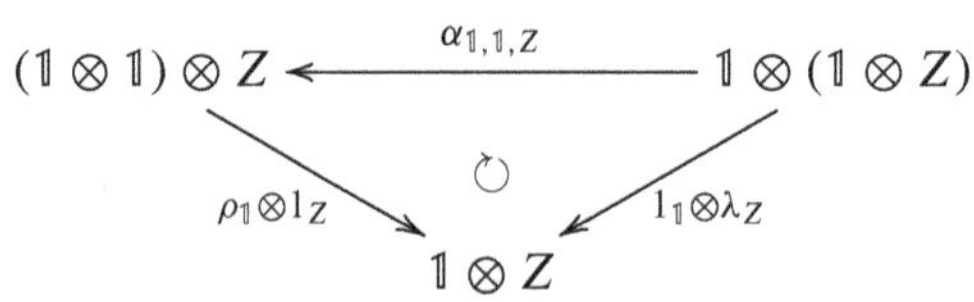

by the functor $- \otimes W$. Therefore, the triangle ③ commutes.

The quadrilateral ④ can be seen as a naturality of

$$\alpha_{\mathbb{1},-,W} : \mathbb{1} \otimes (- \otimes W) \to (\mathbb{1} \otimes -) \otimes W$$

at the morphism $\lambda_Z : \mathbb{1} \otimes Z \to Z$ as follows:

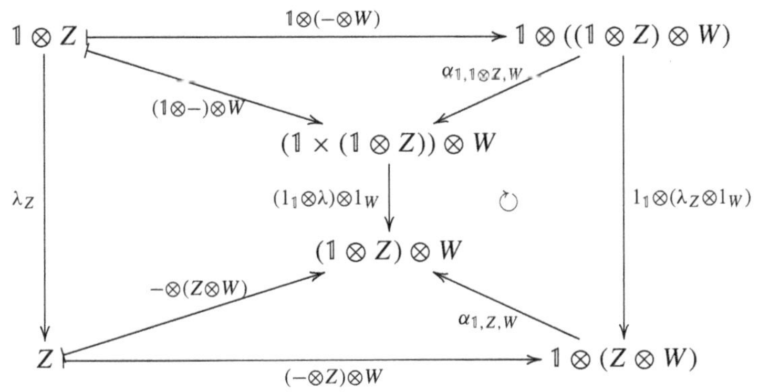

The trapezoid at right is the quadrilateral ④.

Now, we proceed to see if the triangle ⑤ commutes. Using commutativity we have seen so far, we can transform the path from left to right in the following sequence. Notice that the pentagon coherence is used in the step from the third to fourth.

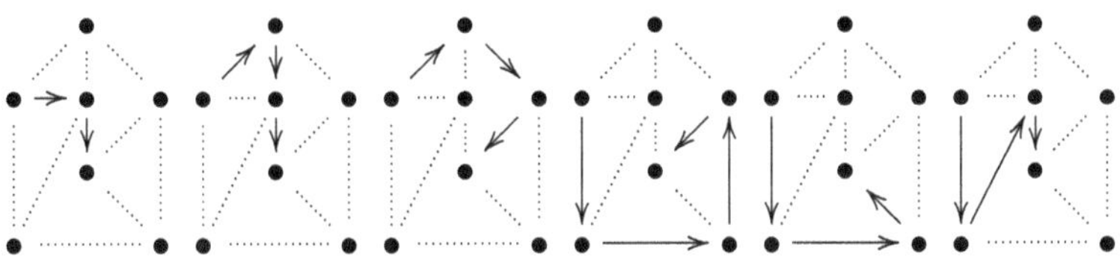

Thus, we have the following.

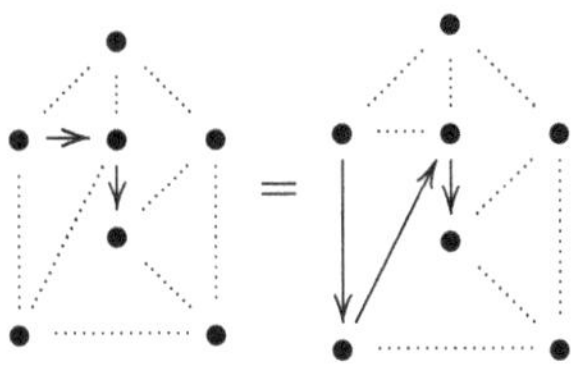

The two paths share the same arrow $\alpha_{\mathbb{1},Z,W}$ at their ends. Since it is an isomorphism, the triangle ⑤ commutes.

We have come halfway. Let us study the following diagram:

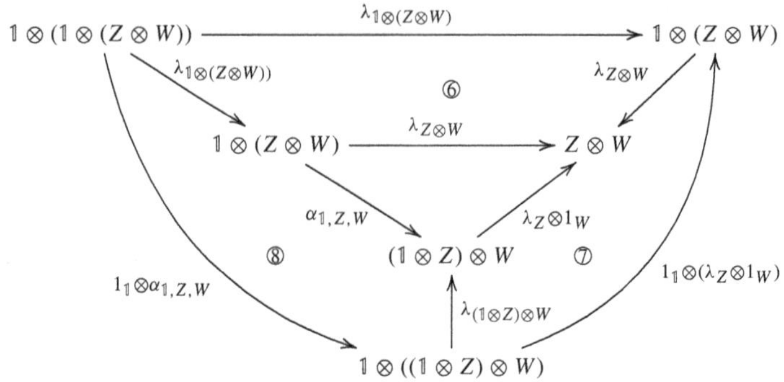

The outermost triangle is ⑤, which we have just proved commutative. The commutativity of the inner triangle is our goal.

We show that the three quadrilaterals surrounding the inner triangle are all commutative. Let us begin with the upper quadrilateral ⑥.

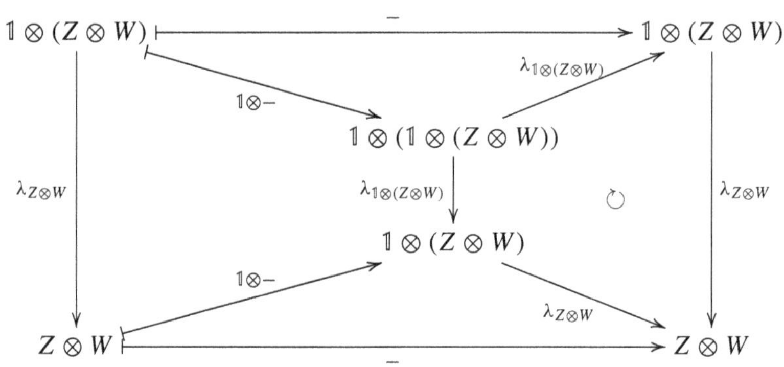

By the naturality of λ at the morphism $\lambda_{Z\otimes W} : \mathbb{1} \otimes (Z \otimes W) \to Z \otimes W$, we see the trapezoid at right in the above diagram, namely, the quadrilateral ⑥ commute.

Next, we see the quadrilateral ⑦.

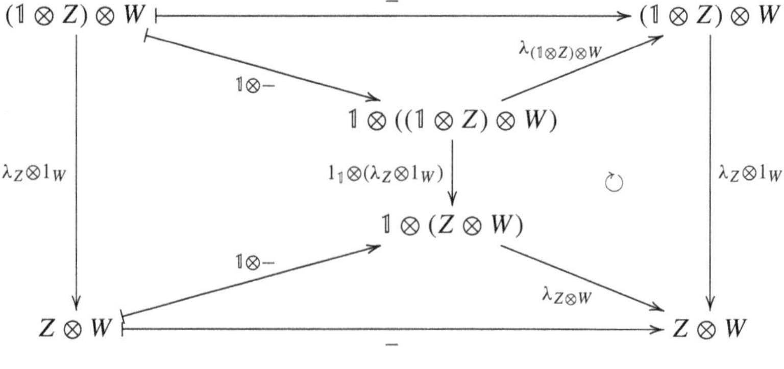

By the naturality of λ at the morphism $\lambda_Z \otimes 1_W : (\mathbb{1} \otimes Z) \otimes W \to Z \otimes W$, we see the trapezoid at right in the above diagram, namely, the quadrilateral ⑦ commute.

Finally, we see the quadrilateral ⑧.

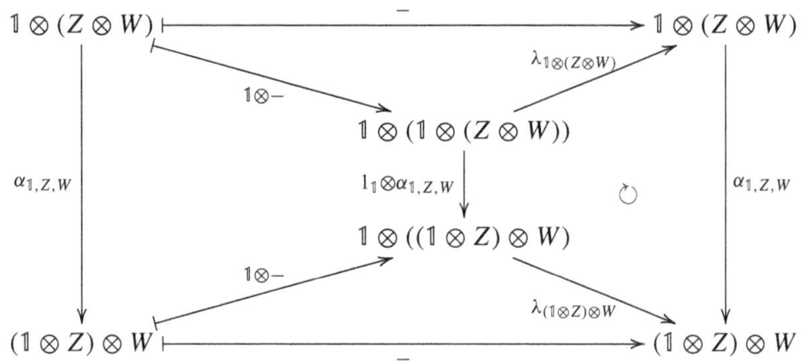

By the naturality of λ at the morphism $\alpha_{\mathbb{1},Z,W} : \mathbb{1} \otimes (Z \otimes W) \to (\mathbb{1} \otimes Z) \otimes W$, we see the trapezoid at right in the above diagram, namely, the quadrilateral ⑧ commute.

Using the commutative quadrilaterals ⑥, ⑦, ⑧, and the large triangle, we can transform the first path to the last as follows. Trace it from upper left to lower right.

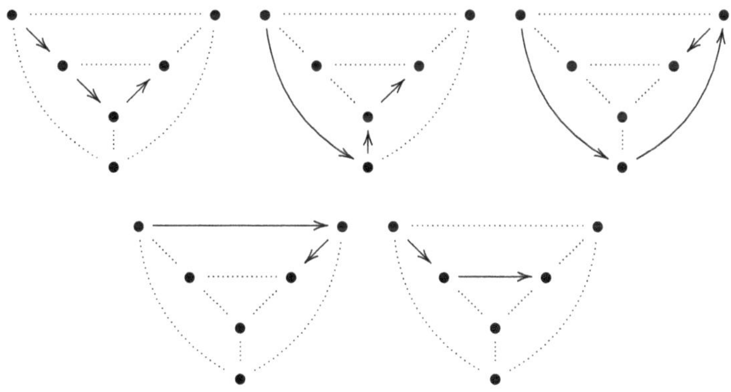

The first path and the last one are identical morphisms. They share one arrow of isomorphism, which can be canceled to obtain the commutativity of the inner triangle.

This establishes the commutativity of (8.5). □

Proof (Second half) We prove the triangle (8.6) commutes. Since most arguments go in the same way, we only show the outline.

If we substitute Z and W in the pentagon diagram by $\mathbb{1}$ adding two instances of the unitor coherence triangle, we obtain the following diagram:

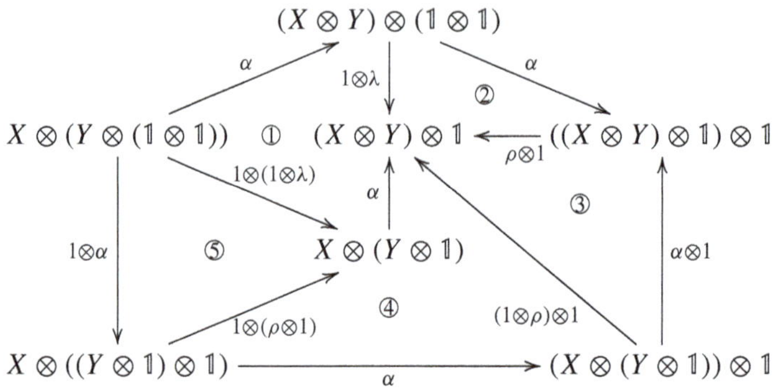

The quadrilateral ① commutes by the naturality of α. To be more specific, the naturality of

$$\alpha_{X,Y,-} : X \otimes (Y \otimes -) \to (X \otimes Y) \otimes -$$

at the morphism $\lambda_{\mathbb{1}} : \mathbb{1} \otimes \mathbb{1} \to \mathbb{1}$.

The triangle ② is an instance of the unitor coherence.

The quadrilateral ④ commutes by the naturality of α. To be more specific, the naturality of

$$\alpha_{X,-,\mathbb{1}} : X \otimes (- \otimes \mathbb{1}) \to (X \otimes -) \otimes \mathbb{1}$$

at the morphism $\rho_Y : Y \otimes \mathbb{1} \to Y$.

Applying the functor $X \otimes -$ to an instance of the unitor coherence, we get the triangle ⑤.

So, ①, ②, ④, and ⑤ all commute leaving ③ unknown at present.

We can transform the first path to the last as follows. Read from left to right.

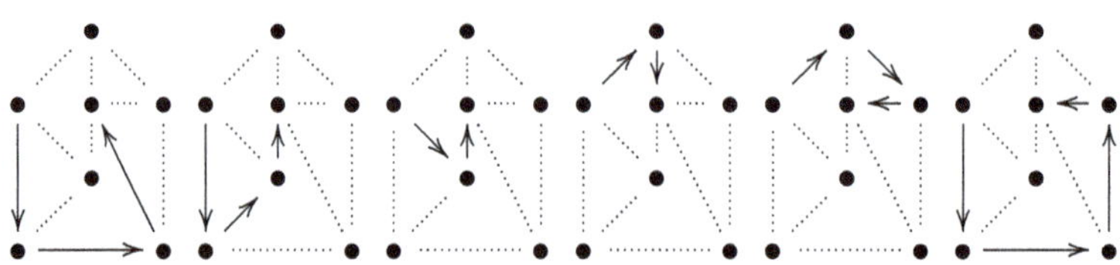

We used the pentagon coherence at the last step.

The first path and the last one share two arrows, which are isomorphisms. By canceling these out, we conclude the triangle ③ is commutative.

Let us study the following diagram. We have just seen that the outer triangle is commutative.

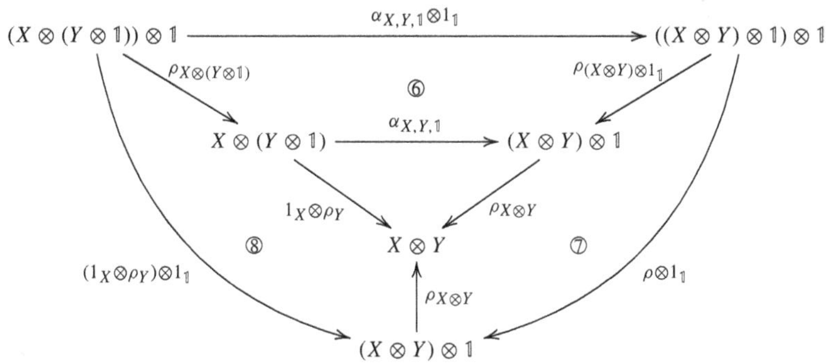

The quadrilateral ⑥ commutes by the naturality of $\rho : - \otimes \mathbb{1} \to -$ at the morphism $\alpha_{X,Y,\mathbb{1}} : X \otimes (Y \otimes \mathbb{1}) \to (X \otimes Y) \otimes \mathbb{1}$.

The quadrilateral ⑦ commutes by the naturality of $\rho : - \otimes \mathbb{1} \to -$ at the morphism $\rho_{X \otimes Y}$.

The quadrilateral ⑧ commutes by the naturality of $\rho : - \otimes \mathbb{1} \to -$ at the morphism $1_X \otimes \rho_Y$.

Using the commutative quadrilaterals ⑥, ⑦, ⑧, and the large triangle, we can transform the first path to the last as follows:

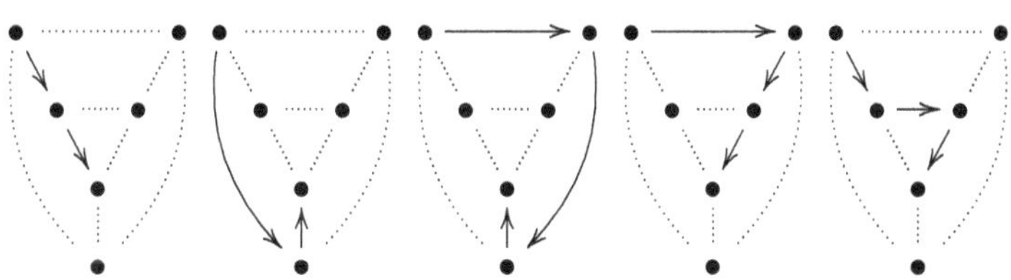

Since the first path and the last one share one isomorphism arrow, we see the inner triangle commutes. □

8.3 Monoidal Functors

Monoidal categories have more structures than vanilla categories. So, we would like functors between monoidal categories to preserve these extra structures in some way.

Definition 8.3 A *strict monoidal functor* from a monoidal category $(\mathscr{C}, \otimes, \mathbb{1}, \alpha, \lambda, \rho)$ to a monoidal category $(\mathscr{C}', \otimes', \mathbb{1}', \alpha', \lambda', \rho')$ is a functor $T : \mathscr{C} \to \mathscr{C}'$ such that, for any $X, Y, Z \in \mathrm{Obj}(\mathscr{C})$ and $f, g \in \mathrm{Mor}(\mathscr{C})$,

$$T(X \otimes Y) = TX \otimes TY, \qquad T(f \otimes) = Tf \otimes Tg, \qquad T\mathbb{1} = \mathbb{1}',$$

$$T\alpha_{X,Y,Z} = \alpha'_{TX,TY,TZ}, \qquad T\lambda_X = \lambda'_{TX}, \qquad T\rho_X = \rho'_{TX}.$$

Strict monoidal functors play the central role in the next section. However, many working examples show strictness is over-requirement. We see one such an example below.

Example 8.3 Let $\Bbbk$ be a field. Let us consider the category $\mathrm{Vect}_{\Bbbk}$ of finite dimensional $\Bbbk$-vector spaces. We know the usual tensor product $\otimes_{\Bbbk}$. With some elaboration based on the basis or purely depending on universality, we can write down a reasonable associator and unitors that satisfy the coherence conditions. But, let us believe all of these are rigorously done.

Let U be the forgetful functor from $(\mathrm{Vect}_{\mathbb{C}}, \otimes_{\mathbb{C}}, \ldots)$ to $(\mathrm{Vect}_{\mathbb{R}}, \otimes_{\mathbb{R}}, \ldots)$, where associators and unitors are omitted in the presentation. "Forgetful" means, in this case, is to regard every one-dimensional complex vector space as a two-dimensional real vector space with basis 1 and $\sqrt{-1}$. Let V_1 and V_2 be any complex vector spaces of dimension n_1 and n_2, respectively. While $\dim_{\mathbb{C}} V_1 \otimes_{\mathbb{C}} V_2 = n_1 \times n_2$ and $\dim_{\mathbb{R}} U(V_1 \otimes_{\mathbb{C}} V_2) = 2 \times n_1 \times n_2$, we see $\dim_{\mathbb{R}} U(V_1) \otimes_{\mathbb{R}} U(V_2) = 4 \times n_1 \times n_2$. So, $U(V_1 \otimes_{\mathbb{C}} V_2)$ and $U(V_1) \otimes_{\mathbb{R}} U(V_2)$ are neither equal nor isomorphic.

Since this kind of forgetful functors play important roles in many places, we have to relax equality to isomorphicity or naturality in various ways.

8.4 Coherence—Part Two

In the second half of the discussion of coherence, we construct some higher order structures out of a monoidal category. This kind of activity is found everywhere in mathematics but especially often in category theory.

We first introduce a category of binary words.

Definition 8.4 We introduce the notion of binary words defined recursively as follows:

- The empty word e is a binary word of length 0.
- The symbol a is a binary word of length 1.
- If v and w are binary words of length m and n, respectively, then $(v) \otimes (w)$ is a binary word of length $m + n$.

The objects of the category $\mathscr{W}$ are all binary words. For any pair of binary words v and w of the same length, there is exactly one morphism $v \to w$. Thus, $\mathscr{W}$ is a preorder with every morphism is invertible. The category $\mathscr{W}$ is a strict monoidal category. The unit object is e. The associator and unitors are uniquely determined since $\mathscr{W}$ is a preorder.

Remark 8.1 A preorder is a category in which, for any two objects x and y, there is at most one morphism from x to y. The preorderedness makes all the diagrams in $\mathscr{W}$ commute. Consider any diagram including an object $\bullet$ below:

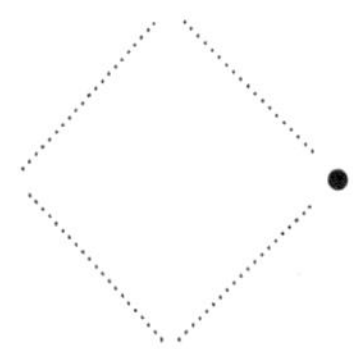

Since all the morphisms in the category $\mathscr{W}$ are invertible, we do not care the directions in the diagram. The diagram above can be seen as a directed path from $\bullet$ to $\bullet$ and $1_\bullet$ is the only morphism allowed. Thus, the path is the identity morphism of the object $\bullet$, which means that the diagram commutes.

Example 8.4 Let $v_1, \ldots, v_5, v_2', v_4' \in \mathrm{Obj}(\mathscr{W})$ as follows:

$$
\begin{aligned}
v_1 &= a \otimes (a \otimes (a \otimes a)) \\
v_2 &= (a \otimes a) \otimes (a \otimes a) \\
v_3 &= ((a \otimes a) \otimes a) \otimes a \\
v_4 &= a \otimes ((a \otimes a) \otimes a) \\
v_5 &= (a \otimes (a \otimes a)) \otimes a \\
v_2' &= (a \otimes e) \otimes (e \otimes a) \\
v_4' &= e \otimes ((a \otimes a) \otimes a)
\end{aligned}
$$

The words $v_1, \ldots, v_5$ all have length 4. The length of the word v_2' is 2. The word v_4' has length 3. Recall that the length of a binary word is the number of a's in the word. Pictorial representations of $v_1, v_2, \ldots$, and v_5 are shown below:

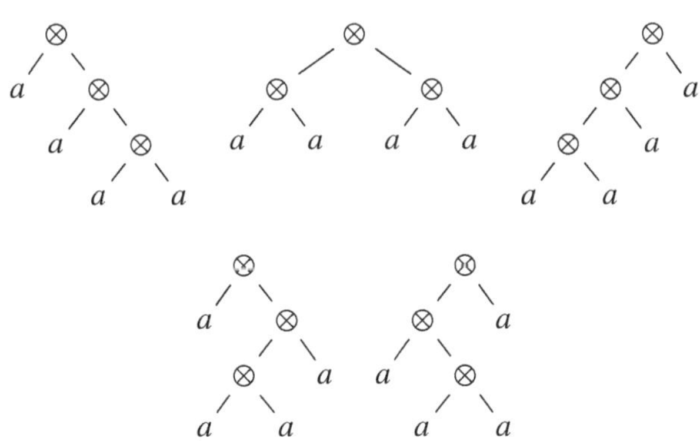

Notice that these five binary trees correspond to vertices in the pentagon diagram (8.1). We will deal with this kind of binary trees with Haskell code later.

Coherence is compactly formulated in the following theorem.

Theorem 8.1 *For any monoidal category* $\mathscr{C}$ *and* $X \in \mathrm{Obj}(\mathscr{C})$*, there is a unique strict monoidal functor* $\mathscr{W} \to \mathscr{C}$ *with* $a \mapsto X$*.*

If there exists a strict monoidal functor as the theorem claims, the function on objects must send each binary word $w \in \mathrm{Obj}(\mathscr{W})$ to the corresponding object $w_X \in \mathrm{Obj}(\mathscr{C})$ by replacing every occurrence of a in w with X, and every occurrence of e with $\mathbb{1}$. Let v and w be two binary words of the same length. There can be several paths consisting of composite morphisms from v to w, which must be all identical in a preorder such as $\mathscr{W}$. The existence of a strict monoidal functor means that all functor images of paths connecting one object to another must commute. In other words, we are facing a problem of well-definedness.

Before diving into the proof, let us play with Haskell code.

Listing 8.1 BinaryWords.hs

```haskell
data Tree a = Pure a | Node (Tree a) (Tree a) deriving Eq

instance Show a => Show (Tree a) where
  show (Pure a)
    = show a
  show (Node (Pure a) (Pure b))
    = show a ++ "." ++ show b
  show (Node (Pure a) t)
    = show a ++ ".(" ++ show t ++ ")"
  show (Node t (Pure a))
    = "(" ++ show t ++ ")." ++ show a
  show (Node t1 t2)
    = "(" ++ show t1 ++ ").(" ++ show t2 ++ ")"


data BWord = A | E deriving Eq

instance Show BWord where
  show A = "a"
  show E = "e"

-- test data
v1 = Node (Pure A) (Node (Pure A) (Node (Pure A) (Pure A)))
v2 = Node (Node (Pure A) (Pure A)) (Node (Pure A) (Pure A))
v3 = Node (Node (Node (Pure A) (Pure A)) (Pure A)) (Pure A)
v4 = Node (Pure A) (Node (Node (Pure A) (Pure A)) (Pure A))
v5 = Node (Node (Pure A) (Node (Pure A) (Pure A))) (Pure A)
v2' = Node (Node (Pure A) (Pure E)) (Node (Pure E) (Pure A))
v4' = Node (Pure E) (Node (Node (Pure A) (Pure A)) (Pure A))

-- associator
alpha :: Tree a -> Tree a
alpha (Pure a) = Pure a
alpha (Node t1 (Node t2 t3)) = Node (Node t1 t2) t3
alpha x = x
```

```

-- utilities
-- generates right heavy trees
rgenerate :: [a] -> Tree a
rgenerate [x] = Pure x
rgenerate (x:xs) = Node (Pure x) (rgenerate xs)

-- generates left heavy trees
lgenerate :: [a] -> Tree a
lgenerate = iter . reverse
              where
                iter [x] = Pure x
                iter (x:xs) = Node (iter xs) (Pure x)

-- repeatedly apply alpha at the top node
-- util it sees Pure at right
alpha' :: Eq a => Tree a -> Tree a
alpha' t@(Node t1 (Pure x)) = t
alpha' t@(Node t1 t2)       = alpha' (alpha t)

-- transforms to the canonical tree
ialpha :: Eq a => Tree a -> Tree a
ialpha (Pure x)              = Pure x
ialpha (Node t1 (Pure x)) = Node (ialpha t1) (Pure x)
ialpha t                     = ialpha (alpha' t)

{-
ghci> map (lgenerate [1..5]==) $ map ialpha $ mkwords [1..5]
[True,True,True,True,True,True,True,True,True,True,True,True,True,True]
-}

-- the len function below can not be used for words
-- that include (Pure E).
len :: Tree a -> Int
len (Pure _)     = 1
len (Node t1 t2) = len t1 + len t2

rank :: Tree a -> Int
rank (Pure a) = 0
rank (Node t1 t2) = rank t1 + rank t2 + len t2 - 1

{-
ghci> rank $ rgenerate [1..5]
6
ghci> rank $ lgenerate [1..5]
0
ghci> map rank $ take 4 $ iterate alpha $ rgenerate [1..5]
[6,3,1,0]
-}

split :: [a] -> [([a],[a])]
split [x] = []
split (x:xs) = [([x],xs)] ++ map (\z -> (x: fst z, snd z)) (split xs)
```

```
89
90 {-
91 ghci> split [1..5]
92 [([1],[2,3,4,5]),([1,2],[3,4,5]),([1,2,3],[4,5])
       ,([1,2,3,4],[5])]
93 -}
94
95 mkwords :: [a] -> [Tree a]
96 mkwords [x] = [Pure x]
97 mkwords xs =
98   [Node t1 t2 |
99     (x1,x2)<- split xs, t1 <- mkwords x1, t2 <- mkwords x2]
```

Line 1 defines a binary tree structure. Leaf nodes are created by the `Pure` constructor. Lines 3–13 define a `String` representation for our binary trees, where the `Tree` constructor is presented by an infix operator dot instead of $\otimes$ and the branching structure by parentheses.

Line 16 defines an alphabet, the set of symbols, for the binary words, the symbols "e" and "a" are defined as the value constructors `E` and `A`, respectively. Lines 18–20 define a `String` representation for these symbols.

Lines 23–29 define test data that appeared in Example 8.4.

Lines 32–35 define the associator α. It can be used as follows:

```
ghci> v1
a.(a.(a.a))
ghci> alpha v1
(a.a).(a.a)
```

We can present this more pictorially.

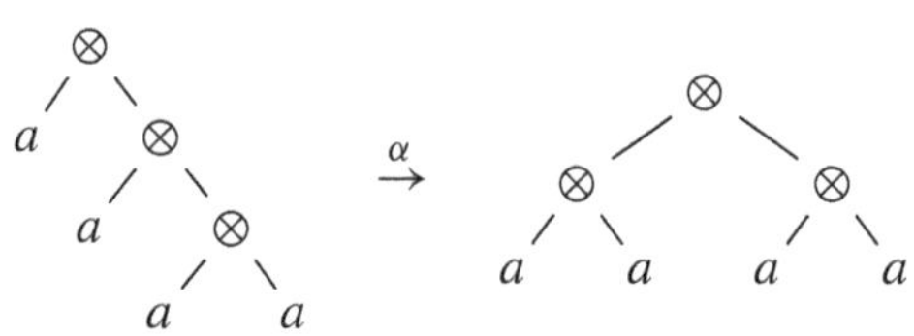

Lines 39–48 define utility functions `rgenerate` and `lgenerate`. The former generates a most right heavy tree and the latter most left. Our `Tree` type is polymorphic. So, we can use it with various types, for example, the `Integer` type as follows:

```
ghci> rgenerate [1..5]
1.(2.(3.(4.5)))
ghci> lgenerate [1..5]
(((1.2).3).4).5
```

Let us give a pictorial representation.

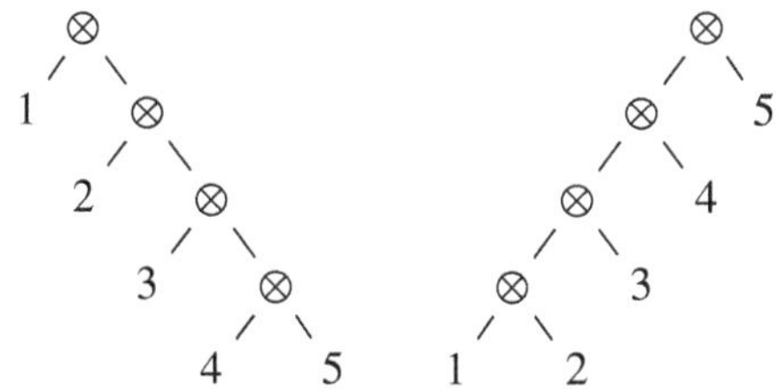

Lines 52–54 define the `alpha'` function that applies the `alpha` function repeatedly at the top level node to finally obtain a tree whose left branch is `Pure`.

Lines 57–60 define the `ialpha` function that recursively downward making all the right branches `Pure`. Thus, `ialpha` transforms any binary trees to its *canonical tree*, that is, the most left heavy one.

```
ghci> rgenerate [1..5]
1.(2.(3.(4.5)))
ghci> ialpha $ rgenerate [1..5]
(((1.2).3).4).5
```

In some cases, repeated application of α at the top node is enough to obtain the canonical tree. For example, see below:

```
ghci> take 4 $ iterate alpha $ rgenerate [1..5]
[1.(2.(3.(4.5))),(1.2).(3.(4.5)),((1.2).3).(4.5),(((1.2).3).4).5]
```

Let us give a pictorial representation.

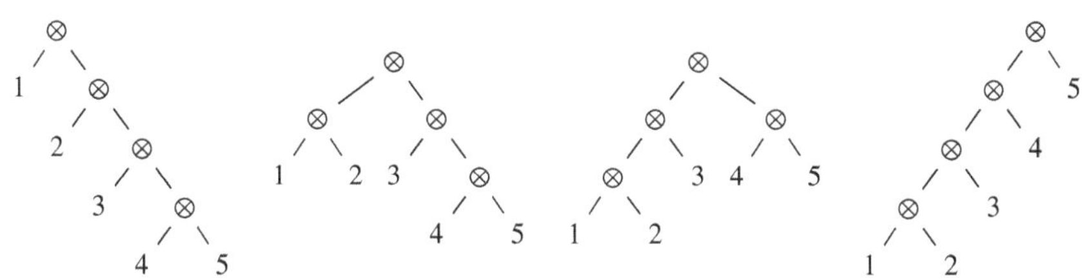

However, this is not always the case.

```
ghci> v5
(a.(a.a)).a
ghci> alpha v5
(a.(a.a)).a
```

This situation is solved by our `ialpha` function as follows:

```
ghci> ialpha v5
((a.a).a).a
```

Lines 69–71 define the `len` function that implements the length function in Example 8.4. Note that the `len` function cannot be applied to binary words that contain the symbol e. However, this is enough for our purpose.

Lines 73–75 define the `rank` function that measures how left heavy the tree is. For example, the ranks of $v_1, \ldots, v_5$ in Example 8.4 can be calculated as follows:

```
ghci> map rank [v1,v2,v3,v4,v5]
[3,1,0,2,1]
```

Lines 86–88 define the `split` function that generates all possible binary division of a given list without disturbing sequence.

```
ghci> split [1..5]
[([1],[2,3,4,5]),([1,2],[3,4,5]),([1,2,3],[4,5]),([1,2,3,4],[5])]
```

Lines 95–99 define the `mkwords` function that generates all the binary trees whose leaves are taken from a given list from left to right.

```
ghci> mkwords [1..4]
[1.(2.(3.4)),1.((2.3).4),(1.2).(3.4),(1.(2.3)).4,((1.2).3).4]
```

We can see that the elements of the resulting list correspond to the vertices of the pentagon coherence (8.1).

The number of all possible binary trees of length 5 is obtained as follows.

```
ghci> length $ mkwords [1..5]
14
```

A more convincing test for the function `ialpha` is shown below:

```
ghci> and $ map (lgenerate [1..7]==) $ map ialpha $ mkwords [1..7]
True
```

Now, we return to Theorem 8.1.

Definition 8.5 Let $\mathscr{W}_0$ be a subcategory of $\mathscr{W}$ by excluding all the binary words and morphisms containing e.

Let $\mathscr{C}$ be a monoidal category and $X \in \mathrm{Obj}(\mathscr{C})$. For the moment, we will work within the category $\mathscr{W}_0$. Let $G_{n,X}$ be a directed graph whose vertices are all binary words in $\mathscr{W}_0$, and whose edge $v \to w$ is a certain morphism, called a basic arrow defined shortly, from v_X to w_X in $\mathscr{C}$.

We define basic arrows recursively:

- $\alpha_{u_X, v_X, w_X} : u_X \otimes (v_X \otimes w_X) \to u_X \otimes (v_X \otimes w_X)$ and its inverse $\alpha^{-1}_{u_X, v_X, w_X}$ is basic,
- for β already defined basic, $1 \otimes \beta$ and $\beta \otimes 1$ are basic.

When the object X is fixed in the discussion, we simply write $\alpha_{u,v,w}$ instead of α_{u_X, v_X, w_X}. A basic arrow is a morphism obtained by tensoring one α or α^{-1} with several 1's. A basic arrow contains exactly one α or α^{-1} in its $\otimes$ factors. We call the former case a *forward* basic arrow, the latter *backward* basic arrow. A path obtained by composing only basic forward arrows is called a *forward path.* A path obtained by composing only basic backward arrows is called a *backward path.*

In the graph $G_{n,X}$ the paths from v to w are a composition of basic arrows, which starts from v_X and ends at w_X.

We want to show that any two paths from v to w yield the same arrow in $\mathscr{C}$. In other words, the graph $G_{n,X}$ is a commutative diagram in $\mathscr{C}$.

Definition 8.6 For any $w \in \mathrm{Obj}(\mathscr{W})$, we assign a integer $\mathrm{rank}(w)$ recursively as follows:

- $\mathrm{rank}(e) = \mathrm{rank}(a) = 0$, and
- $\mathrm{rank}(v \otimes w) = \mathrm{rank}(v) + \mathrm{rank}(w) + \mathrm{len}(w) - 1$.

We have already seen the implementation in Haskell. See Listing 8.1.

Let $w^{(n)}$ be the most left heavy binary word in $\mathscr{W}_0$. Or in other words, the word $w^{(n)}$ is a binary tree all of whose recursive right branches are leaf nodes. See the binary tree representation of $w^{(n)}$ below:

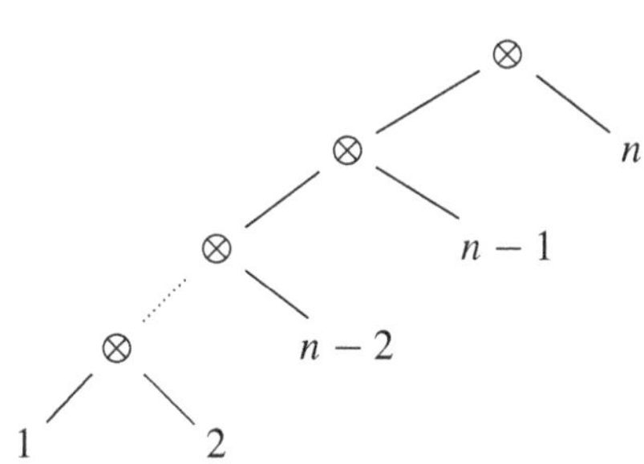

where leaf nodes are numbered from 1 to n.

Lemma 8.2 *For any* $w \in \mathrm{Obj}(\mathscr{W}_0)$*, we have* $\mathrm{len}(w) \geq 0$.

Proof The lemma restricts its claim to words in $\mathscr{W}_0$. So, the rank function is non-decreasing with respect to the $\otimes$ operation. To be specific, we have

$$\mathrm{rank}(v \otimes w) = \mathrm{rank}(v) + \mathrm{rank}(w) + \mathrm{len}(w) - 1 \geq \mathrm{rank}(v) + \mathrm{rank}(w)\ .$$

□

Lemma 8.3 *Let* $w \in \mathrm{Obj}(\mathscr{W}_0)$*. If* $\mathrm{rank}(w) = 0$*, then* $w = w^{(n)}$.

Proof We prove by induction on the $\otimes$ constructs. If $w = a$, that is no $\otimes$ appears, then $w = w^{(1)}$ by definition, which is the base case.

Let $w = w_1 \otimes w_2$. Since we know that the rank function is non-negative and non-decreasing with respect to $\otimes$, we must have $\mathrm{rank}(w_1) = \mathrm{rank}(w_2) = 0$ and $\mathrm{len}(w_2) = 1$. By induction hypothesis, the left branch is $w_1 = w^{(n-1)}$. The right branch w_2 must be a leaf. This completes the proof. □

Lemma 8.4 *Forward basic arrows always decrease the rank of a word.*

Proof Let $v \otimes (w_1 \otimes w_2)$ be a subword of a binary word in question. If we apply α_{u,w_1,w_2} to this part of the binary word, the rank of this subword changes while the length being unchanged. So, all we have to do is to compare the ranks of a subword before and after the operation.

$$\mathrm{rank}(v \otimes (w_1 \otimes w_2)) = \mathrm{rank}(v) + \mathrm{rank}(w_1) + \mathrm{rank}(w_2) + \mathrm{len}(w_1) + 2 \times \mathrm{len}(w_2) - 2$$
$$\mathrm{rank}((v \otimes w_1) \otimes w_2) = \mathrm{rank}(v) + \mathrm{rank}(w_1) + \mathrm{rank}(w_2) + \mathrm{len}(w_1) + \mathrm{len}(w_2) - 2$$

Since the length of w_2 is not zero, the lemma holds. □

Lemma 8.5 *For any* $w \in \mathrm{Obj}(\mathscr{W}_0)$ *of length* n*, there is a forward basic path from* w *to* $w^{(n)}$ *in* $G_{n,X}$.

The algorithm to find a forward path is implemented in Listing 8.1. There are other algorithms to reach this goal. Any way, we have one.

Lemma 8.6 *Let* $v, w \in \mathrm{Obj}(\mathscr{W}_0)$ *be of the same length* n*. If there are two forward paths from* v *to* w *in* $G_{n,X}$*, then they are identical morphisms in* $\mathscr{C}$.

To prove the above lemma, we need the following diamond lemma.

Lemma 8.7 (The Diamond Lemma) *Let* $v \in \mathrm{Obj}(\mathscr{W}_0)$*. If there are forward paths* $\beta : v \to v'$ *and* $\gamma : v \to v''$*, then there are forward paths* $v' \to z$ *and* $v'' \to z$ *that make the following diagram commute:*

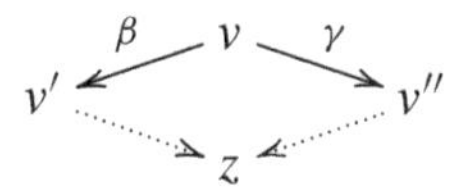

Before proving this, we return to the previous lemma and give it a proof using this diamond lemma.

Proof (Lemma 8.6) We prove by induction on the rank of w. Suppose the lemma holds for all w of lower ranks.

Let $\beta = \beta_k \circ \cdots \circ \beta_1 : w \to w^{(n)}$ and $\gamma = \gamma_\ell \circ \cdots \circ \gamma_1 : w \to w^{(n)}$ are two forward paths, where β_i's and γ_i's are basic forward arrows. We have the following diagram with the commuting diamond:

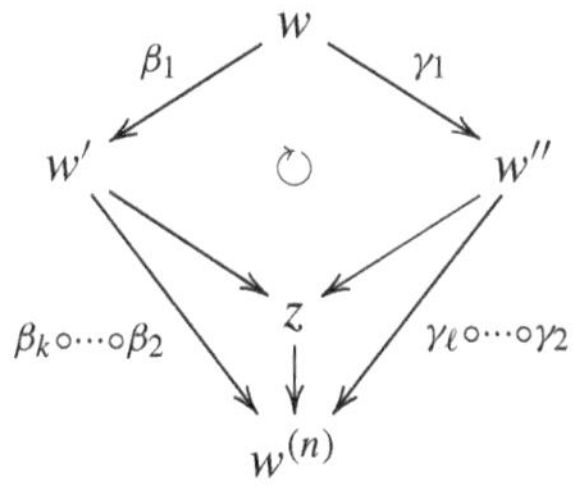

Since w', w'', and z have a lower rank, the two triangles are commutative by the induction hypothesis. This shows that the outer quadrilateral commutes. □

Let us prove the diamond lemma.

Proof (Lemma 8.7) We prove by induction on the rank of v. If $\text{rank}(v) = 0$, then $v = w^{(n)}$. We have nothing to do in the base case since there can be no forward paths from v other than the identity.

Let us suppose the lemma holds for all v of smaller ranks. Let $v = u \otimes w$. The morphism β may appear in three forms:

- $\beta = \beta' \otimes 1_w$:β acts inside u.
- $\beta = 1_u \otimes \beta''$:β acts inside w.
- $\beta = \alpha_{u,s,t}$ where $w = s \otimes t$.

The path γ has the same case division. If both β and γ act inside u or w, we can additionally apply induction on the length of v. We omit this branch of proof.

(Case 1) Let $\beta = \beta' \otimes 1_w$ acts inside u and $\gamma : 1_u \otimes \gamma'$ inside w. The following diagram commutes since $\otimes$ is a bifunctor, namely, component morphisms act parallelly and independently.

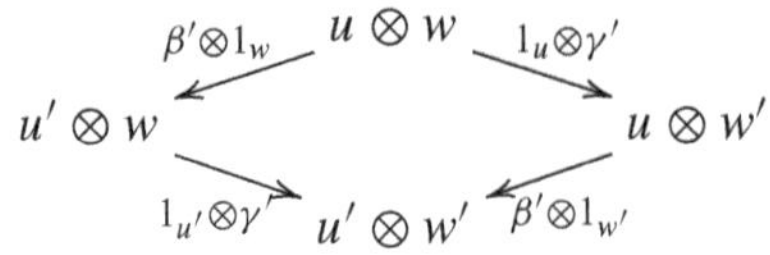

This is a desired diamond.

(Case 2) Let $v = u \otimes (s \otimes t)$. Let $\beta = \alpha_{u,s,t}$ and $\gamma = \gamma' \otimes 1_{s \otimes t}$. We have the following diagram:

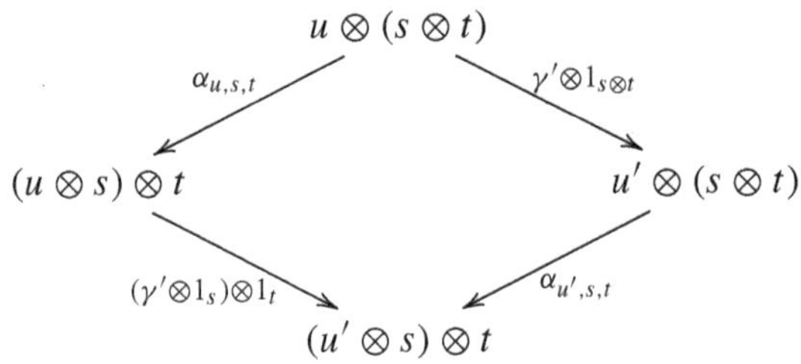

This diagram commutes since α is a natural isomorphism between the functors $- \otimes (- \otimes -)$ and $(- \otimes -) \otimes -$ as follows:

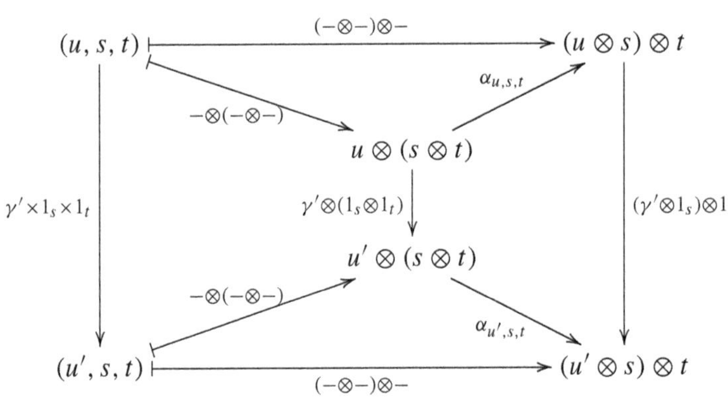

The trapezoid at right must commute, which is identical to the desired diamond.

(Case 3) Let $v = u \otimes (s \otimes t)$. Let $\beta = \alpha_{u,s,t}$ and $\gamma = 1_u \otimes (\gamma' \otimes 1_t)$. We have the following diagram:

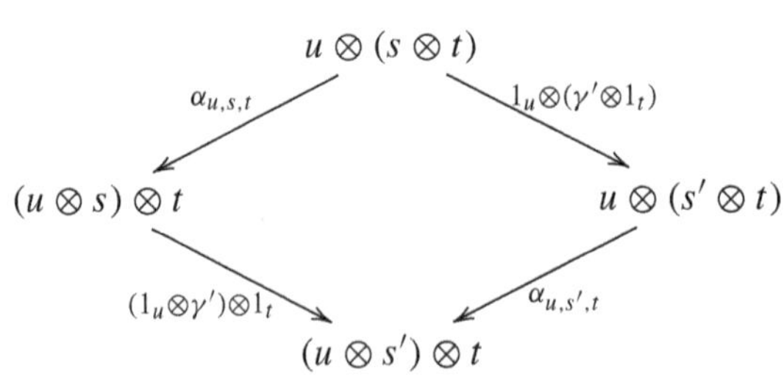

Naturality ensures the trapezoid at right in the following diagram commutes, which is identical to the desired diamond.

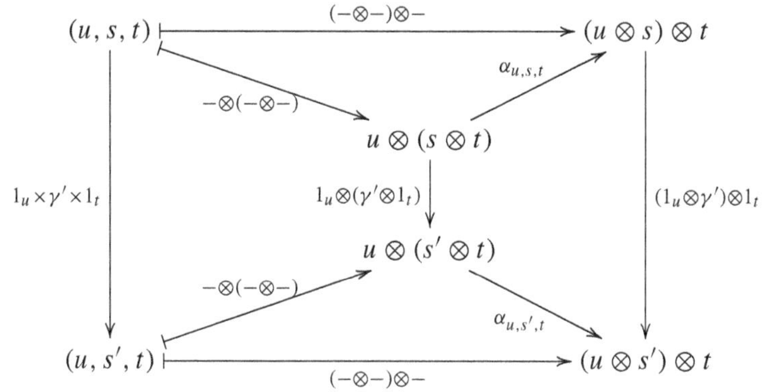

(Case 4) Let $v = u \otimes (s \otimes (p \otimes q))$. Let $\beta = \alpha_{u \otimes s, p, q}$ and $\gamma = 1_u \otimes \alpha_{s,p,q}$. We have the following diagram:

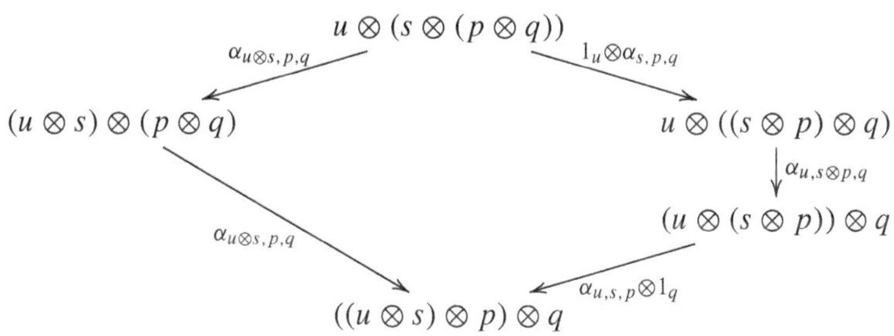

The pentagon coherence assures that this diagram commutes. So, arrows β and γ can be forwardly extended to join at the bottom vertex.

We can omit other cases considering symmetries among them. □

Now, we proceed to one of our main results. The reader might want to revisit Definition 8.5 beforehand.

Lemma 8.8 *The graph $G_{n,X}$ commutes. In other words, for any binary words v and w in $G_{n,X}$, any paths from v to w are identical.*

Proof Along the path from v to w, forward subpaths and backward ones appear alternately. The following diagram illustrates the situation:

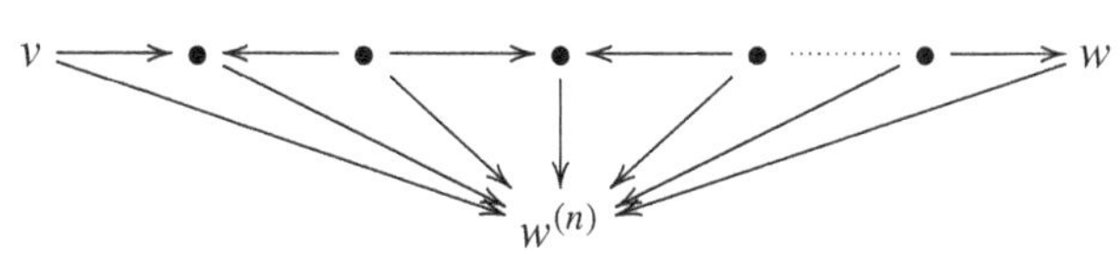

It does not matter whether the path start with a forward subpath or not. The argument will be completely the same. In any case, all the triangles commute due to

Lemma 8.6. This means we can collapse these triangles one by one to obtain the following equation:

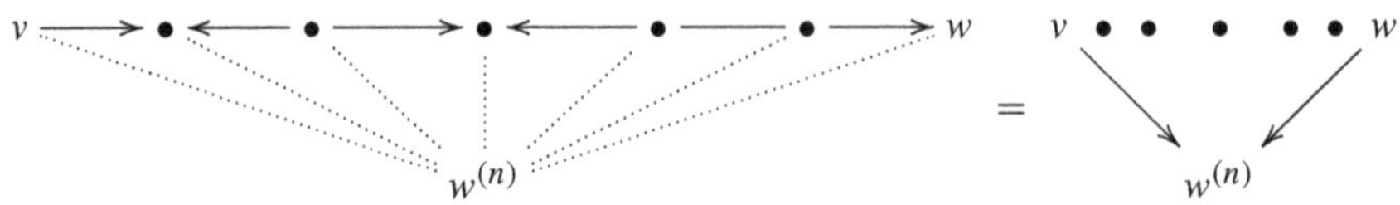

The RHS is determined solely by v and w, which means that all the paths from v to w are identical. □

We have shown that given a monoidal category $\mathscr{C}$ the correspondence of arrows from $\mathscr{W}_0$ to $\mathscr{C}$ can be consistently defined. However Theorem 8.1 claims the unique existence of a strict monoidal functor $\mathscr{W} \to \mathscr{C}$. So, we have to additionally consider all words possibly containing the symbol e.

Definition 8.7 Let $\bar{G}_{n,X}$ be the graph whose vertices are all words in $\mathscr{W}$ of length n. Edges are constructed recursively as follows:

- All basic arrows defined for $G_{n,X}$.
- Previously constructed arrows tensored from left or right by λ and ρ.

The graph $G_{n,X}$ is a subgraph of $\bar{G}_{n,X}$. Arbitrary number of e can be embedded in a word without affecting the length of the word. Therefore, the graph $\bar{G}_{n,X}$ infinite.

Lemma 8.9 *The graph $\bar{G}_{n,X}$ commutes. In other words, for any binary words v and w in $\bar{G}_{n,X}$, any paths from v to w are identical.*

We will reduce the problem to $G_{n,X}$. If we observe the fact that elimination of e from a word by the unitors can be safely done without affecting other parts of the word. Let $v = u \otimes (e \otimes w)$. The unitor coherence

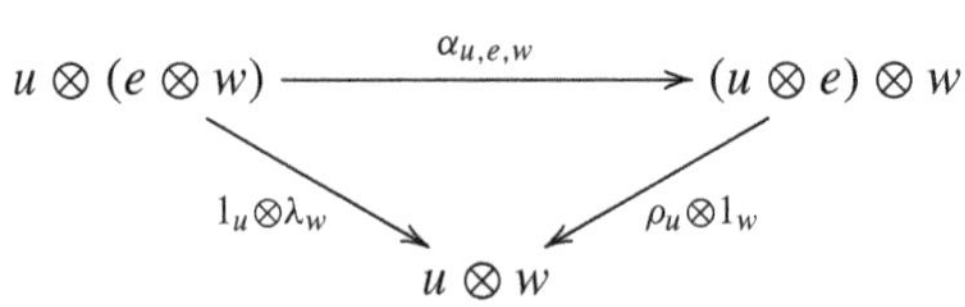

means that elimination of e can be done before or after the action of the associator α, yielding the same result. With the help of Lemma 8.1, we can eliminate e's at an arbitrary step without affecting associator operations. Thus, we can simply suppose that there are no e's in the word.

After a long way of discussing series of lemmas, we got to the complete proof of Theorem 8.1.

We nearly get to the final stage.

Definition 8.8 Let $\mathscr{C}$ be a monoidal category. For any binary word $w \in \mathrm{Obj}(\mathscr{W})$ of length n, we define a functor $w_{\mathscr{C}} : \mathscr{C}^n \to \mathscr{C}$ by replacing all "a"s by the identity functor of $\mathscr{C}$.

- If the length of w is 0, namely $w = e$, then $e_{\mathscr{C}} : \mathscr{C}^0 = \mathbf{1} \to \mathscr{C}$ is the constant functor $\mathrm{Const}_{\mathbb{1}} : \bullet \mapsto \mathrm{Id}_{\mathscr{C}}$, where $\bullet$ is the only object of $\mathbf{1}$ and $\mathrm{Id}_{\mathscr{C}} : \mathscr{C} \to \mathscr{C}$ is the identity functor.
- Let w and w' are words of length n and n', respectively. If $w_{\mathscr{C}}$ and $w'_{\mathscr{C}}$ are already defined, then $(w \otimes w')_{\mathscr{C}}$ is given by

$$\mathscr{C}^{n+n'} = \mathscr{C}^n \times \mathscr{C}^{n'} \xrightarrow{w_{\mathscr{C}} \times w'_{\mathscr{C}}} \mathscr{C} \times \mathscr{C} \xrightarrow{\otimes} \mathscr{C}$$

Example 8.5 Let $w = ((a \otimes a) \otimes a) \otimes (a \otimes a)$. The corresponding functor is given by

$$w_{\mathscr{C}} = ((\mathrm{Id}_{\mathscr{C}} \otimes \mathrm{Id}_{\mathscr{C}}) \otimes \mathrm{Id}_{\mathscr{C}}) \otimes (\mathrm{Id}_{\mathscr{C}} \otimes \mathrm{Id}_{\mathscr{C}}) \ .$$

Let $(X_1, X_2, X_3, X_4, X_5) \in \mathrm{Obj}(\mathscr{C}^5)$. Applying the functor $w_{\mathscr{C}}$, we get

$$((X_1 \otimes X_2) \otimes X_3) \otimes (X_4 \otimes X_5) \ .$$

Definition 8.9 Let $\mathscr{C}$ be a monoidal category. We construct another monoidal category $\mathrm{It}(\mathscr{C})$. Its objects are all pairs (n, T), where $T : \mathscr{C}^n \to \mathscr{C}$ is a functor. Its morphisms from (n, T) to (n', T'), if $n = n'$, are all natural transformations $\gamma : T \to T'$. There is no morphism from (n, T) to (n, T') if $n \neq n'$. We often write (n, T) as T for short when there can be no confusion.

The tensor product on objects is defined as $(n, T) \otimes (n', T') = (n + n', T \otimes T')$, where

$$T \otimes T' : \mathscr{C}^{m+n} \simeq \mathscr{C}^m \times \mathscr{C}^n \xrightarrow{T \times T'} \mathscr{C} \times \mathscr{C} \xrightarrow{\otimes} \mathscr{C} \ .$$

Note that we can safely identify $\mathscr{C}^{m+n} = \mathscr{C}^m \times \mathscr{C}^n$. The unit $\mathbb{1}_{\mathrm{It}(\mathscr{C})}$ is the constant functor

$$\mathscr{C}^0 \to \mathscr{C} : \bullet \mapsto \mathbb{1}_{\mathscr{C}}$$
$$1_\bullet \mapsto 1_{\mathbb{1}} \ ,$$

where $\mathscr{C}^0 = \mathbf{1}$ is the category with one object $\bullet$ and no morphisms other than the identity. We denote this unit $\mathbb{1}_{\mathrm{It}(\mathscr{C})}$ by I in what follows.

We need the associator $\alpha : - \otimes (- \otimes \ ) \to (- \otimes -) \otimes -$. For any $(n, T), (n', T'), (n'', T'')$,

$$\alpha_{T,T',T''} : T \otimes (T' \otimes T'') \to (T \otimes T') \otimes T'' \ ,$$

is defined componentwise as

$$(\alpha_{T,T',T''})_{X,X',X''} = \alpha_{TX,T'X',T''X''} \ ,$$

where X, X', and X'' denote

$$X = (X_1, X_2, \cdots, X_n) \in \mathrm{Obj}(\mathscr{C}^n),$$
$$X' = (X'_1, X'_2, \cdots, X'_{n'}) \in \mathrm{Obj}(\mathscr{C}^{n'}), \text{ and}$$
$$X'' = (X''_1, X''_2, \cdots, X''_{n''}) \in \mathrm{Obj}(\mathscr{C}^{n''}),$$

respectively.

The left unitor $\lambda_T : I \otimes T \to T$ is given as follows. For any $T : \mathscr{C}^n \to \mathscr{C}$ and $X = (X_1, X_2, \ldots, X_n) \in \mathrm{Obj}(\mathscr{C}^n)$, using the left unitor of $\mathscr{C}$, we set

$\lambda_{TX} : \mathbb{1}_{\mathscr{C}} \otimes TX \to TX$. Likewise the right unitor $\rho_T : T \otimes I \to T$ is given by $\rho_{TX} : TX \otimes \mathbb{1}_{\mathscr{C}} \to TX$.

Naturality will be shown later by Lemma 8.10. The pentagon coherence and unitor coherence are consequences of componentwise definitions.

Lemma 8.10 *The associator α, unitors λ and ρ in Definition 8.9 are a natural transformation, in fact a natural isomorphism.*

Proof Let us begin with the associator α. For any triplet of natural transformations $\gamma : T \to S$, $\gamma' : T' \to S'$, $\gamma'' : T'' \to S''$, we have to show the trapezoid at right in the following diagram commute:

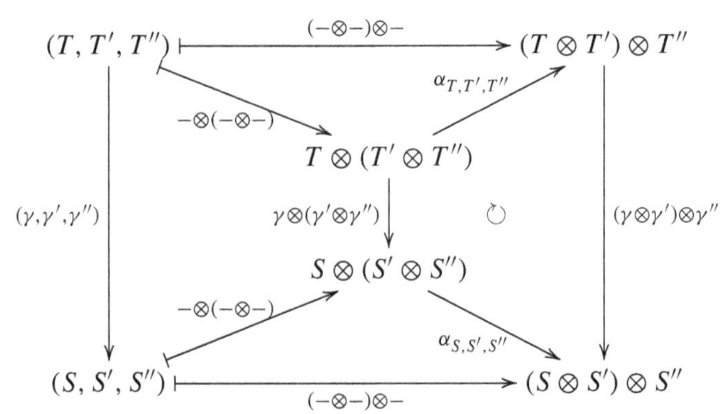

To prove this, we just have to confirm commutativity of the component at an arbitrary object (X, X', X''), where X, X', and X'' denote

$$X = (X_1, X_2, \cdots, X_n) \in \mathrm{Obj}(\mathscr{C}^n),$$
$$X' = (X'_1, X'_2, \cdots, X'_{n'}) \in \mathrm{Obj}(\mathscr{C}^{n'}), \text{ and}$$
$$X'' = (X''_1, X''_2, \cdots, X''_{n''}) \in \mathrm{Obj}(\mathscr{C}^{n''}),$$

respectively. The diagram at this component is as follows:

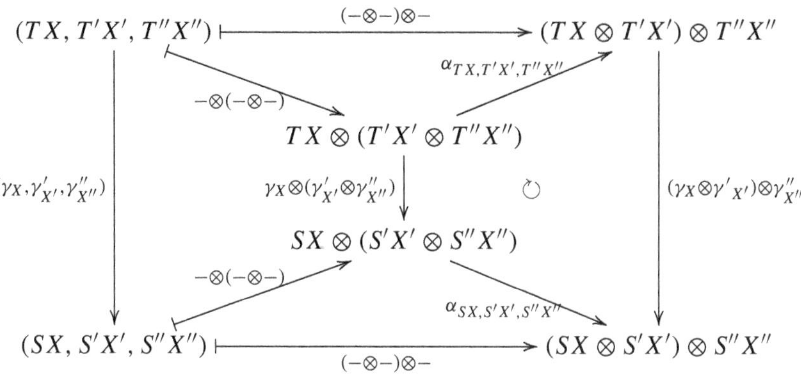

The associator appearing in this diagram is that of the category $\mathscr{C}$, which guarantees that the trapezoid at right commutes. Note that all the involved natural transformations are isomorphisms.

The naturality of the unitor λ requires that, for any (n, T), (n, S), and a natural transformation $\gamma : T \to S$, the trapezoid at right in the diagram below commutes:

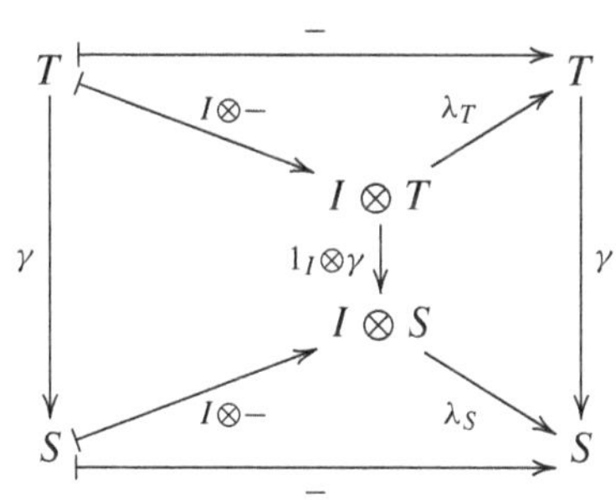

To prove this, we just have to confirm commutativity of the component at an arbitrary object

$$X = (X_1, X_2, \cdots, X_n) \in \mathrm{Obj}(\mathscr{C}^n) .$$

Namely,

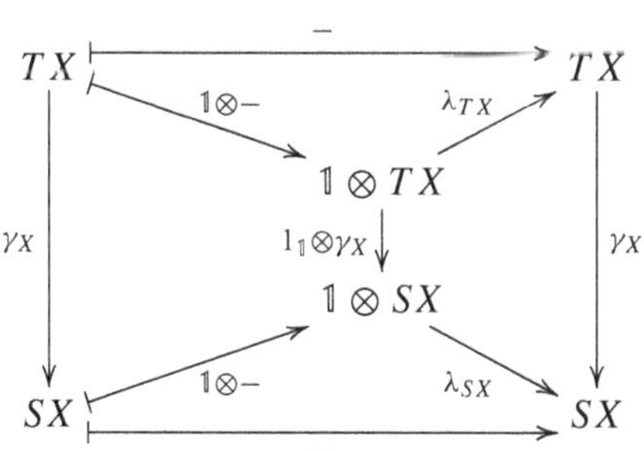

The unitor λ appearing in this diagram is that of the category $\mathscr{C}$, which guarantees that the trapezoid at right commutes. All the morphisms in this diagram are isomorphisms.

The discussion on the unitor ρ goes completely in the same way. □

Using the above construction $\mathrm{It}(\mathscr{C})$, we conclude the discussion on coherence. We give the last theorem.

We need the notion of canonical maps.

Definition 8.10 Let $\mathscr{C}$ be a monoidal category. Let v and w be binary words of the same length. We construct natural isomorphisms from $v_{\mathscr{C}} \to w_{\mathscr{C}}$ called a *canonical map* recursively as follows:

- If $v = w = e$, namely, both are empty word, we defined $v_{\mathscr{C}} = w_{\mathscr{C}} = \mathrm{Const}_{\mathbb{1}}$. The identity natural isomorphism of from the constant functor to itself

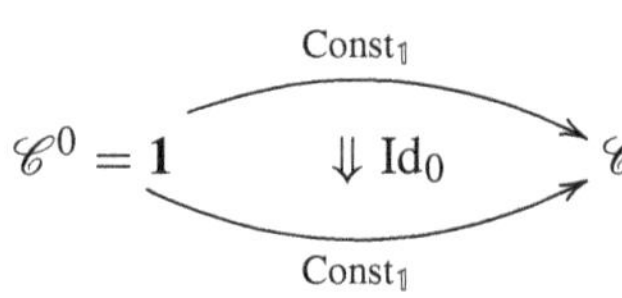

  is canonical.
- If $v = w = a$, then both $v_{\mathscr{C}}$ and $w_{\mathscr{C}}$ equal the identity functor $1_{\mathscr{C}}$. The identity isomorphism of from the identity functor to itself

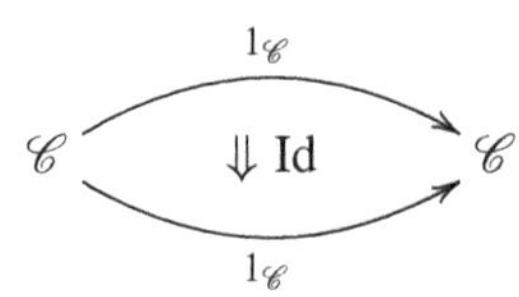

  is canonical.
- The associator, unitors, and their inverses, namely, α, α^{-1}, λ, λ^{-1}, ρ, and ρ^{-1} are canonical. For example, if $v = a \otimes (a \otimes a)$ and $w = (a \otimes a) \otimes a$, then $v_{\mathscr{C}} = - \otimes (- \otimes -)$ and $w_{\mathscr{C}} = (- \otimes -) \otimes -$.

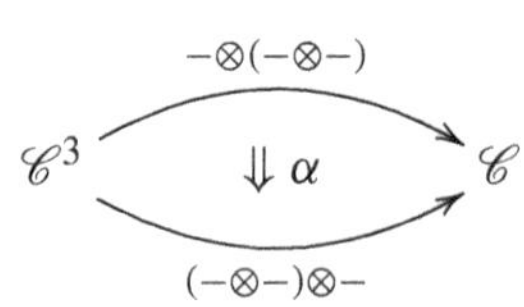

For another example, if $v = e \otimes a$ and $w = a$, then $v_{\mathscr{C}} = \mathbb{1} \otimes -$ and $w_{\mathscr{C}} = -$. Recall that v and w have the same length.

$$\mathscr{C} \simeq \mathbf{1} \times \mathscr{C} \quad \overset{\mathbb{1}\otimes -}{\underset{-}{\Downarrow \lambda}} \quad \mathscr{C}$$

is canonical. Note that $\mathscr{C}$ and $\mathbf{1} \times \mathscr{C}$ can be safely identified.

- The composites and $\otimes$ products of all items above are canonical. For example, if

$$\begin{aligned} v &= a \otimes (\underline{(a \otimes (a \otimes a))} \otimes e) \\ w &= a \otimes (\underline{((a \otimes a) \otimes a)} \otimes e) \end{aligned}$$

then

$$\mathscr{C}^4 \simeq \mathscr{C}^4 \times \mathbf{1} \quad \overset{-\otimes((-\otimes(-\otimes-))\otimes\mathbb{1})}{\underset{-\otimes(((-\otimes-)\otimes-)\otimes\mathbb{1})}{\Downarrow \mathrm{Id} \otimes (\alpha \otimes \mathrm{Const}_{\mathbb{1}})}} \quad \mathscr{C}$$

where we can safely identify $\mathscr{C}^4$ with $\mathscr{C}^4 \times \mathbf{1}$.

The following theorem is a corollary of Theorem 8.1.

Theorem 8.2 *Let $\mathscr{C}$ be a monoidal category. There is a function that assigns to each pair of words v and w of the same length n a unique natural isomorphism*

$$\mathscr{C}^n \quad \overset{v_{\mathscr{C}}}{\underset{w_{\mathscr{C}}}{\Downarrow \mathrm{can}_{\mathscr{C}}(v, w)}} \quad \mathscr{C}$$

from the functor $v_{\mathscr{C}}$ to the functor $w_{\mathscr{C}}$. We call this a canonical map.

Just write down the $(X_1, X_2, \ldots, X_n)$ component of $\mathrm{can}_{\mathscr{C}}(v, w)$, then you will see that this theorem says all about coherence.

References

1. Coecke B, Kissinger A (2017) Picturing quantum processes. Cambridge University Press
2. Turaev V, Virelizier A (2017) Monoidal categories and topological field theory. Birkhäuser
3. Heunen C, Sadrzadeh M, Grefenstette E (eds) (2013) Quantum physics and liguistics. Oxford University Press
4. Mac Lane S (1998) Categories for the working mathematician, Graduate texts in mathematics, vol 5, 2nd edn. Springer

Epilogue

I wonder if the readers enjoyed a long journey to Yoneda's lemma or the coherence problem in monoidal categories. It was possible to take a shortcut to these goals. However, since this book is intended as an introductory book on category theory, readers are supposed to experience various examples of naturality and universality to really appreciate the beauty of these concepts.

Introducing Moggi's ideas with sample code became too lengthy, so I moved most of the correctness proofs to an appendix.

Most of the diagrams in this book were scribbled on my notebooks while I was seated on the commuter train over the past 10 years or so. I drew thousands of diagrams on the notebooks, but I discarded most of them. I selected some of the surviving diagrams that looked good to include in a publication. Criticism from the readers is welcome.

The Haskell code has been used in class on category theory. The code pieces are far from systematically designed. Rather, each piece of code is designed as independently as possible, so the reader can read and test it on the spot.

I had a plan to include the following topics.

- Kan extension
- Ends and coends
- Enriched categories

However, I gave up the original plan because of the objectives of this book as I stated above. In addition, I found the quality of the code for these topics is not good enough for publication. I am looking forward to including these in the near future.

I would like to thank all the readers who have read this book.

S. Yukita, *Category Theory Using Haskell*, Computer Science Foundations and Applied Logic, https://doi.org/10.1007/978-3-031-68538-5

Appendix A
Sets and Universe

The descriptions in this appendix are based on the literature [1], but with less rigor to suit the purpose of this book.

A famous paradox in set theory arises from the construction of sets using the comprehension principle. In the Naive set theory, given a property $\varphi(x)$ of x, we can create a set $\{x \mid \varphi(x)\}$ consisting of all x that satisfies this property. We know that this cannot be done freely without restrictions. Here, when we say the property $\varphi(x)$ of x, we are referring to the function φ that returns true or false for x. The *restricted comprehension principle* allows us to use comprehension only in the form as follows:

$$\{x \mid x \in u, \varphi(x)\},$$

where u is already known to be a set. This proves to give us a good theoretical and practical framework for working in standard mathematics if it observes conditions described below.

In order to handle sets without contradiction, the rules for constructing new sets should be clearly defined. The construction of sets that deviate from the rules should be prohibited. We use a tool called a *universe* to handle "the set of all sets" or "a set of all sets with a given property" without contradiction. It was invented by Grothendieck to handle large categories with a minimum theoretical setting.

We assume the existence of a universe. A *universe* is a set U with the following properties:

1. $x \in y$ and $y \in U$ implies $x \in U$.
2. $x \in U$ and $y \in U$ imply $\{x, y\}$, $\langle x, y \rangle$, and $u \times v \in U$.
3. $x \in U$ implies $\mathscr{P}x \in U$ and $\cup x \in U$.
4. $\omega = \{0, 1, 2, \ldots\} \in U$.
5. if $f : a \to b$ is surjective with $a \in U$ and $b \subset U$, then $b \in U$.

These properties ensure that U is closed under all standard operations in set theory.

S. Yukita, *Category Theory Using Haskell*, Computer Science Foundations and Applied Logic, https://doi.org/10.1007/978-3-031-68538-5

A universe may not be unique (recall that we assumed its existence). So, we take one of them, and fix the universe U, and call any set $u \in U$ a *small set*. Therefore, we may say that the universe U is the set of all small sets.

We say function $f : a \to b$ is *small* if a and b are small sets. In fact, function f can be identified with its graph $\{\langle x, y\rangle \mid \langle x, y\rangle \in a \times b, \langle x, y\rangle \in f\}$. So, by the rules above with the limited comprehension principle, we know that f is a small set.

Given small sets a and b, we can form the small set of functions from a to b. Let

$$\varphi(f) \stackrel{\text{def}}{=} \forall x \in a. \exists! y.\ (x, y) \in f,$$

which is the requirement for a correspondence f to be a *function*. We set

$$\mathrm{Hom}(a, b) = \{f \mid f \in a \times b, \varphi(f)\},$$

which is also a small set.

Now, we can define the category **Set**. Its set of objects is U and its set of morphisms from a to b is $\mathrm{Hom}(a, b)$ as defined above.

Appendix B
Calculations with Output

In this appendix, we will prove all the equations that are required to establish instances of Functor, Applicative, and Monad type classes in the sample code 6.7 `CalcWith-Output2.hs`.

B.1 Functor Instance

We prove equations for Fuctor instances.

Proposition B.1 *The following equality holds.*

```
fmap id = id
```

Proof We rewrite the LHS step by step.

```
fmap id x
      -- apply the definition of fmap
  = T (\s -> let (a,s2) = (unT x) s in (a,s2) )
      -- evaluate the let expression
  = T (\s -> unT x s)
      -- eta conversion
  = T (unT x)
      -- cancel the T/unT pair
  = x
```

□

S. Yukita, *Category Theory Using Haskell*, Computer Science Foundations and Applied Logic, https://doi.org/10.1007/978-3-031-68538-5

Proposition B.2 *The following equality holds.*
```
fmap (g.f) = fmap g.fmap f
```

Proof We rewrite the RHS step by step.

```
fmap g (fmap f ta)
      -- evaluate (fmap f)
  = fmap g (T (\s -> let (a,s2)=unT ta s
                     in (f a,s2) ) )
      -- evaluate (fmap g)
  = T (\t->let
               (b,t2)=unT (T (\s->let (a,s2)=unT ta s
                                  in  (f a,s2))
                     ) t
            in (g b,t2) )
      -- cancel the T/unT pair
  = T (\t->let
               (b,t2)=(\s -> let (a,s2)=unT ta s
                             in  (f a,s2)
                     ) t
            in (g b,t2) )
      -- beta reduction
  = T (\t->let
               (b,t2)=(let
                           (a,s2)=unT ta t
                        in (f a,s2))
            in (g b,t2) )
      -- combine two let-expressions: b = f a, t2 = s2
  = T (\t->let
               (a,s2)=unT ta t
               (b,t2)=(f a,s2)
            in (g b,t2) )
      -- combine substitutions
  = T (\t->let
               (a,t2)=unT ta t
            in ((g.f) a),t2) )
      -- definition of (fmap)
  = fmap (g.f) ta
```

□

B.2 Applicative Instance

Proposition B.3 (Identity) *The following equality holds.*

```
pure id <*> ta = ta
```

Proof We rewrite the LHS step by step.

```
pure id <*> ta
      -- by definition of pure and <*>
  = T (\s ->
            let (h,s2) = unT (T (\s->(id,s)) s
                (a,s3) = unT ta s2
            in  (h a,s3) )
      -- cancel the T/unT pair
  = T (\s ->
            let (h,s2) = (\s->(id,s)) s
                (a,s3) = unT ta s2
            in  (h a,s3) )
      -- beta reduction
  = T (\s ->
            let (h,s2) = (id,s)
                (a,s3) = unT ta s2
            in  (h a,s3) )
      -- do part of the let expression
  = T (\s -> let (a,s3) = unT ta s
            in  (a,s3) )
      -- do another part of the let expression
  = T (\s -> unT ta s)
      -- eta conversion
  = T (unT ta)
      -- cancel the T/unT pair
  = ta
```

□

Proposition B.4 (Composition) *The following equality holds.*

```
pure (.) <*> tu <*> tv <*> tw = tu <*> (tv <*> tw)
```

Proof We rewrite the LHS from left.

```
pure (.)
      -- definition of pure
  = T (\s -> ((.),s) )
      -- Recall that (.) u v w = (u.v) w = u (v w).
  = T (\s -> ((\g -> \f -> \x -> g(f x)),s) )
```

```
shift to right
  pure (.) <*> tu
    = T (\s -> ((\g -> \f -> \x -> g(f x)),s) ) <*> tu
         -- definition of <*>
    = T (\ss->let (h,s2) = unT (T(\s->((\g->\f->\x->g(f x)),s) )
                               ) ss
                  (u,s3) = unT tu s2
              in  (h u,s3) )
         -- cancel the T/unT pair
    = T (\ss->let (h,s2) = (\s->((\g->\f->\x->g(f x)),s) ss
                  (u,s3) = unT tu s2
              in  (h u,s3) )
         -- beta reduction
    = T (\ss->let (h,s2) = ((\g->\f->\x->g(f x)),ss)
                  (u,s3) = unT tu s2
              in  (h a,s3) )
         -- streamline the let expression
    = T (\ss->let h=(\g->\f->\x->g(f x))
                  (u,s3) = unT tu ss
              in  (h u,s3) )
         -- further streamline the let expression
    = T (\s-> let (u,s3) = unT tu s
              in  ((\g->\f->\x->g(f x)) u , s3) )
         -- beta reduction in the "in" clause
    = T (\s-> let (u,s3) = unT tu s
              in  ((\f->\x->u(f x)) , s3) )
         -- rename local a variable
    = T (\s-> let (u,s2) = unT tu s
              in  ((\f->\x->u(f x)) , s2) )

shift further right
  pure (.) <*> tu <*> tv
         -- apply the definition of <*> to the second instance
    = T (\ss->let (h,s3) =
                     unT (T (\s-> let (u,s2) = unT tu s
                                  in  ((\f->\x->u(f x)), s2) )
                     ) ss
                  (v,s4) = unT tv s3
              in  (h v,s4)
      )
         -- cancel the T/unT pair
    = T (\ss->let (h,s3) = (\s-> let (u,s2) = unT tu s
                                 in  ((\f->\x->u(f x)) , s2) )
                           ) ss
                  (v,s4) = unT tv s3
              in  (h v,s4)
      )
         -- beta reduction
    = T (\ss->let (h,s3) = let (u,s2) = unT tu ss
```

```
                        in  ((\f->\x->u(f x)) , s2)
                  (v,s4) = unT tv s3
              in  (h v,s4)
      )
        -- let s3=s2 and h = (\f->\x->u(f x))
        -- beautify
    = T (\ss->let (u,s2) = unT tu ss
                  (v,s3) = unT tv s2
              in  ((\f->\x->u(f x)) v,s3)
      )
        -- beta reduction in the left part on the (in) expression
    = T (\ss->let (u,s2) = unT tu ss
                  (v,s3) = unT tv s2
              in  ((\x->u(v x)),s3)
      )

Complete the calculation of the LHS to the rightmost
  pure (.) <*> tu <*> tv <*> tw
        -- definition of the rightmost <*>
    = T (\s->let (h,s4)=unT (T (\ss->let (u,s2) = unT tu ss
                                         (v,s3) = unT tv s2
                                     in  ((\x->u(v x)),s3)
                           ) s
                 (w,s5)=unT tw s4
             in  (h w,s5)
      )
        -- cancel the T/unT pair
    = T (\s->let (h,s4)=(\ss->let (u,s2) = unT tu ss
                                  (v,s3) = unT tv s2
                              in  ((\x->u(v x)),s3)
                        ) s
                 (w,s5)=unT tw s4
             in  (h w,s5)
      )
        -- beta reduction
    = T (\s->let (h,s4)=let (u,s2) = unT tu s
                            (v,s3) = unT tv s2
                        in  ((\x->u(v x)),s3)
                 (w,s5)=unT tw s4
             in  (h w,s5)
      )
        -- knowing that s4 = s3 and h = (\x->u(v x))
    = T (\s->let (u,s2) = unT tu s
                 (v,s3) = unT tv s2
                 (w,s4) = unT tw s3
             in  ((\x->u(v x)) w,s4)
      )
        -- beta reduction in the (in) expression
    = T (\s->let (u,s2) = unT tu s
```

```
                (v,s3) = unT tv s2
                (w,s4) = unT tw s3
            in  (u(v w),s4)
    )
```

We compute the RHS from the innermost of parentheses.

```
  tv <*> tw
        -- definition of <*>
    = T (\s -> let (v,s2)=unT tv s
                   (w,s3)=unT tw s2
               in  (v w,s3)          )
```

go one level up

```
  tu <*> (tv <*> tw)
    = T (\t -> let (u,s4)=unT tu
                   (x,s5)=unT (T (\s -> let (v,s2)=unT tv s
                                            (w,s3)=unT tw s2
                                        in  (v w,s3) )
                              ) s4
               in  (u x,s5)
        -- rename s4 to s1
    = T (\t -> let (u,s1)=unT tu t
                   (x,s5)=(\s -> let (v,s2)=unT tv s
                                     (w,s3)=unT tw s2
                                 in  (v w,s3)
                          ) s1
               in  (u x,s5)
        )
        -- beta reduction
    = T (\t -> let (u,s1)=unT tu t
                   (x,s5)=let (v,s2)=unT tv s1
                              (w,s3)=unT tw s2
                          in  (v w,s3)
               in  (u x,s5)
        )
        -- streamline the let expresion
    = T (\t -> let (u,s1)=unT tu t
                   (v,s2)=unT tv s1
                   (w,s3)=unT tw s2
                   (x,s5)=(v w,s3)
               in  (u x,s5)
        )
        -- collect
    = T (\s -> let (u,s1)=unT tu s
                   (v,s2)=unT tv s1
                   (w,s3)=unT tw s2
               in  (u (v w),s3)
        )
        -- rename local variables
```

```
    = T (\s -> let (u,s2)=unT tu s
                   (v,s3)=unT tv s2
                   (w,s4)=unT tw s3
               in  (u (v w),s4)
      )
The LHS and RHS gives the same result.
```

□

Proposition B.5 (Homomorphism) *The following equality holds.*
`pure f <*> pure x = pure (f x)`

Proof We rewrite the LHS step by step.

```
pure f <*> pure x
      -- definition of <*>
  = T (s -> let (ff,s2) = unT (T (\t -> (f,t)) ) s
                (xx,s3) = unT (T (\t -> (x,t)) ) s2
            in  (ff xx,s3) )
      -- cancel the T/unT pair
  = T (s -> let (ff,s2) = (\t -> (f,t)) s
                (xx,s3) = (\t -> (x,t)) s2
            in  (ff xx,s3) )
      -- beta reduction at two places
  = T (s -> let (ff,s2) = (f,s)
                (xx,s3) = (x,s2)
            in  (ff xx,s3) )
      -- streamline the let expression
  = T (s -> ((f x),s) )
      -- definition of pure
  = pure (f x)
```

□

Proposition B.6 (Commutativity) *The following equality holds.*
`u <*> pure y = pure ($ y) <*> u`

Proof We rewrite the LHS step by step.

```
u <*> pure y
      -- definitions of <*> and pure
  = T (s -> let (uu,s2) = unT u s
                (yy,s3) = unT (T (\t -> (y,t)) ) s2
            in  (uu yy,s3) )
      -- cancel the T/unT pair
```

```
    = T (s -> let (uu,s2) = unT u s
                  (yy,s3) = (\t -> (y,t)) s2
              in  (uu yy,s3) )
        -- beta reduction
    = T (s -> let (uu,s2) = unT u s
                  (yy,s3) = (y,s2)
              in  (uu yy,s3) )
        -- streamline substitutions
    = T (s -> let (uu,s2) = unT u s
              in  (uu y,s2) )
        -- simplify local variables
    = T (s -> let (f,s2) = unT u s
              in  (f y,s2) )

We compute the RHS.
  pure ($ y) <*> u
        -- eta conversion ($ y) = \f -> f y
    = pure (\f -> f y) <*> u
        -- definition of pure
    = T (\s -> ((\f->f y),s) ) <*> u
        -- definition of <*>
    = T (\t -> let (g,t2) = unT (T (\s -> ((\f->f y),s) )) t
                   (a,t3) = unT u t2
               in  (g a,t3) )
        -- cancel the T/unT pair
    = T (\t -> let (g,t2) = (\s -> ((\f->f y),s) ) t
                   (a,t3) = unT u t2
               in  (g a,t3) )
        -- beta reduction
    = T (\t -> let (g,t2) = ((\f->f y),t)
                   (a,t3) = unT u t2
               in  (g a,t3) )
        -- streamline the let expression
    = T (\t -> let (a,t3) = unT u t
               in  ((\f->f y) a,t3) )
        -- beta reduction
    = T (\t -> let (a,t3) = unT u t
               in  (a y,t3) )
        -- rename local variables
    = T (\s -> let (f,s2) = unT u s
               in  (f y,s2) )
```

□

B.3 Monad Instance

We prove equations for functor instances.

Proposition B.7 (left unit) *The following equality holds.*
`return x >>= f = f x`

Proof We rewrite the LHS step by step.

```
return x >>= f
      -- definition of return
  = T (\s -> (x,s)) >>= f
      -- definition of (>>=)
  = sharp f $ T (\s -> (x,s))
  = T (\st1 -> let (y,st2) = unT T (\s -> (x,s)) s
               in  unT (f y) st2 )
      -- cancel the T/unT pair
  = T (\st1 -> let (y,st2) = (\s -> (x,s)) st1
               in  unT (f y) st2 )
      -- beta reduction
  = T (\st1 -> let (y,st2) = (x,st1)
               in  unT (f y) st2 )
      -- streamline the let expression
  = T (\st1 -> unT (f x) st1 )
      -- eta conversion
  = T (unT (f x))
      -- cancel the T/unT pair
  = f x
```

□

Proposition B.8 (right unit) *The following equality holds.*
`m >>= return = m`

Proof

```
m >>= return
      -- definition of (>>=)
  = sharp return m
  = T (\st1 -> let (y,st2) = unT m st1
               in  unT (return y) st2 )
      -- definition of return
  = T (\st1 -> let (y,st2) = unT m st1
               in  unT (T (\s->(y,s) ) st2 )
      -- cancel the T/unT pair
  = T (\st1 -> let (y,st2) = unT m st1
```

```
                   in   (\s->(y,s) ) st2 )
            -- beta reduction
        = T (\st1 -> let (y,st2) = unT m st1
                   in   (y,st2) )
            -- streamline the let expression
        = T (\st1 -> unT m st1)
            -- eta conversion
        = T (unT m)
            -- cancel the T/unT pair
        = m
```

□

Proposition B.9 (associativity) *The following equality holds.*
(m >>= f) >>= g = m >> = (\x → f x >>= g)

Proof We first calculate a subexpression of the LHS.

```
   m >>= f
        -- definition of (>>=)
     = sharp f m
     = T (\st1 -> let (y,st2) = unT m st1
                  in  unT (f y) st2 )

Next, we calculate the whole LHS.
   (m >>= f) >>= g
        -- replace (>>= g) with sharp
     = sharp g $ T (\st1 -> let (y,st2) = unT m st1
                            in  (f y) st2 )
        -- definition of sharp
     = T (\s1 -> let (p,s2) =
                    unT (T (\st1 -> let (y,st2) = unT m st1
                                    in  unT (f y) st2 ))
                    s1
                 in  unT (g p) s2 )
        -- cancel the T/unT pair
     = T (\s1 -> let (p,s2) = (\st1 -> let (y,st2) = unT m st1
                                       in  unT (f y) st2 )
                              s1
                 in  unT (g p) s2 )
        -- beta reduction
     = T (\s1 -> let (p,s2) = let (y,st2) = unT m s1
                              in  unT (f y) st2
                 in  unT (g p) s2 )
        -- streamline the let expression
     = T (\s1 -> let (y,st2) = unT m s1
                     (z,st3) = unT (f y) st2
                 in  unT (g z) st3 )
        -- rename local variables
     = T (\s1 -> let (v2,s2) = unT m s1
```

```
                  (v3,s3) = unT (f v2) s2
              in  unT (g v3) s3 )

We calculate a subexpression of the RHS.
  f x >>= g
        -- replace (>>= g) with sharp
    = sharp g (f x)
    = T (\st1 -> let (y,st2) = unT (f x) st1
                 in  unT (g y) st2 )
We calculate a subexpression that contains this.
  \x -> (f x >>= g)
        -- use the result above
    = \x -> T (\st1 -> let (y,st2) = unT (f x) st1
                       in  unT (g y) st2 )
We calculate the whole RHS.
  m >>= (\x -> f x >>= g)
        -- replace (>>=) with sharp
    = sharp (\x -> T (\st1 -> let (y,st2) = unT (f x) st1
                              in  unT (g y) st2 )  )  m
        -- definition of sharp
    = T (\s1 -> let (p,s2) = unT m s1
                in  unT ((\x -> T (\st1 -> let (y,st2) = unT (f x) st1
                                           in  unT (g y) st2 ) )
                         p)
                    s2
        -- beta reduction
    = T (\s1 -> let (p,s2) = unT m s1
                in  unT (T (\st1 -> let (y,st2) = unT (f p) st1
                                    in  unT (g y) st2) )
                    s2
        -- cancel the T/unT pair
    = T (\s1 -> let (p,s2) = unT m s1
                in  (\st1 -> let (y,st2) = unT (f p) st1
                             in  unT (g y) st2)
                    s2
        -- beta reduction
    = T (\s1 -> let (p,s2) = unT m s1
                in  let (y,st2) = unT (f p) s2
                    in  unT (g y) st2 )
        -- streamline the let expression
    = T (\s1 -> let (p,s2) = unT m s1
                    (y,st2) = unT (f p) s2
                in  unT (g y) st2 )
        -- change the names of local variables
    = T (\s1 -> let (v2,s2) = unT m s1
                    (v3,s3) = unT (f v2) s2
                in  unT (g v3) s3 )
```

□

Proposition B.10 (compatibility of Applicative and Monad) *The following equality holds.*

```
mf <*> mx = mf >> = \f  → mx >> = \x  → return (f x)
```

Proof We rewrite the LHS step by step.

```
  mf <*> mx
        -- definition of <*>
    = T (\s1 -> let (f,s2) = unT mf s1
                    (x,s3) = unT mx s2
                in  (f x,s3) )
We calculate subexpressions of the RHS from the innermost.
  return (f x)
    = T (\s -> (f x,s))

  mx >>= \x -> return (f x)
        -- replace (>>=) with sharp
    = sharp (\x -> T (\s->(f x,s))) mx
        -- definition of sharp
    = T (\t1 -> let (y,t2) = unT mx t1
                in  unT ((\x -> T (\s->(f x,s)))
                         y)
                    t2 )
        -- beta reduction
    = T (\t1 -> let (y,t2) = unT mx t1
                in  unT (T (\s->(f y,s)))
                    t2 )
        -- cancel the T/unT pair
    = T (\t1 -> let (y,t2) = unT mx t1
                in  (\s->(f y,s))
                    t2 )
        -- beta reduction
    = T (\t1 -> let (y,t2) = unT mx t1
                in  (f y,t2) )

  mf >>= \f -> mx >>= \x -> return (f x)
    = mf >>= \f -> T (\t1 -> let (y,t2) = unT mx t1
                             in  (f y,t2) )
        -- replace (>>=) with sharp
    = sharp (\f -> T (\t1 -> let (y,t2) = unT mx t1
                             in  (f y,t2) )) mf
        -- definition of sharp
    = T (\s1 -> let (v1,s2) = unT mf s1
                in  unT ((\f -> T (\t1 -> let (y,t2) = unT mx t1
                                          in  (f y,t2) ))
                         v1)
```

```
                s2 )
    -- beta reduction
  = T (\s1 -> let (v1,s2) = unT mf s1
              in  unT (T (\t1 -> let (y,t2) = unT mx t1
                                 in  (v1 y,t2) ) )
                s2 )
    -- cancel the T/unT pair
  = T (\s1 -> let (v1,s2) = unT mf s1
              in (\t1 -> let (y,t2) = unT mx t1
                         in  (v1 y,t2) )
               s2 )
    -- beta reduction
  = T (\s1 -> let (v1,s2) = unT mf s1
                  let (y,t2) = unT mx s2
                  in  (v1 y,t2) )
    -- streamline the let expression
  = T (\s1 -> let (v1,s2) = unT mf s1
                  (y,t2) = unT mx s2
              in  (v1 y,t2) )
    -- change the names of local variables
  = T (\s1 -> let (f,s2) = unT mf s1
                  (x,s3) = unT mx s2
              in  (f x,s3) )
```

□

Appendix C
Continuation Passing

In this appendix, we will prove all the equations that are required to establish instances of Functor, Applicative, and Monad type classes in the sample code 6.10 `Continuation2.hs`.

C.1 Functor Instance

Proposition C.1 *The following equality holds.*

```
fmap id = id
```

Proof We will show that the LHS and RHS produce the same result when applied to any h.

```
        -- definition of fmap
  T (\b' -> unT h (b'.id))
        -- omit id
    = T (\b' -> unT h b')
        -- eta conversion
    = T $ unT h
        -- cancel the T/unT pair
    = h
```

□

Proposition C.2 *The following equality holds.*

```
fmap (g.f) = fmap g . fmap f
```


S. Yukita, *Category Theory Using Haskell*, Computer Science Foundations and Applied Logic, https://doi.org/10.1007/978-3-031-68538-5

Proof We will show that the LHS and RHS produce the same result when applied to any h.
Applying the LHS to h gives the following.

```
fmap (g.f) h = T (\c' -> unT h (c'.g.f))
```

Applying the RHS to h gives the following.

```
fmap g (fmap f h)
      -- apply definition to the inner fmap
  = fmap g (T (\b' -> unT h (b'.f))
      -- apply definition to the remaining fmap
  = T (\c' -> unT (T (\b' -> unT h (b'.f))) (c'.g))
      -- cancel the T/unT pair
  = T (\c' -> (\b' -> unT h (b'.f)) (c'.g))
      -- beta reduction
  = T (\c' -> unT h (c'.g.f))
```

□

C.2 Applicative Instance

Proposition C.3 *The following equality holds.*

```
pure id <*> ta = ta
```

Proof We rewrite the LHS step by step.

```
pure id <*> ta
      -- replace pure by eta
  = eta id <*> ta
      -- definition of eta
  = T (\aa' -> aa' id) <*> ta
      -- definition of <*>
  = T (\b' -> unT (T (\aa' -> aa' id)) $ \f -> unT ta (b'.f))
      -- cancel the T/unT pair
  = T (\b' -> (\aa' -> aa' id) $ \f -> unT ta (b'.f))
      -- beta reduction
  = T (\b' -> (\f -> unT ta (b'.f)) id)
      -- beta reduction
  = T (\b' -> unT ta (b'.id))
      -- omit id
  = T (\b' -> unT ta b')
      -- eta conversion
  = T (unT ta)
      -- cancel the T/unT pair
  = ta
```

□

Proposition C.4 *The following equality holds.*

```
pure (.) <*> tu <*> tv <*> tw = tu <*> (tv <*> tw)
```

Proof We give general useful remarks.

```
Remark 1.  Function composition operator (.) can be defined as
follows.
  (.) :: (b->c) -> (a->b) -> (a->c)
  (.) = \g -> \f -> \x -> g (f x)

Remark 2.  Function composition sometimes takes the form as follows.
   f . (\x -> y) = \x -> f y

We calculate the LHS.
Use a local variable bcabac'.
  -- bcabac'  :: ((b->c) -> (a->b) -> (a->c)) -> v
  -- pure (.) :: T v ((b->c) -> (a->b) -> (a->c))
We get the following.
  pure (.) = T (\bcabac' -> bcabac' (.))

Use local variables tu and abac'.
  -- tu              :: T v (b->c)
  -- unT tu          :: ((b->c)->v) -> v
  -- pure (.) <*> tu :: T v ((a->b) -> (a->c))
  -- abac'           :: ((a->b) -> (a->c)) -> v
  -- abac'.z         :: (b->c)->v
  -- z               :: (b->c) -> ((a->b) -> (a->c))
We get the following.
  pure (.) <*> tu
    = T (\bcabac' -> bcabac' (.) ) <*> tu
        -- definition of <*>
    = T (\abac' -> unT (T (\bcabac' -> bcabac' (.) ))
                   \z -> unT tu
                         (abac'. z)
        -- cancel the T/unT pair
    = T (\abac' -> (\bcabac' -> bcabac' (.))
                   \z -> unT tu
                         (abac'. z)
        -- beta reduction
    = T (\abac' -> (\z -> unT tu
                          (abac'. z)
                          (.) )
        -- beta reduction
    = T (\abac' -> unT tu
                   (abac'. (.) ) )
        -- Remark 1
    = T (\abac' -> unT tu
                   (abac'. (\u -> \v -> \x -> u (v x) ) )
        -- Remark 2
```

```
    = T (\abac' -> unT tu
                   \u -> abac'
                         \v -> \x -> u (v x) )

Use local variables tu, tv, and g.
  -- tv               :: T v (a->b)
  -- unT tv           :: ((a->b)->v) -> v
  -- pure (.) <*> tu <*> tv :: T v (a->c)
  -- g                :: b->c
We get the following.
  pure (.) <*> tu <*> tv
    = T ( \abac' -> unT tu
                    \u -> abac'
                          \v -> \x -> u (v x) )
     <*> tv
    = T ( \ac' -> (\abac' -> unT tu
                             \u -> abac'
                                   \v -> \x -> u (v x) )
                  \g -> unT tv (ac'. g)  )
        -- beta reduction
    = T ( \ac' -> unT tu
                  \u -> (\g -> unT tv (ac'. g))
                        \v -> \x -> u (v x) )
        -- beta reduction
    = T ( \ac' -> unT tu
                  \u -> unT tv
                        ac'. (\v -> \x -> u (v x)) )

Use local variables tu, tv, and tw.
  -- tw                            :: T v a
  -- pure (.) <*> tu <*> tv <*> tw :: T v c
We get the following.
  pure (.) <*> tu <*> tv <*> tw
    = T (c' -> \ac' -> unT tu
                       \u -> unT tv
                             ac'. (\v -> \x -> u (v x)) )
               f -> unT tw (c'. f)
        -- beta reduction
    = T (c' -> unT tu
               \u -> unT tv
                     (f -> unT tw (c'.f)) . (\v -> \x -> u (v x)) )
        -- Remark 2
    = T (c' -> unT tu
               \u -> unT tv
                     \v -> unT tw $ c' . (\x -> u (v x))  )
        -- Remark 2
    = T (c' -> unT tu
               \u -> unT tv
                     \v -> unT tw
```

```
                        \x -> c'
                              u (v x)   )
        -- beta reduction
     = T (c' -> unT tu
                 \u -> unT tv
                       \v -> unT tw
                             c'. u. v)

We calculate the RHS.
Use local variables tv, tw, b', and v.
  -- tv          :: T v (a->b)
  -- unT tv     :: ((a->b)->v) -> v
  -- tw          :: T v a
  -- unT tw     :: (a->v) -> v
  -- tv <*> tw :: T v b
  -- b'          :: b -> v
  -- v           :: a -> b
We get the following
  tv <*> tw
    = T (\b' -> unT tv
                 \v -> unT tw (b'. v))

Use local variables tu, tv, tw, c', b', and u.
  -- tu <*> (tv <*> tw) :: T v c
  -- c'                  :: c -> v
  -- b'                  :: b -> v
  -- u                   :: b -> c
We get the following.
  tu <*> (tv <*> tw)
    = T (\c' -> unT tu
                 \u -> unT (T (\b' -> unT tv
                                       \v -> unT tw (b'. v))
                       (c'. u) )
        -- cancel the T/unT pair
    = T (\c' -> unT tu
                 \u -> (\b' -> unT tv
                               \v -> unT tw (b'.v))
                       (c'. u) )
        -- beta reduction
    = T (\c' -> unT tu
                 \u -> unT tv
                       \v -> unT tw
                             (c' . u . v) )
```

□

Proposition C.5 (homomorphism) *The following equality holds.*

```
pure f <*> pure x == pure (f x)
```

Proof Use local variables f, x, ab', b', a', and g.

```
-- f          :: a -> b
-- x          :: a
-- pure f     :: T v (a->b)
-- pure x     :: T v a
-- pure (f x) :: T v b
-- ab'        :: (a->b) -> v
-- b'         :: b -> v
-- a'         :: a -> v
-- g          :: a -> b
```

We get the following.

```
pure f <*> pure x
      -- definition of pure
  = T (\ab'->ab' f) <*> T (\a'->a' x)
      -- definition of <*>
  = T (b' -> unT (T (\ab'->ab' f))
             \g -> unT (T (\a'->a' x))
                   (b'.g) )
      -- cancel the T/unT pair
  = T (b' -> (\ab'->ab' f)
             \g -> (\a'->a' x)
                   (b'.g) )
      -- beta reduction
  = T (b' -> (\ab'->ab' f)
             \g -> (b'.g) x )
      -- beta reduction
  = T (b' -> (\g -> (b'.g) x ) f )
      -- beta reduction
  = T (b' -> (b'.f) x )
  = T (b' -> b' (f x) )
      -- definition of pure
  = pure (f x)
```

□

Proposition C.6 (commutativity) *The following equality holds.*

```
u <*> pure y == pure ($ y) <*> u
```

Proof Use local variables u, y, a', b', and g.

```
-- u              :: T v (a->b)
-- y              :: a
-- pure y         :: T v a
-- u <*> pure y   :: T v b
-- a'             :: a -> v
-- b'             :: b -> v
```

```
  -- g           :: a -> b

We calculate the LHS.
  u <*> pure y
        -- definition of pure
    = u <*> T (\a'->a' y)
        -- definition of <*>
    = T (\b' -> unT u
                \g -> unT (T (\a'->a' y))
                      (b'.g) )
        -- cancel the T/unT pair
    = T (\b' -> unT u
                \g -> (\a'->a' y)
                      (b'.g) )
        -- beta reduction
    = T (\b' -> unT u
                \g -> (b'.g) y )
    = T (\b' -> unT u
                \g -> b' (g y) )

We calculate the RHS.
  pure ($ y) <*> u
        -- recall ($ y) = \f -> f y
    = pure (\f -> f y) <*> u
        -- definition of pure
    = T (\g -> g (\f -> f y)) <*> u
        -- definition of <*>
    = T (\b' -> unT (T (\g -> g (\f -> f y)))
                \h -> unT u
                      (b'.h) )
        -- cancel the T/unT pair
    = T (\b' -> (\g -> g (\f -> f y))
                \h -> unT u (b'.h) )
        -- beta reduction
    = T (\b' -> (\h -> unT u (b'.h)) (\f -> f y) )
        -- beta reduction
    = T (\b' -> unT u
                b'. (\f -> f y) )
        -- recall Remark 2 on composition   b' . (\f -> f y)
    = T (\b' -> unT u
                \f -> b' (f y) )
        -- change the name of local variable
    = T (\b' -> unT u
                \g -> b' (g y) )
```

□

C.3 Monad Instance

Proposition C.7 (left unit) *The following equality holds.*

```
return x >>= f = f x
```

Proof Use local variables f and b'.

```
f  :: a -> T v b
b' :: b -> v
```

We calculate the LHS.

```
return x >>= f
        -- definition of return
  = T (\a' -> a' x) >>= f
        -- replace (>>=) with sharp
  = sharp f $ T (\a' -> a' x)
        -- definition of sharp
  = T (\b' -> unT (T (\a' -> a' x))
              \a -> unT (f a) b' )
        -- cancel the T/unT pair
  = T (\b' -> (\a' -> a' x)
              \a -> unT (f a) b' )
        -- beta reduction
  = T (\b' -> unT (f x) b')
        -- eta conversion
  = T $ unT (f x)
  = f x
```

□

Proposition C.8 (right unit) *The following equality holds.*

```
m >>= return = m
```

Proof Use a local variable m.

```
m      :: T v a
return :: a -> T v a
```

We calculate the LHS.

```
m >>= return
        -- replace (>>=) with sharp
  = sharp return m
        -- definition of sharp
  = T (\a' -> unT m
              \a -> unT (return a) a' )
        -- definition of return
  = T (\a' -> unT m
```

```
                 \a -> unT (T (\f -> f a)) a' )
          -- cancel the T/unT pair
      = T (\a' -> unT m
                 \a -> (\f -> f a)
                         a' )
          -- beta reduction
      = T (\a' -> unT m
                 \a -> a' a )
          -- eta conversion
      = T (\a' -> unT m a')
          -- eta conversion
      = T (unT m)
          -- cancel the T/unT pair
      = m
```

□

Proposition C.9 (associativity) *The following equality holds.*

```
(m >>= f) >>= g = m >>= (\x -> f x >>= g)
```

Proof Use local variables m, f, and g.

```
m   :: T v a
f   :: a -> T v b
g   :: b -> T v c
```

We calculate the LHS.

```
  (m >>= f) >>= g
      -- replace the first (>>=) with sharp
    = (sharp f m) >>= g
      -- replace the remaining (>>=) with sharp
    = sharp g (sharp f m)
      -- apply the definition to the second appearance of sharp
    = sharp g (T (\b' -> unT m
                         \a -> unT (f a) b'))
      -- definition of sharp
    = T (\c' -> (unT (T (\b' -> unT m
                              \a -> unT (f a)
                                      b')))
               \b -> unT (g b) c' )
       -- cancel the T/unT pair
    = T (\c' -> (\b' -> unT m
                        \a -> unT (f a)
                               b')
               \b -> unT (g b) c' )
       -- beta reduction
    = T (\c' -> unT m
               \a -> unT (f a)
```

```
                    \b -> unT (g b) c' )
```

```
We calculate the RHS. Use local variables x and g.
  -- x   :: a
  -- f x :: T v b
  -- g   :: b -> T v c

Begin with the subexpression in the parentheses.
  f x >>= g
      -- replace (>>=) with  sharp
    = sharp g (f x)
      -- definition of sharp
    = T (\c' -> unT (f x)
                (\b -> unT (g b) c') )

We calculate the whole RHS. Use local variables m, c', and d'.
  -- (\x -> f x >>= g) :: a -> T v c
  -- m                 :: T v a
  -- c', d'            :: c -> v
  m >>= (\x -> f x >>= g)
       -- replace (>>=) with sharp
    = sharp (\x -> T (\c' -> unT (f x)
                             (\b -> unT (g b) c') ))
      m
       -- definition of sharp
    = T (\d' -> unT m
                \a -> unT ( T (\c' -> unT (f a)
                                      (\b -> unT (g b)
                                                 c') ) )
                     d')
       -- cancel the T/unT pair
    = T (\d' -> unT m
                \a -> (\c' -> unT (f a)
                              (\b -> unT (g b)
                                         c') )
                     d')
       -- beta reduction
    = T (\d' -> unT m
                \a -> (unT (f a)
                      (\b -> unT (g b)
                                 d') )
       -- change the name of a local variable
    = T (\c' -> unT m
                \a -> unT (f a)
                     \b -> unT (g b) c' )
```

□

Reference

1. Mac Lane S (1998) Categories for the working mathematician, Graduate texts in mathematics, vol 5, 2nd edn. Springer

Index

A
Ab, 9
$\mathscr{A}^{\text{op}}$, 10
$[\mathscr{A}, \mathscr{B}]$, 35
ε_A, 122
$F \dashv U$, 120
Grp, 9
H^A, 197
$Hask$, 10
H_A, 202
H^-, 201
1, 9
Set, 9
Top, 9
2, 10
η_X, 122
Vect_k, 9
0, 9
adjoint equivalence, 141
adjunct, 120
apex, 54, 61
Applicative, 175, 183
arrow, 7
associator, 227

B
backward path, 253
bifunctor, 229

C
canonical map, 263
category, 7
category of all monoids, 192
category with coproducts, 59
category with products, 55
characteristic function, 203
class, 7
cocone, 69
codomain, 7
coequalizer, 77
colimit, 70
colimits in **Set**, 101
composite, 7
computation that may fail, 163
cone, 61
Cones, 62
constant functor, 25
continuation passing, 159
contravariant functor, 16
coones, 70
coproduct, 59, 79
coprojection, 58, 69
cospan, 58
counit, 122
covariant functor, 16

D
diagonal functor, 148
diagram of shape J, 61
direct product, 67
direct sum, 57, 79
discrete category, 67
disjoint union, 57, 101
disjoint union with gluing, 73
domain, 7

S. Yukita, *Category Theory Using Haskell*, Computer Science Foundations and Applied Logic, https://doi.org/10.1007/978-3-031-68538-5

E
endofunctor, 152
epimorphism, 138
equalizer, 66
equivalence of categories, 39
essentially surjective, 42
evaluation functor, 207

F
faithful functor, 18
fiber product, 63
forward path, 253
free monoid, 192
free object, 135
full and faithful functor, 18
full functor, 18
function, 1
function on morphisms, 13
function on objects, 13
Functor, 175
functor category, 35

G
generator, 11
graph of a relation, 1

H
have colimits, 80
have limits, 80

I
identity morphism, 8
index category, 61
initial object, 51
isomorphism, 8
isomorphism of categories, 39

K
Kleisli triple, 152

L
large category, 8
left adjoint, 120
limit, 62
limiting cocone, 70
limiting cone, 62
limits in **Set**, 91
List functor, 14
locally small category, 8

M
mapping, 1
Maybe functor, 14
mediating morphism, 62, 70
Monad, 165, 175
monoid, 192
monoid homomorphism, 192
monoidal category, 227
monoidal functor, 246
monomorphsim, 138
morphism, 7

N
natural, 120
natural isomorphism, 36
natural transformation, 21
non-deterministic, 154

O
opposite category, 10

P
partial function, 2
pentagon coherence, 228
poset, 10
preserve colimits, 104
preserve limits, 104
principle of duality, 10
product, 55, 67
product category, 12
projection, 61
pullback, 63
pushout, 71

R
representable, 198, 203
Representation of contravariant functors, 202
Representation of covariant functors, 197
retract, 139
retraction, 139
right adjoint, 120

S
section, 139
small category, 8
small set, 7, 268
span, 54
split epimorphism, 139
split monomorphism, 139
strict monoidal functor, 246
system of generators, 12

T
tagged union, 101
terminal object, 51
transposition, 120
Tree functor, 15
triangle coherence, 228

U
underlying set, 9, 192
unique up to isomorphisms, 56
unit, 122
unit of monoid, 192
unitality constraint, 228
unitor, 228
universal cocone, 70
universal cone, 62
universe, 7, 267

V
vertical composition, 27

W
with output, 156

Y
The Yoneda Lemma, 204
Yoneda embedding, 204

www.ingramcontent.com/pod-product-compliance
Lightning Source LLC
Chambersburg PA
CBHW070048060125
19957CB00003B/72

9 783031 685378